Causation, Physics, and the Constitution of Reality

Causation, Physics, and the Constitution of Reality

Russell's Republic Revisited

EDITED BY
Huw Price and Richard Corry

CLARENDON PRESS · OXFORD

This book has been printed digitally and produced in a standard specification in order to ensure its continuing availability

UNIVERSITY PRESS

Great Clarendon Street, Oxford OX2 6DP

Oxford University Press is a department of the University of Oxford.
It furthers the University's objective of excellence in research, scholarship,
and education by publishing worldwide in

Oxford New York

Auckland   Cape Town   Dar es Salaam   Hong Kong   Karachi
Kuala Lumpur   Madrid   Melbourne   Mexico City   Nairobi
New Delhi   Shanghai   Taipei   Toronto
With offices in
Argentina   Austria   Brazil   Chile   Czech Republic   France   Greece
Guatemala   Hungary   Italy   Japan   South Korea   Poland   Portugal
Singapore   Switzerland   Thailand   Turkey   Ukraine   Vietnam

Oxford is a registered trade mark of Oxford University Press
in the UK and in certain other countries

Published in the United States
by Oxford University Press Inc., New York

© the Several Contributors 2007

The moral rights of the author have been asserted

Database right Oxford University Press (maker)

Reprinted 2009

All rights reserved. No part of this publication may be reproduced,
stored in a retrieval system, or transmitted, in any form or by any means,
without the prior permission in writing of Oxford University Press,
or as expressly permitted by law, or under terms agreed with the appropriate
reprographics rights organization. Enquiries concerning reproduction
outside the scope of the above should be sent to the Rights Department,
Oxford University Press, at the address above

You must not circulate this book in any other binding or cover
And you must impose this same condition on any acquirer

ISBN 978-0-19-927819-0

Preface

This collection has its origins in a conference on 'Causal Republicanism', organized by the Centre for Time of the University of Sydney, in July 2003. However, the final volume takes advantage of the fact that for an edited collection, unlike a conference, there is no need to assemble the contributors at a single point in spacetime. Less constrained in physical space, collections can be much more focused in conceptual space. Accordingly, while some of the present chapters descend from papers presented at the conference, this is not a volume of conference proceedings. Not all the conference speakers are represented here, and, some who are have written on new topics; and several of the present authors did not attend the conference.

We would like to acknowledge the generous financial support of the Australian Research Council, and of the University of Sydney. This support contributed greatly to the viability and success of the original conference, and to the research and collaborations of which this volume is a product. We are also greatly indebted to two anonymous readers for Oxford University Press, for extensive comments on all the chapters in the volume.

H. P. and R. C.

Contents

Notes on Contributors	ix
1. A Case for Causal Republicanism? Huw Price and Richard Corry	1
2. Causation as Folk Science John D. Norton	11
3. What Russell Got Right Christopher Hitchcock	45
4. Causation with a Human Face Jim Woodward	66
5. Isolation and Folk Physics Adam Elga	106
6. Agency and Causation Arif Ahmed	120
7. Pragmatic Causation Antony Eagle	156
8. Causation in Context Peter Menzies	191
9. Hume on Causation: The Projectivist Interpretation Helen Beebee	224
10. Causal Perspectivalism Huw Price	250
11. Counterfactuals and the Second Law Barry Loewer	293
12. The Physical Foundations of Causation Douglas Kutach	327
13. Causation, Counterfactuals, and Entropy Mathias Frisch	351
Index	397

Notes on Contributors

ARIF AHMED is Lecturer in Philosophy at the University of Cambridge

HELEN BEEBEE is Professor of Philosophy at the University of Birmingham

RICHARD CORRY is Associate Lecturer in Philosophy at the University of Tasmania

ANTONY EAGLE is Lecturer in Philosophy at the University of Oxford and a Fellow of Exeter College

ADAM ELGA is Assistant Professor of Philosophy at Princeton University

MATHIAS FRISCH is Associate Professor of Philosophy at the University of Maryland, College Park

CHRISTOPHER HITCHCOCK is Professor of Philosophy at the California Institute of Technology

DOUGLAS KUTACH is Assistant Professor of Philosophy at Brown University

BARRY LOEWER is Professor of Philosophy at Rutgers University

PETER MENZIES is Professor of Philosophy at Macquarie University

JOHN D. NORTON is Professor of History and Philosophy of Science at the University of Pittsburgh

HUW PRICE is ARC Federation and Challis Professor of Philosophy at the University of Sydney

JIM WOODWARD is the J. O. and Juliette Koepfli Professor of the Humanities at the California Institute of Technology

1
A Case for Causal Republicanism?

HUW PRICE AND RICHARD CORRY

In 1912, in the lull between *Principia Mathematica* and the Great War, Bertrand Russell turned a famously iconoclastic eye on the problem of causation:

All philosophers, of every school, imagine that causation is one of the fundamental axioms or postulates of science, yet, oddly enough, in advanced sciences such as gravitational astronomy, the word 'cause' never occurs... The law of causality, I believe, like much that passes muster among philosophers, is a relic of a bygone age, surviving, like the monarchy, only because it is erroneously supposed to do no harm.   (Russell 1913)

Ninety years later, both targets seem to have survived Russell's attack. The monarchy in question remains firmly in place (Russell having proved one of the lesser trials of a troubled century). And causation still cuts the mustard in philosophy, apparently, despite further threats from new revolutions in physics.

Causation thrives in contemporary philosophy in two senses, in fact. First, the philosophy *of* causation remains a popular topic—if Russell was right in thinking that causality is a relic of a bygone age, many contemporary philosophers are still living in the past. Secondly, and perhaps more importantly, causation has come to play an increasingly central role in the toolkit of contemporary philosophy, invoked in the foundations of popular approaches to a very wide range of other philosophical topics. To mention a few examples, it is crucial to popular arguments for the identity of mind and brain, central to discussions of free will, action, and agency, and often invoked in criteria for realism about particular topics; and it underpins some accounts of epistemic reliability and argument to best explanation, and grounds well-known accounts of many other matters, such as reference, perception, memory, and the direction of time.

On the other hand, the issues Russell raised about causation remain unresolved and profoundly puzzling. In particular, there is still no satisfactory account of the asymmetries of causation—the difference between cause and effect, and the fact that cause–effect pairs are (almost?) always aligned from past to future. There has been considerable progress in our understanding of the *physics* of temporal asymmetry, but to date, at any rate, this has not yielded an explanation of how to ground the causal asymmetries in a physical world of the kind we inhabit.

Thus the issue of the place of causation in the constitution of the kind of reality revealed to us by physics remains both highly problematic and highly important. Taking its point of departure from Russell's famous article, this volume explores this issue in several directions.

One key theme of the volume turns on the possibility that in presenting philosophy with a stark choice between finding causation in physics and rejecting it altogether, Russell missed an important range of intermediate views. In particular, he missed what, by a natural extension of his own constitutional analogy, we may call *the republican option*. In the political case, rejecting the view that political authority is vested in our rulers by God leaves us with two choices: we may reject the notion of political authority altogether; or we may regard it, with republicans, as vested in our rulers by us. Arguably, the republican option exists in metaphysics, too. Causal republicanism is thus the view that although the notion of causation is useful, perhaps indispensable, in our dealings with the world, it is a category provided neither by God nor by physics, but rather constructed by us. (From this republican standpoint, then, thinking of eliminativism about causality as the sole alternative to full-blown realism is like thinking of anarchy as the sole alternative to the divine right of kings.)

Several chapters in the volume explore and develop this republican view of causation. The common thought, roughly, is that in looking for causation in fundamental physics, we may be looking for the wrong thing. Causation may be important, both in science and in everyday life, and yet not the sort of thing we should expect to find in physics.

The first four chapters (Norton, Hitchcock, Woodward, and Elga) take on Russell's argument explicitly. Each of these chapters endorses, to some extent, Russell's claim that there is a tension between our usual notions of causation on the one hand, and our theories of fundamental physics on the other. Yet each chapter rejects Russell's conclusion that the concept of causation is of no use.

John Norton begins to clear the way for a republican view of causation by arguing that causal notions should be regarded as part of a false, but approximately true, folk theory. Norton joins Russell in arguing against 'causal fundamentalism': the claim that causality is a fundamental part of nature. He argues that unless causal fundamentalism places some constraint on the factual content of scientific theories, it is an empty position. But a quick survey of the history of science, he says, is enough to show that all plausible candidates for such causal constraints (the existence of final causes, no action at a distance, determinism, determination of probabilities) have been violated. To make this point graphic, Norton provides a beautiful example of the failure of determinism even in simple Newtonian mechanics. Norton concludes that the claim that causes play a fundamental role in nature has been falsified by science. But the falsity of a theory, he goes on to point out, does not mean it is useless. A false theory can be very useful in certain domains if, within these domains, the world behaves as if the theory were true. Such is the case with the Newtonian theory of gravitation, or the caloric theory of heat, and such is the case, Norton suggests, with the folk theory of causation.

Christopher Hitchcock, Jim Woodward, and Adam Elga can be seen as adding flesh to Norton's suggestion by investigating the kinds of physical domains in which the notion of causation is useful, and why it is that causation is useful in those domains. Interestingly, there is a lot of general agreement among these authors. In particular they all agree on the following:

(1) All three seem to favour (though Elga less explicitly than the other two) an 'agency' or 'interventionist' approach to causation, which holds that notions of agency or manipulation play an essential role in a philosophical elucidation of the notion of causation. On this view, as Woodward puts it elsewhere, 'causes are to be regarded as handles or devices for manipulating effects' (2001, p. 1)—and it is the role that the concept of causation plays in means–end reasoning that explains its usefulness, even though the concept may have no place in fundamental physics.

(2) Based on this interventionist approach, all three of these authors argue that the concept of causation only makes sense when applied to the kinds of system that Pearl (2000) calls 'small worlds'—systems that are relatively closed and autonomous, and yet are embedded in a larger world so that interventions into the system are possible.

(3) All three suggest that the variables under consideration must be extremely coarse grained, or imprecise, from the point of view of fundamental physics. It is this coarse graining that allows us to divide the world into discrete causal relations, and to identify something less than the entire backward light-cone as causally relevant to an event.

Despite this level of agreement, these three chapters develop their arguments in different directions. Hitchcock is concerned with the implications this picture has for the kinds of questions that we can sensibly ask about causal situations. He argues that careful attention must be paid here if we are not to be led astray by the 'misleading associations' suggested by Russell. Woodward focuses on developing a detailed interventionist analysis of causation and showing how this concept of causation is useful in common sense and the 'upper sciences'. Elga, on the other hand, looks at what it is about the physics of our world that makes possible the kinds of systems in which causation is useful. In particular, Elga argues that the existence of stable coarse-grained systems relies on a certain probability distribution over the initial conditions of the universe—the very same distribution that is responsible for the temporal asymmetry of thermodynamics. Thus, as we shall see, Elga's article provides a bridge between the chapters in the first and last sections of this volume.

As we have already mentioned, the interventionist, or agency, approach to causation is an important element of many chapters in this volume. However, opponents of such approaches accuse the view of circularity, on the grounds that agency itself is a causal notion. In his chapter (ch. 6), Arif Ahmed turns his attention to this charge of circularity. He reviews a well-known response by Menzies and Price (1993), argues that it is unsuccessful, and then proposes an alternative response, based on the phenomenology of decision.

Consideration of the interventionist approach also leads nicely to a second key theme of this volume, namely the extent to which our causal claims essentially involve a projection onto the world of features of our perspective as deliberative agents. While our first four authors focus on the objective physical conditions that allow causal reasoning to take place, they are all motivated by the need to explain the usefulness of causation for agents involved in means–end reasoning—thus the need for the interventionist approach. Now, some writers (e.g. Hausman 1998) take intervention to be a thoroughly objective notion. (In this form, the view has close links to that of Pearl 2000. See also Woodward 2001). But other proponents of the agency approach (e.g. Menzies and Price 1993) take it to imply that causation is anthropocentric, or a secondary quality. So although our first four authors seem to view causality as an objective notion, their discussion raises the question of the role of the agent's perspective.

This issue of objectivity is taken up first by Antony Eagle, who develops a counterfactual analysis of causation that is broadly compatible with the common elements of Norton, Hitchcock, Woodward, and Elga. However, Eagle points out that since (as Russell suggests) causation is not *reducible* to fundamental physics, one might argue that we should not accept causal relations into our ontology. In response, Eagle argues that fundamental physics is not the final arbiter of what ontology we should accept. The acceptance of theories (and hence their ontologies), says Eagle, is a pragmatic matter; theories are accepted for a purpose and from a particular perspective. From the perspective of limited agents such as ourselves, with the aim of planning effective strategies, Eagle claims, the notion of causation is extremely useful and can be legitimately accepted. From a different perspective, fundamental physics is the most acceptable theory. What we must not do, says Eagle, is make the mistake of thinking that the concepts and ontology of one perspective will replace those of the other.

The chapters by Peter Menzies and Helen Beebee (chs. 8 and 9) further develop the idea that the agent's perspective is essential to the concept of causation, though they take very different directions.

Menzies begins by arguing that the truth of causal judgements is context sensitive. Looking at a single situation from different contexts, he says, we may legitimately make different judgements about what causes what in that situation. This is true, he argues, even if we do not change our mind about the objective structure of physical events and relations in that situation. In support of his claim, Menzies develops an analysis of context-dependent causal claims based on the theory of causal modeling put forward by

Pearl (2000) and argues that this analysis correctly deals with numerous examples of apparent context-variability. Given that the truth of causal claims is not determined by the objective physical features of a situation, Menzies concludes, with Russell, that causal facts cannot supervene on objective physical facts. However, like Eagle, Menzies cautions us against hastily following Russell to the conclusion that causation has no place in an objective account of reality. Menzies hints at another option, which he calls *perspectival realism*. The perspectival realist accepts the perspectival nature of causal claims yet holds that causal relations are a genuine part of reality nonetheless.

Beebee gives the volume a historical twist by suggesting that Hume is best interpreted as something like a causal republican. Following a suggestion made by Simon Blackburn, Beebee advocates an interpretation according to which Hume could be described as holding a quasi-realist projectivist position on causation. The position Beebee ascribes to Hume is characterized by three claims: (i) statements about causal connections do not have representational content, rather they are expressions of our inductive commitments; (ii) nonetheless, because of the role that these inductive commitments play in our cognitive lives, it is quite legitimate to treat causal statements as propositions that can be objects of belief, knowledge, truth, and falsity; (iii) once we acquire these inductive commitments, they change the nature of our perceptual experience, so that we really do perceive the world as containing causal connections. Thus, as well as turning the standard interpretation of Hume on its head, Beebee presents us with an alternative way of seeing causation as something that is projected onto the world by deliberative agents.

Huw Price (ch. 10) also takes up the question of objectivity, but considers, in particular, the objectivity of the temporal asymmetries that seem to be associated with causation. Thus Price leads us into the third major theme of this volume. This theme concerns the connections between three clusters of temporal asymmetries: (i) the modal asymmetries of cause and effect and of counterfactual dependence; (ii) the decision-theoretic asymmetries of knowledge and action; and (iii) the physical asymmetries associated with the second law of thermodynamics. There are good reasons for thinking that these three clusters of asymmetries are intimately related, but wide differences of opinion about the nature of the connections.

Price argues that the modal asymmetries are perspectival—features of the world as it looks from the viewpoint of creatures characterized by the decision-theoretic asymmetries, rather than of the world in itself. The role of the physical, thermodynamic, asymmetries is to make possible the existence of such creatures. On this view, the temporal orientation of causation reflects that of agents (here connecting again with some ideas in Russell's paper), and this provides one clear sense in which causation is less than fully objective—differently oriented creatures, in a region of the universe in which the thermodynamic asymmetry had the opposite orientation, would regard it as having the opposite direction, and neither view is objectively correct. Price appeals to this argument, among others, to argue that only the perspectival view makes good sense of the key role of intervention in an account of causation.

In opposition to Price's position is the view that the modal asymmetries of causation are objective, and reducible to (or explicable in terms of) the physical asymmetries—which, either via this route or directly, perhaps also supports the decision-theoretic asymmetries. This view has affinities with that of David Lewis (1979), and is also related to those of Papineau (1985), Ehring (1982), and Hausman (1998). However, it is vulnerable to criticisms of the kind raised by Price (1992a), Field (2003), Elga (2000), Frisch and others. The view has recently surfaced in a new form, claimed by its proponents to evade the problems facing Lewis' account. The new variant, drawing on work by Albert (2000), and developed most thoroughly by Kutach (2002), accords a crucial role to fact that entropy is very low in the past. This objectivist view is developed in this volume in Chapters 11, 12 by Barry Loewer and Douglas Kutach, and criticized in this chapter.

Based on considerations of statistical mechanics, Loewer argues that our generalizations about the macroscopic evolution of the world presuppose both that the early universe was in a state of very low entropy and that there is a uniform probability distribution over the microstates that could realize a given macrostate (here we see the close connection to Chapter 5 by Elga). To be consistent with our generalizations about the macroscopic world, then, we should also hold fixed these two presuppositions when evaluating counterfactual claims. If we do this, Loewer argues, Lewis' analysis will indeed give rise to an asymmetry in counterfactual dependence.

Kutach also identifies the assumption of a low entropy state of the early universe (which is not balanced by the symmetric assumption that the late universe is also in a low entropy state) as the source of causal and counterfactual asymmetries. In this sense, Kutach is in agreement with Loewer, and in opposition to Price. Yet Kutach points out that asymmetry is only one aspect of our notion of causation. Another important aspect is the idea that causes determine, or necessitate, their effects. However, he argues, determination is only a feature of the objective world at the micro-scale, and at this level of description, the entropic asymmetry plays no role. Thus our concept of causation relies on mixing up the micro- and macro-levels of description, and does not correspond to any objective reality. In the end, then, Kutach seems to agree with the more general republican claim that our concept of causation can only be justified pragmatically.

Frisch argues that attempts, like those of Loewer and Kutach, to base causal asymmetries upon thermodynamic asymmetries are unlikely to succeed. He has no quarrel here with the claim that we must hold fixed the hypothesis that the early universe was in a state of very low entropy. However, Frisch questions whether this constraint does actually lead to an asymmetry of counterfactual dependence between the past and the future. On the one hand, Frisch argues that the constraint is not strong enough to generate the asymmetry of counterfactual reasoning. On the other hand, he argues that if the constraint were strong enough it would lead to the counter-intuitive claim that records of the past are more reliable than any inference we might draw from the present macro-state of the world together with the laws of nature.

The chapters in this volume thus exhibit a considerable degree of agreement. Almost all our authors agree with Russell that causation is not to be found in fundamental physics. Yet all disagree with Russell's conclusion that the concept of causation is therefore useless. There is also some consensus that the way to reconcile these two positions is to explain how it is that causal concepts are useful in the deliberative lives of agents like us (hence the prominence of interventionist approaches). Thus, in some sense, most of the chapters in this volume have republican sentiments. Where our authors disagree is on some of the details. What exactly are the objective facts (about the world, and about us) that account for the usefulness of our causal reasoning? And accordingly, which features of our concept of causation reflect objective features of the world, and which are merely projected onto the world (even if on the basis of objective features of ourselves)? In particular, how are we to understand the temporal characteristics of causation, in this framework?

Price notes that in treating our modal categories as products of our perspective as knowers and agents, the perspectival view is neo-Kantian or pragmatist in character. In other words, it is a republican view, in the sense outlined earlier. Given the apparent centrality of causal and counterfactual reasoning in science and everyday life, a successful defence of this view, informed by contemporary understanding of the physics of time asymmetry, would be an important victory for a neo-Kantian metaphysics.

In view of the centrality of causation in contemporary metaphysics, moreover, the battle over causation cannot remain a regional skirmish. A republican victory here would be a real revolution in philosophy—even if only Kant's Copernican revolution, rediscovered and reinvigorated by our new understanding of our place in the temporal world revealed to us by physics. Hence—we think—the great importance of the issues discussed in this volume. Few issues in contemporary metaphysics are so central and so timely.

References

Albert, D. (2000). *Time and Chance*. Cambridge MA: Harvard University Press.
Cartwright, N. (1979). 'Causal Laws and Effective Strategies', in *Noûs*, 13 (4) Special Issue on Counterfactuals and Laws: 419–37.
Ehring, D. (1982). 'Causal Asymmetry', in *The Journal of Philosophy*, 79 (12): 761–774.
Elga, A. (2000). 'Statistical Mechanics and the Asymmetry of Counterfactual Dependence', in *Philosophy of Science*, Suppl. 68: s313–s324.
Field, H. (2003). 'Causation in a Physical World', in M. Loux and D. Zimmerman (eds), *Oxford Handbook of Metaphysics*. Oxford: Oxford University Press.
Hausman, D. (1998). *Causal Asymmetries*. Cambridge: Cambridge University Press.
Kutach, D. (2002). 'The Entropy Theory of Counterfactuals', in *Philosophy of Science*, 69 (1): 82–104.
Lewis, D. (1979). 'Counterfactual Dependence and Time's Arrow', in *Noûs*, 13 (4) Special Issue on Counterfactuals and Laws: 455–76.
Menzies, P. and Price, H. (1993). 'Causation as a Secondary Quality', in *British Journal for the Philosophy of Science*, 44: 187–203.

Papineau, D. (1985). 'Causal Asymmetry', in *British Journal for the Philosophy of Science*, 36: 273–89.

Pearl, J. (2000). *Causality: Models, Reasoning, and Inference.* Cambridge: Cambridge University Press.

Price, H. (1992). 'Agency and Causal Asymmetry', in *Mind*, 101: 501–20.

Russell, B. (1913). 'On the Notion of Cause', *Proceedings of the Aristotelian Society*, 13: 1–26.

Woodward, J. (2001) 'Causation and Manipulability' in Zalta, E (ed.) *The Stanford Encyclopedia of Philosophy (Fall 2001 Edition).* http://plato.stanford.edu/archives/fall2001/entries/causation-mani/

2
Causation as Folk Science

JOHN D. NORTON

2.1 Introduction

Each of the individual sciences seeks to comprehend the processes of the natural world in some narrow domain—chemistry, the chemical processes, biology; living processes, and so on. It is widely held, however, that all the sciences are unified at a deeper level in that natural processes are governed, at least in significant measure, by cause and effect. Their presence is routinely asserted in a law of causation or principle of causality—roughly that every effect is produced through lawful necessity by a cause—and our accounts of the natural world are expected to conform to it.[1]

This chapter was previously published in *Philosophers' Imprint,* vol. 3, no. 4: http://www.philosophersimprint.org/003004/ with the following abstract: I deny that the world is fundamentally causal, deriving the skepticism on non-Humean grounds from our enduring failures to find a contingent, universal principle of causality that holds true of our science. I explain the prevalence and fertility of causal notions in science by arguing that a causal character for many sciences can be recovered, when they are restricted to appropriately hospitable domains. There they conform to loose and varying collections of causal notions that form folk sciences of causation. This recovery of causation exploits the same generative power of reduction relations that allows us to recover gravity as a force from Einstein's general relativity and heat as a conserved fluid, the caloric, from modern thermal physics, when each theory is restricted to appropriate domains. Causes are real in science to the same degree as caloric and gravitational forces.

I am grateful to Holly Anderson, Gordon Belot, Jim Bogen, Anjan Chakravartty, Kevin Davey, John Earman, Sam Floyd, Doreen Fraser, Brian Hepburn, Francis Longworth, Sandra Mitchell, and two referees of the journal *Philosophers' Imprint* for helpful discussion, although they are in no way complicit (excepting Earman).

[1] Some versions are: Kant (1787: 218) 'All alterations take place in conformity with the law of the connection of cause and effect'; 'Everything that happens, that is, begins to be, presupposes something upon which it follows according to a rule.' Mill (1872, bk. III, ch. 5, §2): 'The law of causation, the recognition of which is the main pillar of inductive science, is but the familiar truth that invariability of succession is found by observation to obtain between every fact in nature and some other fact which has preceded it, independently of all considerations respecting the ultimate mode of production of

My purpose in this chapter is to take issue with this view of causation as the underlying principle of all natural processes. I have a negative and a positive thesis.

In the negative thesis I urge that the concepts of cause and effect are not the fundamental concepts of our science and that science is not governed by a law or principle of causality. This is not to say that causal talk is meaningless or useless—far from it. Such talk remains a most helpful way of conceiving the world and I will shortly try to explain how that is possible. What I do deny is that the task of science is to find the particular expressions of some fundamental causal principle in the domain of each of the sciences. My argument will be that centuries of failed attempts to formulate a principle of causality, robustly true under the introduction of new scientific theories, have left the notion of causation so plastic that virtually any new science can be made to conform to it. Such a plastic notion fails to restrict possibility and is physically empty. This form of causal skepticism is not the traditional Humean or positivistic variety. It is not motivated by an austere epistemology that balks at any inference to metaphysics. It is motivated by taking the content of our mature scientific theories seriously.

Mature sciences, I maintain, are adequate to account for their realms without need of supplement by causal notions and principles. The latter belong to earlier efforts to understand our natural world or to simplified reformulations of our mature theories, intended to trade precision for intelligibility. In this sense I will characterize causal notions as belonging to a kind of folk science, a crude and poorly grounded imitation of more developed sciences. More precisely, there are many folk sciences of causation corresponding to different views of causation over time and across the present discipline. While these folk sciences are something less than our best science, I by no means intend to portray them as pure fiction. Rather I will seek to establish how their content can be licensed by our best science, without the causal notions becoming fundamental.

In the positive thesis, I will urge that ordinary scientific theories can conform to a folk science of causation when they are restricted to appropriate, hospitable processes; and the way they do this exploits the generative

---

phenomena and of every other question regarding the nature of "things in themselves".' For a short survey, see Nagel (1961, ch. 10, sect. v).

power of reduction relations, a power usually used to recover older theories from newer ones in special cases.

This generative power is important and familiar. It allows Einstein's general theory of relativity to return gravity to us as a Newtonian force in our solar system, even though Einstein's theory assures us that gravity is fundamentally not a force at all. And it explains why, as long as no processes interchange heat and work, heat will behave like a conserved fluid, as caloric theorists urged. In both domains it can be heuristically enormously helpful to treat gravity as a force or heat as a fluid and we can do so on the authority of our best sciences. My positive thesis is that causes and causal principles are recovered from science in the same way and have the same status: they are heuristically useful notions, licensed by our best sciences, but we should not mistake them for the fundamental principles of nature. Indeed we may say that causes are real to the same degree that we are willing to say that caloric or gravitational forces are real.

The view developed here is not an unalloyed causal skepticism. It has a negative (skeptical) and a positive (constructive) thesis and I urge readers to consider them in concert. They are motivated by the same idea. If the world is causal, that is a physical fact to be recovered from our science. So far our science has failed to support the idea of principle of causality at the fundamental level (negative thesis); but a causal character can be recovered from the science as looser, folk sciences that obtain in restricted domains (positive thesis).

2.1.1 To Come

In Section 2.2, I will describe the causal skepticism I call 'anti-fundamentalism' and lay out the case for the negative thesis in the form of a dilemma. In Section 2.3, in support of the arguments of Section 2.2, I will give an illustration of how even our simplest physical theories can prove hostile to causation. In Section 2.4, I will begin development of the positive thesis by outlining the generative power of reduction relations. In Section 2.5, I will describe one type of the possible folk theories of causation in order to illustrate the sorts of causal structure that can be recovered from the generative power of reduction relations. Section 2.6 has examples of this folk theory used to identify first and final causes and to display the domain dependence of the recovery. Section 2.7 has a brief conclusion.

2.2 The Causal Fundamentalist's Dilemma

2.2.1 *The Dispensability of Causes*

Russell (1917, p. 132) got it right in his much celebrated riposte:

> All philosophers, of every school, imagine that causation is one of the fundamental axioms or postulates of science, yet, oddly enough, in advanced sciences such as gravitational astronomy, the word 'cause' never occurs... The law of causality, I believe, like much that passes muster among philosophers, is a relic of a bygone age, surviving like the monarchy, only because it is erroneously supposed to do no harm.

When they need to be precise, fundamental sciences do not talk of causes, but of gravitational forces, or voltages, or temperature differences, or electrochemical potentials, or a myriad of other carefully devised, central terms. Nonetheless they are still supposed to be all about causes. Perhaps the analogy is to an account of a bank robbery. It can be described in the most minute detail—the picking of the lock, the bagging of the cash—without ever actually mentioning 'theft' or 'robbery'. If one thinks cause might have a similar surreptitious role in science, it is sobering to compare the case of causation with that of energy. Many sciences deal with a common entity—energy, which manifests itself quite directly throughout the science. Sometimes it appears by name—kinetic energy, potential energy, field energy, elastic energy—and other times it appears as a synonym: heat, work, or the Hamiltonian. However there is little doubt that each of the sciences is dealing with the very same thing. In each science, the energies can be measured on the same scale, so many Joules, for example, and there are innumerable processes that convert the energy of one science into the energy of another, affirming that it is all the same stuff. The term is not decorative; it is central to each theory.

2.2.2 *Causal Fundamentalism*

If one believes that the notions of cause and effect serve more than a decorative function in science, one must find some manifest basis for their importance. It is clearly too severe to demand that causes all be measurable on some common scale like energies. We can afford to be a little more forgiving. However we must find some basis; taking cash is theft because of an identifiable body of criminal law. What should that basis be in the case of causes? In it, the notion of cause must betoken some factual property of natural processes; otherwise its use is no more than an exercise in labeling. And the notion must be the same or similar in the various sciences; otherwise the use of the same term in many places would be no more than a pun. I believe this basis to be broadly accepted and to energize much of the philosophical literature on causation. I shall call it:

Causal fundamentalism: Nature is governed by cause and effect; and the burden of individual sciences is to find the particular expressions of the general notion in the realm of their specialized subject matter.

My goal in this section is to refute this view. In brief, I regard causal fundamentalism as a kind of *a priori* science that tries to legislate in advance how the world must be. These efforts have failed so far. Our present theories have proven hard enough to find and their content is quite surprising. They have not obliged us by conforming to causal stereotypes that were set out in advance and there is little reason to expect present causal stereotypes to fare any better. The difficulty for causal fundamentalism is made precise in:

Causal fundamentalist's dilemma: EITHER conforming a science to cause and effect places a restriction on the factual content of a science; OR it does not.

In either case, we face problems that defeat the notion of cause as fundamental to science. In the first horn, we must find some restriction on factual content that can be properly applied to all sciences; but no appropriate restriction is forthcoming. In the second horn, since the imposition of the causal framework makes no difference to the factual content of the sciences, it is revealed as an empty honorific.

2.2.3 The First Horn

Discerning how causation restricts the possibilities has been the subject of a long tradition of accounts of the nature of cause and effect and of the law or principle of causality. One clear lesson is learned from the history of these traditions. Any substantial restriction that they try to place on a science eventually fails. There is no shortage of candidates for the factual restriction of the first horn. The trouble is, none works. Let us take a brief tour.

Aristotle described four notions of cause: the material, efficient, final and formal; with the efficient and final conforming most closely to the sorts of things we would now count as a cause. The final cause, the goal towards which a process moves, was clearly modeled on the analogy between animate processes and the process of interest. In the seventeenth century, with the rise of the mechanical philosophy, it was deemed that final causes simply did not have the fundamental status of efficient causes and that all science was to be reconstructed using efficient causes alone. (De Angelis 1973) Although talk of final causes lingers on, this is a blow from which final causes have never properly recovered.

The efficient cause, the agent that brings about the process, provided its share of befuddlement. Newton (1692/3; third letter) pulled no punches in his denunciation of gravity as causal action at a distance:

> that one body may act upon another at a distance through a vacuum, without the mediation of anything else, by and through which their action and force may be conveyed from one to another, is to me so great an absurdity, that I believe no man, who has in philosophical matters a competent faculty of thinking, can ever fall into it. Gravity must be caused by an agent acting constantly according to certain laws

Causes cannot act where they are not. Nonetheless several centuries of failed attempts to find a mechanism or even finite speed for the propagation of gravity brought a grudging acceptance in the nineteenth century that this particular cause could indeed act where it was not.

In the same century, causes were pressed to the forefront as science came to be characterized as the systematic search for causes, as in Mill's *System of Logic*. At the same time, an enlightened, skeptical view sought to strip the notion of causation of its unnecessary metaphysical and scholastic decorations. While it might be customary to distinguish in causal processes between agent and patient, that which acts and that which is acted upon, Mill (1872, bk III, ch. 5 §4) urged that the distinction is merely a convenience. Or, he urged, the continued existence of the cause is not needed after all for the persistence of the effect (§7). All that remained was the notion that the cause is simply the unconditional, invariant antecedent: 'For every event there exists some combination of objects or events, some given concurrence of circumstances, positive and negative, the occurrence of which is always followed by that phenomenon' (§2).

Causation had been reduced to determinism: fix the present conditions sufficiently expansively and the future course is thereby fixed. Thus the nineteenth century brought us the enduring image of Laplace's famous calculating intelligence, who could compute the entire past and future history of the universe from the forces prevailing and the present state of things. This great feat was derived directly from the notion that cause implied determinism, as the opening sentence of Laplace's (1825, p. 2) passage avows: 'We ought then to consider the present state of the universe as the effect of its previous state and the cause of that which is to follow.'

This lean and purified notion of causation was ripe for catastrophe, for it inhered in just one fragile notion, determinism. The advent of modern quantum theory in the 1920s brought its downfall. For in the standard approach, the best quantum theory could often deliver were *probabilities* for future occurrences. The most complete specification of the state of the universe now cannot determine whether some particular Radium-221 atom will decay over the next 30 seconds (its half life); the best we say is that there is a chance of 1/2 of decay. A lament for the loss of the law of causality became a fixture in modern physics texts (e.g. Born 1935, p. 102).

While the refutation seemed complete, causation survived, weakly. If causes could not compel their effects, then at least they may raise the probabilities. A new notion of causation was born, probabilistic causation.[2] Quantum theory brought other, profound difficulties for causation. Through its non-separability, quantum theory allows that two particles that once interacted may remain entangled, even though they might travel light years away from each other, so that the behavior of one might still be affected instantly by that of the other. This places severe obstacles in the way of any account of causality that tries to represent causes locally in space and time and seeks to prohibit superluminal causal propagation.

One could be excused for hoping that this enfeebled notion of probabilistic causation might just be weak enough to conform peacefully with our physics. But the much neglected fact is that it never was! All our standard physical theories exhibit one or other form of indeterminism. (See Earman 1986; Alper et al. 2000.)[3] That means, that we can always find circumstances in which the full specification of the present fails to fix the future. In failing to fix the future, the theories do not restrict the range of possibilities probabilistically, designating some as more likely than

---

[2] 'This quantum indeterminacy is, in fact, the most compelling reason for insisting upon the need for probabilistic causation.' (Salmon 1980: n. 19)

[3] Curiously the most likely exception is a no collapse version of quantum theory since it is governed fully by the Schroedinger equation, which is deterministic.

others. They offer no probabilities at all. This failure of determinism is a commonplace for general relativity that derives directly from its complicated spacetime geometries in which different parts of spacetime may be thoroughly isolated from others. For determinism to succeed, it must be possible to select a spatial slice of spacetime that can function as the 'now' and is sufficiently well connected with all future times that all future processes are already manifest in some trace form on it. Very commonly spacetimes of general relativity do not admit such spatial slices. What is less well known is that indeterminism can arise in ordinary Newtonian physics. Sometimes it arises in exotic ways, with 'space invaders' materializing with unbounded speed from infinity and with no trace in earlier times; or it may arise in the interactions of infinitely many masses. In other cases, it arises in such prosaic circumstances that one wonders how it could be overlooked and the myth of determinism in classical physics sustained. A simple example is described in the next section.

With this catalog of failure, it surely requires a little more than naïve optimism to hope that we still might find some contingent principle of causality that can be demanded of all future sciences. In this regard, the most promising of all present views of causation is the process view of Dowe, Salmon, and others (Dowe 1997). In identifying a causal process as one that transmits a conserved quantity through a continuous spatiotemporal pathway, it seeks to answer most responsibly to the content of our mature sciences. Insofar as the theory merely seeks to identify which processes in present science ought to be labeled causal and which are not, it succeeds better than any other account I know. If however, it is intended to provide a factual basis for a universal principle of causality, then it is an attempt at *a priori* science, made all the more fragile by its strong content. If the world is causal according to its strictures, then it must rule out *a priori* the possibility of action at a distance, in contradiction with the standard view of gravitation in science in the nineteenth century and the non-local processes that seem to be emerging from present quantum theory. Similar problems arise in the selection of the conserved quantity. If we restrict the conserved quantity to a few favored ones, such as energy and momentum, we risk refutation by developments in theory. Certain Newtonian systems are already known to violate energy and momentum conservation (Alper et al. 2000) and in general relativity we often cannot define the energy and momentum of an extended system. But if we are permissive in selection of the conserved quantity, we risk trivialization by the construction of artificial conserved quantities specially tailored to make any chosen process come out as causal.

Or do we ask too much in seeking a single, universal principle? Perhaps we should not seek a *universal* principle, but just one that holds in some subdomain of science that is fenced off from the pathologically acausal parts of science. The first problem with this proposal is that we do not know where to put the fence. The common wisdom has been that the fence should lie between the pathologically acausal quantum theory and the causally well-behaved classical physics. Yet some dispute whether quantum theory has shrunk the domain in which the causal principle holds. (Bunge 1979, pp. 346–51; Margenau 1950, pp. 96, 414). And the example of the next section shows that even the simplest classical physics still admits acausal pathologies. The second problem is, if we did find where to put the fence, what confidence can we have of finding a single principle that applies in the causal domain? The proliferation of different accounts of causation and the flourishing literature of counterexamples suggests no general agreement even on what it means to say that something is a cause. So perhaps we should also give up the search for a *single* principle and allow each causally well-behaved science to come with its own, distinct principle of causality.[4] Now the real danger is that we eviscerate the notion of causation of any factual content. For now we can go to each science and find some comfortable sense in which it satisfies its own principle of causality. Since, with only a little creativity, that can be done with essentially any science, real or imagined, the demand of conformity to cause and effect places no restriction on factual content—and we have left the realm of the first horn.

2.2.4 The Second Horn

Let us presume that conforming a science to cause and effect places no restriction on the factual content of the science. The immediate outcome is that any candidate science, no matter how odd, may be conformed to cause and effect; the notion of causation is sufficiently plastic to conform to whatever science may arise. Causal talk now amounts

---

[4] Or we may purchase broad scope by formulating a principle so impoverished that it no longer resembles causation but contradicts no present science. Margenau (1950, §19.5) proposes that causality is the 'temporal invariability of laws': 'Causality holds if the laws of nature (differential equations) governing closed systems do not contain the time variable in explicit form.'

to little more than an earnest hymn of praise to some imaginary idol; it gives great comfort to the believers, but it calls up no forces or powers.

Or is this just too quick and too clever? Even if there is no factual principle of causality in science to underwrite it, might not the concept of cause be somehow indispensable to our science? Perhaps the most familiar and longest lived version of this idea is drawn from the Kantian tradition. It asserts that we must supply a conception of causation if we are to organize our experiences into intelligible coherence. A variant of this is Nagel's (1961, p. 324, his emphasis) proposal that the principle of causality even in vague formulation

is an *analytic consequence* of what is commonly meant by 'theoretical science.' ... it is difficult to understand how it would be possible for modern theoretical science to surrender the general ideal expressed by the principle without becoming thereby transformed into something incomparably different from what that enterprise actually is.

Nagel (1961, p. 320) formulates the principle as a methodological rule of heuristic value which 'bids us to analyze physical processes in such a way that their evolution can be shown to be independent of the particular times and places at which those processes occur'. This version conforms to the second horn since Nagel (1961, p. 320, his emphasis) insists this principle of causality is a '*maxim* for inquiry rather than a statement with definite empirical content.'

Appealing as these approaches may be, they do not defeat this second horn of the dilemma. One could well imagine that a concept of causation might be indispensable or an injunction to find causes heuristically useful, if the conception of causation reflected some factual properties of the world. Then something like causation must arise when we conform our concepts to the world. Or an heuristic principle could exploit those facts to assist discovery. But that is the province of the first horn, where I have already described my reasons for doubting that there are such facts. The presumption of this second horn is that there are no such factual properties of the world. In the context of this second horn, conceptual indispensability or heuristic fertility must derive not from facts in the world but from facts about us, our psychology and our methods. So a supposed indispensability or fertility of the notion of causation is at most telling us something about us and does not establish that the world is governed at some fundamental level by a principle of causality.

2.2.5 Varieties of Causal Skepticism

The form of causal skepticism advocated here is not the more traditional Humean and positivistic skepticism that is based on an austere epistemology and aversion to metaphysics. My anti-fundamentalism is based on an aversion to *a priori* science; it requires that a metaphysics of causation that pertains to the physical character of the world must be recovered from our science. It is worth distinguishing a few varieties of causal skepticism in more detail.

Humean/positivist skepticism. This dominant tradition of causal skepticism in philosophical analysis depends upon an austere epistemology that denies we can infer to entities, causal or otherwise, beyond direct experience. What passes as causation is really just constant conjunction or functional dependence within actual experiences. Hume (1777, section VII, parts I–II) initiated the tradition when he urged that the necessity of causal connection cannot be discerned in the appearances; the latter supply only constant conjunctions. The critique was sustained by the positivists of the late nineteenth and early twentieth centuries as part of their program of elimination of metaphysics. Mach (1960, p. 580, his emphasis) concluded 'There is no cause nor effect in nature; nature has but an individual existence; nature simply *is*.' Where Hume saw constant conjunction, Mach saw functional dependence: 'The concept of cause is replaced...by the concept of function: the determining of the dependence of phenomena on one another, the economic exposition of actual facts...' (Mach 1960, p. 325). Very similar themes are found in Pearson (1911, p. vi, chs. 4, 5). Russell (1903, p. 478; 1917, pp. 142, 150–1) also endorsed a functionalist view akin to Mach's.

Anti-fundamentalism. The skepticism of this paper is grounded in the content of our mature sciences and the history of its development. Skepticism about causal fundamentalism is derived from the failure of that content and history to support a stable, factual notion of causation. Insofar as it is able to take the content of our mature sciences seriously, with that content extending well beyond direct experience, it relies on a fertile epistemology rather than the barren epistemology of Humean and positivist skepticism. I believe this anti-fundamentalist form of causal skepticism is quite broadly spread. What did the most to promote the view was the advent of quantum theory and the resulting demise of determinism. On the basis of the content of the latest science, a generation of physicists and philosophers of science lamented the failure of causation. However I have found it hard to locate expositions in which that lament is systematically developed into a strongly argued version of anti-fundamentalism. It appears to be the position of Campbell (1957, ch. 3). He noted that the relations expressed by many laws of nature cannot be causal, since they do not conform to the characteristic properties of causal relations, which are temporal, asymmetric, and binary. 'So,' he concluded (p. 56), 'far from all laws asserting causal relations, it is doubtful whether any assert them'.

These two forms of skepticism should be distinguished from: *Eliminativism*.[5] In this view causal skepticism is derived from the possibility of formulating our sciences without explicitly causal terms, like cause and effect. Bunge (1979, p. 345) correctly protested that this is a simple verbal trap and not strong enough to support a robust skepticism. However there is a converse trap. Most forms of causal skepticism, including mine, lead to the view that the notion of cause is dispensable. Mach (1894, p. 254) 'hope[d] that the science of the future will discard the idea of cause and effect, as being formally obscure...' But that should then not be mistaken as the basis of their skepticism.

2.3 Acausality in Classical Physics

While exotic theories like quantum mechanics and general relativity violate our common expectations of causation and determinism, one routinely assumes that ordinary Newtonian mechanics will violate these expectations only in extreme circumstances, if at all. That is not so. Even quite simple Newtonian systems can harbor uncaused events and ones for which the theory cannot even supply probabilities. Because of such systems, ordinary Newtonian mechanics cannot license a principle or law of causality. Here is an example of such a system fully in accord with Newtonian mechanics. It is a mass that remains at rest in a physical environment that is completely unchanging for an arbitrary amount of time—a day, a month, an eon.

---

[5] I borrow the term from Schaffer (2003, section 2.1), although I am not sure that we define the term in the same way.

Then, without any external intervention or any change in the physical environment, the mass spontaneously moves off in an arbitrary direction with the theory supplying no probabilities for the time or direction of the motion.

2.3.1 The Mass on the Dome

The dome of Figure 2.1a sits in a downward directed gravitational field, with acceleration due to gravity g. The dome has a radial coordinate r inscribed on its surface and is rotationally symmetric about the origin $r = 0$, which is also the highest point of the dome. The shape of the dome is given by specifying h, how far the dome surface lies below this highest point, as a function of the radial coordinate in the surface, r. For simplicity of the mathematics, we shall set $h = (2/3g)r^{3/2}$. (Many other profiles, though not all, exhibit analogous acausality.)

A point-like unit mass slides frictionlessly over the surface under the action of gravity. The gravitational force can only accelerate the mass along the surface. At any point, the magnitude of the gravitational force tangential to the surface is $F = d(gh)/dr = r^{1/2}$ and is directed radially outward. There is no tangential force at $r = 0$. That is, on the surface the mass experiences a net outward directed force field of magnitude $r^{1/2}$. Newton's second law '$F = ma$' applied to the mass on the surface sets the radial acceleration d^2r/dt^2 equal to the magnitude of the force field:

(1) $d^2r/dt^2 = r^{1/2}$

If the mass is initially located at rest at the apex $r = 0$, then there is one obvious solution of Newton's second law for all times t:

(2) $r(t) = 0$

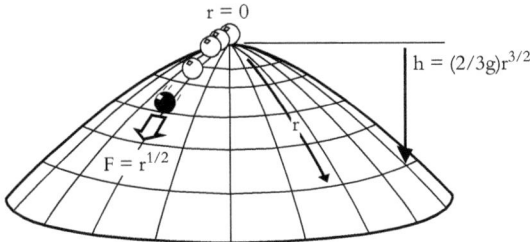

Figure 2.1a. Mass sliding on a dome.

The mass simply remains at rest at the apex for all time. However there is another large class of unexpected solutions. For any radial direction:

(3) $r(t) = (1/144)(t - T)^4$ for $t \geq T$
$ = 0$ for $t \leq T$

where $T \geq 0$ is an arbitrarily chosen constant. One readily confirms that the motion of (3) solves Newton's second law (1).[6]

If we describe the solutions of (3) in words, we see they amount to a violation of the natural expectation that some cause must set the mass in motion. Equation (3) describes a point mass sitting at rest at the apex of the dome, whereupon at an arbitrary time $t = T$ it spontaneously moves off in some arbitrary radial direction.

2.3.2 Properties

Two distinct features of this spontaneous excitation require mention:

No cause. No cause determines when the mass will spontaneously accelerate or the direction of its motion. The physical conditions on the dome are the same for all times t prior to the moment of excitation, $t = T$ and are the same in all directions on the surface.

No probabilities. One might think that at least some probabilistic notion of causation can be preserved insofar as we can assign probabilities to the various possible outcomes. Nothing in the Newtonian physics requires us to assign the probabilities, but we might choose to try to add them for our own conceptual comfort. It can be done as far as the *direction* of the spontaneous motion is concerned. The symmetry of the surface about the apex makes it quite natural for us to add a probability distribution that assigns equal probability to all directions. The complication is that there is no comparable way for us to assign probabilities for the *time* of the spontaneous excitation that respects the physical symmetries of solutions (3). Those solutions treat all candidate excitation times T equally. A probability distribution that tries to make each candidate time equally likely cannot be proper—that is, it cannot assign unit probability to the union of all disjoint outcomes.[7] Or one

---

[6] By direct computation $d^2r/dt^2 = (1/12)(t-T)^2 = [(1/144)(t-T)^4]^{1/2}$ for $t \geq T$ and 0 otherwise; so that $d^2r/dt^2 = r^{1/2}$.

[7] Since all excitation times T would have to be equally probable, the probability that the time is in each of the infinitely many time intervals, (0,1), (1,2), (2,3), (3,4), ... would have to be the same, so that

that is proper can only be defined by inventing extra physical properties, not given by the physical description of the dome and mass, Newton's laws and the laws of gravitation, and grafting them unnaturally onto the physical system.[8]

2.3.3 *What about Newton's First Law?*

The solutions (3) are fully in accord with Newtonian mechanics in that they satisfy Newton's requirement that the net applied force equals mass x acceleration at all times. But, one may still worry that spontaneous acceleration somehow violates Newton's First Law:

In the absence of a net external force, a body remains at rest or in a state of uniform motion in a straight line.

It is natural to visualize 'uniform motion in a straight line' over some time interval, but we will need to apply the law at an instant. At just one instant, the law corresponds to motion with zero acceleration. So the instantaneous form of Newton's First Law is:

In the absence of a net external force, a body is unaccelerated.

Returning to the concern, there is no net force on the mass at $t = T$, so, by this law, shouldn't the mass remain at rest? A more careful analysis shows the motions of (3) are fully in accord with Newton's First Law.

For times $t \leq T$, there is no force applied, since the body is at position $r = 0$, the force free apex; and the mass is unaccelerated.

For times $t > T$, there is a net force applied, since the body is at positions $r > 0$ *not* at the apex, the only force free point on the dome; and the mass accelerates in accord with $F = ma$.

---

zero probability must be assigned to each of these intervals. Summing over all intervals, this distribution entails a zero probability of excitation ever happening.

[8] For example, consider the natural condition that, at any time t, we always have the same probability of *no* excitation occurring over the next (arbitrarily chosen, but fixed) time interval Δt, given that no excitation has occurred by the start of that time interval. This condition uniquely picks out the exponential decay rule $P(t) = \exp(-t/\tau)$ where $P(t)$ is the probability of no excitation over the time interval $(0,t)$ and τ is some positive time constant. (At any time t, the probability of excitation in the ensuing time interval Δt is just $\exp(-(t + \Delta t)/\tau)/\exp(-t/\tau) = \exp(-\Delta t/\tau)$, which is independent of t as required.) The problem is that the dynamics of excitation is governed by the magnitude of the time constant τ, which is the mean time to excitation. A small τ means that we likely will have rapid excitation; a large τ means we will not. Nothing in the physical set-up of the dome and mass enables us to fix a value for τ. We must fix its value by arbitrary stipulation, thereby inventing the new physical property of rate of decay, which is not inherent in the original physical system.

But what of the crucial time $t = T$? The solutions of (3) entail that the acceleration $a(t)$ of the mass is given by

(4) $a(t) = (1/12)(t - T)^2$ for $t \geq T$
    $= 0$ for $t \leq T$

We confirm by substitution into (3) that at $t = T$, the mass is still at the force free apex $r = 0$ and, by substitution into (4), that the mass has an acceleration $a(0)$ of zero. This is just what Newton's first law demands. At $t = T$, there is no force and the mass is unaccelerated. At any $t > T$, there is a non-zero force and the mass is accelerated accordingly.

2.3.4 No First Instant of Motion—No Initiating Cause

Why is it so easy to be confused by this application of Newtonian mechanics? Our natural causal instinct is to seek the first instant at which the mass moves and then look for the cause of the motion at that instant. We are tempted to think of the instant $t = T$ as the first instant at which the mass moves. But that is not so. It is the *last* instant at which the mass does *not* move. There is no first instant at which the mass moves. The mass moves during the interval $t > T$ only and this time interval has no first instant. (Any candidate first instant in $t > T$, say $t = T + \varepsilon$ for any $\varepsilon > 0$, will be preceded by an earlier one, $t = T + \varepsilon/2$, still in $t > T$.) So there is no first instant of motion and thus no first instant at which to seek the initiating cause.

2.3.5 Still Not Happy?

There is a simple way to see that the spontaneous motion of the mass is actually not that strange. Instead of imagining the mass starting at rest at the apex of the dome, we will imagine it starting at the rim and that we give it some initial velocity directed exactly at the apex. If we give it too much initial velocity, it will pass right over the apex to the other side of the dome. So let us give it a smaller initial velocity. We produce the trajectory T_1 of Figure 2.1b.

The mass rises towards the apex, but before it arrives it loses its motion, momentarily halts and then falls back to the rim. So we give it a little more initial velocity to produce trajectory T_2. The mass rises closer to the apex but does not reach it before momentarily halting and falling back. We

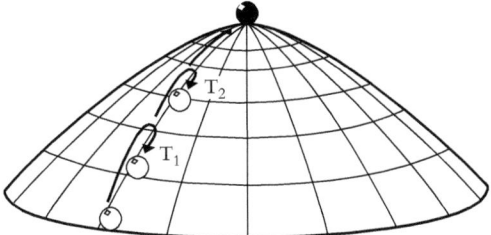

Figure 2.1b. Projecting the mass towards the apex.

continue this process until we give the mass just the right initial velocity so that it rises up and momentarily halts exactly at the apex. In this last case, we have ended up with the mass momentarily at rest at the one force free point on the dome, the one point where, if it is at rest, the mass can (but need not) remain at rest. So let us imagine that it does remain at rest once it arrives. We now have a trajectory in which the mass rises up to the apex, halts there and remains there at rest for any arbitrary time period we care to nominate.[9]

An important feature of Newtonian mechanics is that it is time reversible; or at least that the dynamics of gravitational systems invoked here are time reversible. This means that we can take any motion allowed by Newton's theory and generate another just by imagining that motion run in reverse in time. So let us do that with the motion we have just generated. That reversed motion corresponds to a mass that remains at rest at the apex of the dome for some arbitrary time period and then spontaneously moves off towards the rim. And that is just a qualitative description of one of the solutions of (3).

This time reversal trick is powerful, but we must be cautious not to overrate it. It is best used just to make the acausal behavior plausible, while the proper mathematical analysis of (1), (3), and (4) proves it. The reason is that there is a loophole. The spontaneous motion can happen only on domes of the right shape, such as those of Figure 2.1a. It cannot happen on others such as a hemispherical dome. The time reversal argument fails for these other cases for a reason that is easy to overlook. As we proceed through the trajectories $T_1, T_2, \ldots$ on a hemispherical dome, the time taken

---

[9] In an analogous analysis, we consider trajectories with too much initial velocity so that the mass reaches the apex with some non-zero velocity and passes over it. We reduce the initial velocity until the velocity at the apex is zero and then proceed as in the first analysis.

for the mass to rise to its momentary halt increases without bound. The final trajectory we seek, the one that momentarily halts at the apex, turns out to require infinite time. This means that the mass never actually arrives. Its time reverse displays a mass that has been in motion at all past times, without any spontaneous launches. The corresponding time for the dome of Figure 2.1a, however, is finite, so the analysis does succeed for this case.

2.4 The Generative Capacity of Reduction Relations and its Utility for Causation

2.4.1 *Reduction of Gravitational Force and Particles...*

The negative thesis asserts that science is not based fundamentally on cause and effect. That is *not* to say that notions of cause and effect are purely fictions; that would be too severe. There is a sense in which causes are properly a part of our scientific picture of the natural world and my goal in the positive thesis is to find it. I shall urge that the place of causes in science is closely analogous to the place of superseded theories. In 1900, our picture of the natural world seemed secure. We concluded that the planet earth orbited the sun because of a gravitational force exerted on it by the sun; and matter consisted of many small charged particles, called ions or electrons. All this was supported by an impressive body of observational and experimental evidence. Three decades later, these conclusions had been overturned. Einstein's general theory of relativity assured us that gravitation was not a force after all, but a curvature of spacetime. Quantum theory revealed that our fundamental particles were some mysterious conglomeration of both particle and wavelike properties.

The earlier theories did not disappear; and they could not. The large bodies of evidence amassed by Newton in favor of gravitational forces and by Thomson for electrons as particles needed to be directed to favor the new theories. The simplest way of doing this was to show that the older theories would be returned to us in suitable limiting cases. General relativity tells us gravitation does behave just like a force, as long as we deal only with very weak gravity; and quantum theory tells us we can neglect the wavelike properties of electrons as long as we stay away from circumstances in which interference effects arise. In the right conditions,

CAUSATION AS FOLK SCIENCE 29

the newer theories revert to the older so that evidence for the older could be inherited by the newer.

2.4.2 ... and the Caloric

A simpler and more convenient example is the material theory of heat. In the eighteenth and early nineteenth century, heat was conceived of as a conserved fluid. The temperature measured the density of the fluid and the natural tendency of the fluid to flow from high to low density was manifested as a tendency to flow from high to low temperature. The theory flourished when Lavoisier (1790) included the matter of heat as the element caloric in his treatise that founded modern chemistry; and Carnot (1824) laid the foundations of modern thermodynamics with an analysis of heat engines that still presumed the caloric theory. Around 1850, through the work of Joule, Clausius, Thomson, and others, this material theory of heat fell with the recognition that heat could be converted into other forms of energy. Heat came to be identified with a disorderly distribution of energy over the very many component subsystems of some body; in the case of gases, the heat energy resided in the kinetic energy of the gas molecules, verifying a kinetic theory of heat. The older material theory could still be recovered as long as one considered processes in which there was no conversion between heat energy and other forms of energy such as work. An example would be the conduction of heat along a metal bar. Exactly because heat is a form of energy and energy is conserved, the propagating heat will behave like a conserved fluid. In the newer theory, the temperature is measured by the average energy density. It is a matter of overwhelming probability that energy will pass from regions of higher temperature (higher average energy) to those of lower temperature (lower average energy) with the result that the heat energy distribution moves towards the uniform. This once again replicates a basic result of the caloric theory: heat spontaneously moves from hotter to colder.

2.4.3 Generative Capacity

I call this feature of reduction relations their 'generative capacity'. In returning the older theories, the relations revive a defunct ontology. More precisely, they do not show that heat is a fluid, or gravity is a force, or that electrons are purely a particle; rather they show that in the right domain the world behaves just as if they were. The advantages of this generative capacity are great. It is not just that the newer theories could now inherit the evidential base of the old. It was also that the newer theories were conceptually quite difficult to work with and reverting to the older theories often greatly eases our recovery of important results. Einstein's general relativity does assure as that planets orbit the sun almost exactly in elliptical orbits with the sun at one focus. But a direct demonstration in Einstein's theory is onerous. Since much of the curvature of spacetime plays no significant role in this result, the easiest way to recover it is just to recall that Einstein's theory reverts to Newton's in the weak gravity of the solar system and that the result is a familiar part of Newton's theory. In many cases it is just conceptually easier and quite adequate to imagine that gravity is a force or heat a fluid.

2.4.4 Applied to Causation: Are Causes Real?

The situation is the same, I urge, with causation. We have some idea of what it is to conform to cause and effect, although what that amounts to has changed from epoch to epoch and even person to person. The world does not conform to those causal expectations in the sense that they form the basis of our mature sciences. However in appropriately restricted circumstances our science entails that nature will conform to one or other form of our causal expectations. The restriction to those domains generates the causal properties in the same way that a restriction to our solar system restored gravity as a force within general relativity; or ignoring conversion processes restored heat as a conserved fluid. The causes are not real in the sense of being elements of our fundamental scientific ontology; rather in these restricted domains the world just behaves as if appropriately identified causes were fundamental.

So, are causes real? My best answer is that they are as real as caloric and gravitational forces. And how real are they? That question is the subject of an extensive literature in philosophy of science on the topic of reduction. (For a survey, see Silberstein 2002.) I will leave readers to make up their own minds, but I will map out some options, drawn from the reduction literature, and express an opinion. One could be a fictionalist and insist that causes, caloric and gravitational forces are ultimately just inventions, since they are not present in the fundamental ontology. Or one could be a realist and insist upon the autonomy of the various levels of science. To withhold reality from an entity, one might say, because it does not fall in the fundamental ontology of our most advanced science is to risk an infinite regress that leaves us with no decision at all about the reality of anything in our extant sciences, unless one is confident that our latest science can never be superseded. My own view is an intermediate one: causes, caloric and gravitational forces have a derivative reality. They are not fictions insofar as they are not freely invented by us. Our deeper sciences must have quite particular properties so that these entities are generated in the reduction relation. Whatever reality the entities have subsist in those properties and these properties will persist in some form even if the deeper science is replaced by a yet deeper one. But then they cannot claim the same reality as the fundamental ontology. Heat is, after all, a form of energy and not a conserved fluid. Hence I call the compromise a derivative reality.

2.4.5 *Science Versus Folk Science*

A major difference between causation and the cases of caloric and gravitational force lies in the precision of the theory governing the entities and processes called up by reduction relations. In the case of caloric and gravitational forces, we call up quite precise theories, such as Newton's theory of gravitation. In the case of causes, what we call up is a collection of causal notions that do not comprise a theory as precise as Newton's theory of gravitation. That no correspondingly precise theory is possible or, at least, presently available is implicit in the continuing proliferation of different accounts of causation in the literature. Yet there is some system and regularity to causal notions called up by reduction relations; hence I have labeled that system a 'folk theory', or 'folk science'.

There are comparable cases in science in which reduction relations call up powers that are not governed by a well worked out, but now defunct theory. The simplest example pertains to vacua. We know that in classical physics vacua have no active powers, yet we routinely attribute to them the ability to draw things in—to suck. The appearance of this active power arises in a special, but common case: the vacuum is surrounded by a fluid such as air with some positive pressure. The power of the vacuum is really just that of the pressure of the surrounding fluid according to ordinary continuum mechanics. There is no precise account of the active power of the vacuum in the resulting folk theory, which is always employed qualitatively. Any attempt to make it quantitative almost immediately involves replacing the active power of the vacuum by the active power of a pressure differential, whereby we return to the ordinary theory of continuum mechanics. Nonetheless, at a purely qualitative level, it is very convenient to talk of creating the vacuum and to explain resulting processes in terms of a supposed active power of the vacuum.

Causal talk in science has the same status. In many familiar cases, our best sciences tell us that the world behaves as if it were governed by causes obeying some sort of causal principle, with quantitatively measured physical properties such as forces, chemical potentials and temperature gradients being replaced by qualitative causal powers and tendencies. This proves to be a very convenient way to grasp processes that might otherwise be opaque, just as attributing active powers to a vacuum can greatly simplify explanatory stories. No harm is done as long as we take neither the active powers of the vacuum nor the causal principle too seriously.

2.4.6 *The Multiplicity of Folk Notions of Causation*

There is a second major difference. Newton's theory of gravitational forces or the historical theory of caloric are fixed, so that their recovery from a newer science unequivocally succeeds or fails. Matters are far less clear with causation. The little historical survey of notions of causation in Section 2.2 and the present literature in philosophy of causation shows considerable variation in views on what counts as causal. Therefore we do not have one unambiguous notion that must be generated by the reduction relations, but many possibilities. Moreover the notion seems to vary from domain to domain. The sort of causation we recover in physical systems is not quite the same as the sort we recover in biological domains, for example. Finally our notion of causation evolves in response to developments in the science. May causes act at a distance? Is causation anything more than determinism? The answers depend on who you ask and when you ask; and those differences in part result from developments in the relevant science. In crude analogy, seeking causation in nature is akin to seeking images in the clouds. Different people naturally see different images. And different clouds incline us to seek different images. But once an image has been identified, we generally all see it. Moreover, the image is not a pure fiction. It is grounded in the real shape of the cloud; the nose of the face does correspond to a real lobe in the cloud.

2.5 A Folk Notion of Cause

2.5.1 A Sample of the Folk Theories

When restricting a science to hospitable domains generates cause and effect, just what it is that becomes manifest? We identify a pattern that we label as causal and codify its properties in a folk theory. I have just indicated, however, that there is considerable fluidity in the content of the folk theory and that there are many possible theories appropriate to different times, people and domains. Nonetheless, I do not think that the fitting of causal notions is arbitrary. To illustrate the extent to which this fitting is a systematic activity, in this section, I will to try outline one possible folk theory that I think fairly represents one mainstream view of what it is to be causal. In the next section, I will illustrate how it is applied. The folk theory will be based on a relation and seven properties that may be attributed to it. I am fairly confident that most causal theorists would not want to endorse all the properties at once. For this reason the account below really describes a class of folk theories with the different members of the class arising with different choices of properties. Choosing different subsets of properties, in effect, gives a different folk theory, more amenable to different domains and different views of causation.

My goal is to be distinguished from the accounts of causation that are standard fare in the philosophy literature. Their goal is the one, true account of causation that is sufficiently robust to evade the existing repertoire of ingenious counterexamples and the new ones that critics may devise to harass it. My purpose is more modest. I am not trying to enunciate a fundamental principle that must have a definite and unambiguous character. I merely seek to give a compendium of the sorts of things at least some of us look for when we identify a process as causal, without presuming that the compendium is recoverable from a deeper, principled account of the nature of causation. No doubt the account I offer could be elaborated, but I think little would be gained from the elaboration, because of the imprecision inherent in our current notions of causation. That imprecision supports a multiplicity of distinct theories of causation in the literature and I have nothing to add to their efforts at capturing the true essence of causation.

In giving folk theories of causation this fragile character, I am being a little more pessimistic about the solidity of a folk theory of causation than is evident in the recent philosophical literature on folk psychology, where the notion of a folk theory is most commonly encountered (see Ravenscroft 1997). In the spirit of that literature, Menzies (1996) has also sought to characterize causation through what he calls a folk theory of causation. His account is different from mine in that his motivations are not skepticism and his postulates differ from those given below, depending essentially on a probabilistic notion of causation.[10]

2.5.2 *The Basic Notion*

It has long been recognized that human action is the prototype of cause and effect. At its simplest, we identify processes as causal if they are sufficiently analogous.[11] We push over a pile of stones and they fall; our action causes the effect of the toppling. We build a tall tower that is too weak and gravity pulls it down; the action of gravity causes the effect of the fall. Using human action as a prototype, *we identify terms in the cause and effect relation whenever we have one that brings about or produces the other; and we identify the process of production as the causal process.*

A popular explication relates causation to manipulability. When a cause brings about the effect, we can manipulate the effect through the cause but not vice versa. This falls short of a fully satisfactory definition since the notion of manipulation contains residual anthropomorphism and 'produces' is little more than a synonym for 'causes'. However I do not think it is possible to supply a non-circular definition and, in practice, that does not seem to matter, since, as I shall indicate in a moment, we are able to apply the notion without one.

2.5.3 *Applying the Notion*

It is done as follows. We restrict a science to some hospitable domain. We recover certain processes that are still fully described in the vocabulary of the full science; for example, an acid corrodes holes in a metal foil. We

---

[10] For comparison, his folk theory is based on three 'crucial platitudes': '[T]he causal relation is a relation holding between distinct events.' '[T]he causal relation is an intrinsic relation between events.' 'Aside from cases involving pre-emption and overdetermination, one event causes another event just when the two events are distinct and the first event increases the chance of the second event.'

[11] That is, I do not mean to offer an account of the nature of causation in terms of human action. I am merely making the weaker point that, in a rough and ready way, we identify causal process by their analogy to human action. I do not wish to say that anything in this identification is constitutive of causation.

then compare the restricted science to the folk theory of causation and see if we can set up correspondences between terms in the restricted science and in the folk theory. In this case, the acid is the agent that produces or brings about the holes; so we identify the acid as the cause, the holes as the effect and corrosion as the causal process.

How do we know which terms in the science to associate with the cause and effect? There is no general principle. In practice, however, we have little trouble identifying when some process in science has the relevant productive character that warrants the association. Forces cause the effect of acceleration; or heat causes the effect of thermal expansion; or temperature differences cause the motion of heat by conduction; or concentration gradients cause the diffusion of solutes; or electric currents cause the effect of heating of a resistor; or the cause of a particular electron quantum state produces the effect of a raised probability of a particle detection. The terms in the causal relation may be states at a moment of time; or entities; or properties of entities.

2.5.4 The Blobs and Arrows Diagram

The relation of cause and effect is so often represented by a particular diagram that I believe the diagram itself can be an important part of a folk theory of causation. It is a diagram in which the cause C and effect E are represented by blobs and the asymmetric causal relation between them by an arrow (see Fig. 2.2). It is common to represent a complicated set of causal interactions by a correspondingly complicated diagram (see Fig. 2.3).

Figure 2.2. Cause C produces effect E.

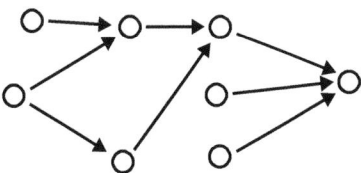

Figure 2.3. A complicated causal nexus.

The particular interpretation of these figures varies by context. In the causal modeling literature, for example, the blobs represent variables that enter into sets of equations (usually linear); the arrows represent the immediate dependencies encoded within the equations. (Spirtes et al. 2000) In other cases, the blobs might represent the presence or absence of some entity or property and whether the relevant term is present at a blob is determined by some Boolean formula (generally specified separately) from the immediately antecedent blobs.

2.5.5 Properties

The blob and arrow diagrams are quite fertile insofar as they suggest properties routinely (though not universally) presumed for causal relations that can be read either directly from the diagram or from simple manipulations of them:

Principle of causality. All states, entities, and properties enter at least as an effect and sometimes also as a cause in causal relations as depicted in Figure 2.2. Each must enter as an effect, or else we would violate the maxim (equivalent to the principle of causality) that *every effect has a cause*. We would have an uncaused state, entity or property. In terms of the blobs and arrows diagrams, this means that there can be no blobs that escape connection with arrows; and that a blobs and arrows diagram is incomplete if it has any blob that is not pointed to by an arrow, that is, one that is not an effect (see Fig. 2.4.) The cause brings about the effect by necessity; this is expressed in the constancy of causation: *the same causes always bring about the same effects*.

Asymmetry. The causal relation is asymmetric as indicated by the arrowhead. Causes bring about effects and not *vice versa*.

Time precedence. The effect cannot precede the cause in time. Insofar as times are associated with the blobs, the arrows point from one blob to another, contemporaneous or later in time.

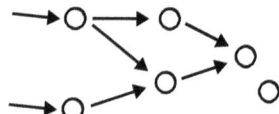

Figure 2.4. A prohibited uncaused event.

CAUSATION AS FOLK SCIENCE   37

Locality. The blobs indicate that at some level of description, causes can be localized. Most commonly they are localized in space and time, but they need not be. For example, in medicine we might identify a particular drug as having some causal effect and portray it as a little blob in a diagram, while the drug is actually spatially distributed throughout the body. The action itself is also presumed local, so that both cause and effect are localized in the same place. If the locality is in space and time, then this requirement prohibits action at a distance; causes here can only produce effects there, if their action is carried by a medium.[12]

Dominant cause. While many entities and properties may enter into the causal process, it is common to identify just one as the dominant cause and the remainder as having a secondary role. This can be represented diagrammatically by 'chunking', the grouping of blobs into bigger blobs or the suppression or absorption of intermediate blobs into the connecting arrow. Chunking as shown in Figure 2.5 allows a complicated causal nexus of the form of Figure 2.3 to be reduced to the simple diagram of Figure 2.2 with a single dominant cause.

First cause. On the model of changes brought about by human action, we expect that every causal process has an initiating first cause. This notion prohibits an infinite causal regress and can be represented by chunking (see Fig. 2.6).

Final cause. In cases in which the end state exercises a controlling influence on the course of a process, the process is governed by a final cause. We are used to explaining away apparent cases of final causation as really produced

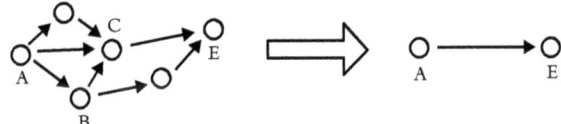

Figure 2.5. Chunking reveals a dominant cause.

Figure 2.6. Chunking reveals a first cause.

---

[12] In his later far less skeptical treatment of causation, Russell (1948, pp. 491–2) makes this requirement a 'postulate of spatio-temporal continuity.'

by efficient (initiating) causes. So the modern tendency is to think of final causes as derivative and efficient causes as fundamental. Since I hold neither to be fundamental, there is no reason to deny final cause a place in this list. As we shall see in the next section, invoking the notion of a final cause can supply the same sorts of heuristic advantages as efficient causes. I do not know a simple way of representing final causes in a blob and arrow diagram.

While all these properties have been invoked often enough to warrant inclusion here, they are by no means universally accepted. For example, asymmetry might well not be accepted by functionalists about causation, that is, those like Russell and Mach who see causation as residing entirely in functional relations on variables. Time precedence would be denied by someone who thinks time travel or backward causation are physically possible—and a growing consensus holds that whether they are possible is a contingent matter to be decided by our science. Locality must be renounced by someone who judges action at a distance theories or quantum theory to be causal. Someone like Mill who essentially equates causation with determinism may not want to single out any particular element in the present determining state as dominant. The demand for a first cause would not be felt by someone who harbors no fear of infinite causal regresses.

Also, because of their antiquarian feel, I have omitted a number of causal principles that can be found in the literature. Some have been conveniently collected by Russell (1917, pp. 138–9): 'Cause and effect must more or less resemble each other.' 'Cause is analogous to volition, since there must be an intelligible *nexus* between cause and effect.' 'A cause cannot operate when it has ceased to exist, because what has ceased to exist is nothing.'

Finally I do not expect that all the properties will be applied in each case. One may well be disinclined to seek first causes in a domain in which final causes are evident; and conversely invoking a first cause may lead us to eschew final causes. In choosing the appropriate subsets of properties, we can generate a variant form of the folk theory specifically adapted to the domain at hand.

2.6 Illustrations

To apply this folk theory to a science, we restrict the science to some suitably hospitable domain. We then associate terms within the restricted science with the central terms of the folk theory; we seek to identify causes

CAUSATION AS FOLK SCIENCE 39

and effects such that they are related by a suitable relation of production. Finally we may seek particular patterns among the causal relations such as those listed on pp. 36–7. The way we make the association; just what counts as a relation of production; and the patterns we may find will depend upon the particular content present. If we are interested in weather systems, for example, we would not ask for dominant causes. Because of its chaotic character, the smallest causes in weather systems may have the largest of effects.

2.6.1 The Dome: A First Cause

As an illustration let us return to the dome considered on p. 23. This will illustrate both how the patterns of the folk theory may be fitted to the science and how successful fitting requires a restricted domain. We shall see that expanding the domain can defeat the fit by embracing an evident failure of causality.

The failure of causality arises specifically at time $t = T$ when the system spontaneously accelerates. Before and after, the system is quite causal insofar as we can map the appropriate causal terms onto the system. To see how this works, let us assume for simplicity that $T = 0$, so that the system spontaneously accelerates at $t = 0$ and consider the sequence of states at $t = 0.5, 0.6, \ldots, 1.0$ in the causal period. Neglecting intermediate times for convenience, we can say that the state at each time is the effect of the state at the earlier time and the force then acting. If we represent the state at time t by the position $r(t)$ and velocity $v(t)$ and the force at t by $F(t)$, we can portray the causal relations in a blobs and arrows diagram. By chunking we can identify the first cause (see Fig. 2.7). If, however, we extend the time period of interest back towards the moment of spontaneous excitation at $t = 0$, we can find an infinite sequence of causes at times, say, $t = 1, 1/2, 1/4, 1/8, \ldots$ for which there is no first cause (see Fig. 2.8).

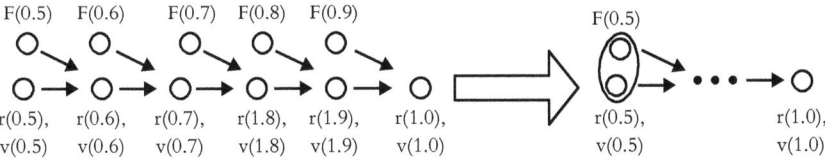

Figure 2.7. A first cause for the causal part of the motion on the dome.

40   JOHN D NORTON

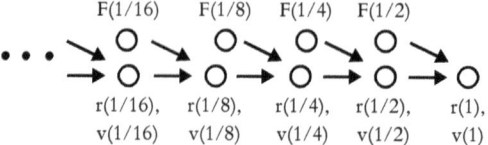

Figure 2.8. An infinite chain of causes with no first cause.

We have already seen the reason for this in Section 2.3. The mass moves during the time interval t > 0 only and there is no first instant of motion in this time interval at which to locate the first cause. (Any candidate first instant t = ε, for ε > 0, is preceded by t = $\varepsilon/2$.) Might we locate the first cause at t = 0, the last instant at which the mass does not move? As we saw in Section 3, nothing in the state at t = 0 is productive of the spontaneous acceleration. One might be tempted to insist nonetheless that there must be something at t = 0 that functions as a cause. The result will be the supposition of a cause whose properties violate the maxim that the same cause always brings about the same effect. For the physical state of the system at t = 0 is identical with the physical states at earlier times t = −1, t = −2, ..., but the state at t = 0 only is (by false supposition) causally effective, where the other identical states are not. The folk theory of causation can only be applied in hospitable domains; this difficulty shows that when we added the instant t = 0, the domain ceases to be hospitable.

Analogous problems arise in the case of big bang cosmology. The universe exists for all cosmic times t > 0 and its state at each time might be represented as the cause of the state at a later time. However, there is no state at t = 0 (loosely, the moment of the big bang) and the demand that there be a first cause for the process must conjure up causes that lie outside the physics.

2.6.2 Dissipative Systems: a Final Cause

There are many processes in physics in which the final state exercises a controlling influence on the course of the process. In thermal physics, processes spontaneously move towards a final state of highest entropy, which is, in microphysical terms, the state of highest probability. Dissipative physical systems are those in which mechanical energy is not conserved; through friction, for example, mechanical energy is lost irreversibly to heat.

CAUSATION AS FOLK SCIENCE 41

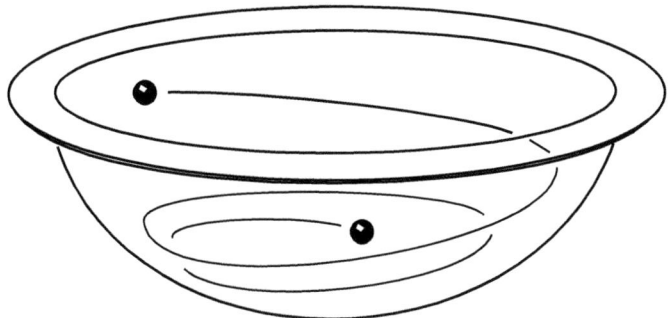

Figure 2.9. A dissipative system.

Such systems can be controlled by their end states in a way that merits the appellation 'final cause.' Consider, for example, a mass sliding *with friction* in a bowl (see Fig. 2.9).

As long as the initial motion of the mass is not so great as to fling the mass out of the bowl, we know what its ultimate fate will be. The mass may slide around inside the bowl in the most complicated trajectory. Throughout the process it will dissipate energy so that its maximum height in the bowl will lessen until the mass finally comes to rest at the lowest point in the bowl. That ultimate state is the final cause of the process. It seems quite natural to assign that final state the status of final cause. At a coarse-grained level of description, the evolution of the process is largely independent of the initial state, but strongly dependent on the final.

One may want to object that it is somehow improper to assign the term 'cause' to this ultimate state; for it exerts no power on the mass in the way, say, that the earth pulls the mass down through a gravitational force. In my view the objection is misplaced. It elevates a gravitational force to the status of a true and fundamental cause, while there are no such things. A gravitational force is a cause in the same way that heat is a material fluid. The generative powers of the reduction relation that confer the character of cause on it can be used equally to confer the character of final cause on the ultimate state of this dissipative process.

2.6.3 Scope

These illustrations, and those earlier in the paper, are drawn largely from the physical sciences, for there it is possible for me to give the most precise account of the extent of viability of causal notions. I intend and hope that the account will be applicable in other sciences, although it is beyond this paper to put my hope to the test. I expect that the character of the application may change. In the physical sciences, an important reason for choosing a restricted domain is to fence off processes that are acausal. A second reason to which I gave less attention is that different domains manifest different sorts of causes. In this section, the first illustration manifests efficient causes; the second, final causes. I expect this latter case to be prevalent in chemistry and non-physical sciences—that the restriction to different domains will divide different types of causation, as opposed to fencing off acausal processes. So chemical potentials might appear as efficient causes in one domain of chemistry and, in another, equilibrium states might appear as final causes. Correspondingly in biology, viruses and bacteria might appear as efficient causes of diseases, while adapted forms in evolutionary biology might appear as final causes.

2.7 Conclusion

The goal of this account of causation in science has been to reconcile two apparently incompatible circumstances. On the one hand, causes play no fundamental role in our mature science. Those sciences are not manifestly about causation and they harbor no universally valid principle of causality. On the other, the actual practice of science is thoroughly permeated with causal talk: science is often glossed as the search for causes; and poor science or superstition is condemned because of its supposed failure to conform to a vaguely specified principle of causality. I have argued that we can have causes in the world of science in the same way as we can retain the caloric. There is no caloric in the world; heat is not a material substance. However, in many circumstances heat behaves just as if it were a material fluid and it can be very useful to think of heat this way. It is the same with cause and effect. At a fundamental level, there are no causes and effects in science and no overarching principle of causality. However, in appropriately restricted domains our science tells us that the world behaves just as if it conformed to some sort of folk theory of causation such as the one already outlined. We should expect these folk theories to be sketchy; we should not expect them to conjure up the exceptionless explication of causation that continues to elude the present philosophical literature. They serve their purpose, however, if they capture what it is we seek then we fit causal conceptions to some process, even if they preserve the vaguenesses of our practice.

References

Alper, J. Bridger, M. Earman, J. Norton, J. (2000). 'What is a Newtonian System? The Failure of Energy Conservation and Determinism in Supertasks', *Synthese*, 124: 281–93.

Born, M. (1935). *Atomic Physics*. Glasgow: Blackie and Sons, 1935; New York: Dover, 1989.

Bunge, M. (1979). *Causality and Modern Science*. New York: Dover.

Campbell, N. R. (1957). *Foundations of Science*. New York: Dover.

Carnot, S. (1824). *Reflections on the Motive Power of Fire*. New York: Dover, 1960.

De Angelis, E. (1973). 'Causation in the Seventeenth Century, Final Causes,' P. P. Wiener (ed.) *Dictionary of the History of Ideas*. New York: Charles Scribner's Sons. Vol. 1: 300–4.

Dowe, P. (1997). 'Causal Processes', *The Stanford Encyclopedia of Philosophy* (Fall 1997 edn), Edward N. Zalta (ed.), http://plato.stanford.edu/archives/fall1997/entries/causation-process/.

Earman, J. (1986). *A Primer on Determinism*. Dordrecht: Reidel.

Hume, D. (1777). *Enquiries Concerning Human Understanding and Concerning the Principles of Morals*. Oxford: Clarendon Press, 1975.

Kant, I. (1787). *Critique of Pure Reason*. N. Kemp Smith, (trans.) London: MacMillan, 1933

Laplace, P. (1825). *Philosophical Essay on Probabilities*. A. I. Dale, (trans.) New York: Springer, 1995.

Lavoisier, A. (1790). *Elements of Chemistry*. New York: Dover, 1993.

Mach, E. (1894). 'On the Principle of Comparison in Physics,' in *Popular Scientific Lectures*. T. J. McCormack (trans.) 5th edn. La Salle, IL: Open Court, 1943: 236–58.

——— (1960). *The Science of Mechanics: A Critical and Historical Account of its Development*. T. J. McCormack (trans.) 6th edn. La Salle, IL: Open Court.

Margenau, H. (1950). *The Nature of Physical Reality: A Philosophy of Modern Physics*. New York: McGraw-Hill.

Menzies, P. (1996). 'Probabilistic Causation and the Preemption Problem,' *Mind*, 105: 85–117.

Mill, J. S. (1872). *A System of Logic: Ratiocinative and Inductive: Being a Connected View of the Principles of Evidence and the Methods of Scientific Investigation.* 8th edn. London: Longman, Green, and Co., 1916.

Nagel, E. (1961). *The Structure of Science.* London: Routledge and Kegan Paul.

Newton, I. (1692/93). 'Four Letters to Richard Bentley,' reprinted in M. K. Munitz, ed., *Theories of the Universe* New York: Free Press, 1957: 211–19.

Pearson, K. (1911). *The Grammar of Science. Part I–Physical.* 3rd edn. London: Macmillan.

Ravenscroft, I. (1997). 'Folk Psychology as a Theory', in *The Stanford Encyclopedia of Philosophy* (Winter 1997 edition), Edward N. Zalta (ed.), <http://plato.stanford.edu/archives/win1997/entries/folkpsych-theory/>.

Russell, B. (1903). *The Principles of Mathematics.* New York: W. W. Norton, 1996.

—— (1917). 'On the Notion of Cause', ch. 9 in *Mysticism and Logic and Other Essays.* London: Unwin, 1917, 1963.

—— (1948). *Human Knowledge: Its Scope and Limits.* New York: Simon and Schuster.

Salmon, W. (1980). 'Probabilistic Causality,' *Pacific Philosophical Quarterly,* 61: 50–74.

Schaffer, J. (2003). 'The Metaphysics of Causation', *The Stanford Encyclopedia of Philosophy* (Spring 2003 edn), Edward N. Zalta (ed.), <http://plato.stanford.edu/archives/spr2003/entries/causation-metaphysics/>

Silberstein, M. (2002). 'Reduction, Emergence and Explanation,' ch. 5 in P. Machamer and M. Silberstein, *The Blackwell Guide to the Philosophy of Science.* Malden, MA: Blackwell.

Spirtes, P., Glymour, C. and Scheines, R. (2000). *Causation, Prediction, and Search.* 2nd edn. Cambridge, MA: MIT Press.

3
What Russell Got Right

CHRISTOPHER HITCHCOCK

3.1 Introduction

Bertrand Russell's 'On the Notion of Cause' is now[1] 90 years old, yet its central claim still provokes. To summarize his argument, I can do no better than to provide an excerpt from his oft-quoted introduction:

> the word 'cause' is so inextricably bound up with misleading associations as to make its complete extrusion from the philosophical vocabulary desirable...
>
> All philosophers, of every school, imagine that causation is one of the fundamental axioms or postulates of science, yet, oddly enough, in advanced sciences such as gravitational astronomy, the word 'cause' never appears. Dr James Ward... makes this a ground of complaint against physics... To me, it seems that... the reason why physics has ceased to look for causes is that, in fact, there are no such things. The law of causality, I believe, like much that passes muster among philosophers, is a relic of a bygone age, surviving, like the monarchy, only because it is erroneously supposed to do no harm. (Russell 1913, p. 1).[2]

The most basic principles of 'advanced sciences' are not couched in the form 'A causes B', but rather in the form of differential equations. Most philosophers have rejected Russell's main conclusion, mostly for good reasons. I wish to argue, however, that Russell's arguments carry important insights that can survive the rejection of his primary conclusion. While

---

For helpful comments I would like to thank audience members at the University of Queensland, the University of Helsinki, and the Causal Republicanism conference in Sydney, especially Peter Menzies, Huw Price, and David Spurrett. I would also like to thank Alan Hájek, Jonathan Schaffer, and Jim Woodward for comments on later drafts of the paper

[1] i.e. at the time of this writing, 2003. By the time you are reading the published version of this paper, it will be several years older than that.

[2] Henceforth, all page references are to this source unless stated otherwise.

there is no need for the word 'cause' to be completely extruded from the philosophical vocabulary, it should be used with much greater care than is usually exercised by philosophers. In developing my case, I will consider two of the most interesting critiques of Russell's argument, due to Nancy Cartwright (1983) and Patrick Suppes (1970). In the final section, I will pull the various threads together to paint a general picture of the nature of causation, and draw some conclusions for the use of causal concepts in philosophical projects.

3.2 Russell's Targets

Even in the short excerpt above, it is possible to identify no fewer than four targets of Russell's attack. In order (starting with the title) they are: (1) the notion of cause; (2) the word 'cause'; (3) the existence of causes; and (4) the 'law of causality'. Correspondingly, Russell seems to be make the following claims:

R1 The notion of cause is incoherent, or fundamentally confused.

R2 The word 'cause' has 'misleading associations', and should be eliminated from philosophical usage. (Presumably this is supposed to apply to cognates such as 'causation' as well.)

R3 There are no causes.

R4 The 'law of causality' is obsolete and misleading.

Russell does not seem to distinguish clearly between these claims, although they are far from being equivalent. R1 does seem to entail R2 through R4, but R2 does not clearly imply *any* of the others. R3 does not imply R1 or R2: there are no 'unicorns', but the notion of a unicorn is not incoherent (although see Hull 1992), nor should 'unicorn' be expunged from the philosophical vocabulary. (Indeed, precisely because there are no unicorns, the word 'unicorn' is useful in formulating philosophical examples like this one.) And R4 appears to be compatible with the denial of any of R1–R3: the notion of cause might be coherent and useful, the word 'cause' might be useful, and causes might exist, even if it is wrong or misleading to say that *every* event is preceded by a cause from which it invariably follows.

Russell's third claim does not seem to reappear beyond the second paragraph, although as we have noted it would seem to follow as a natural consequence of R1. For this reason I will not consider R3 any further. In what follows, I will grant that R4 is basically correct, and argue that R1 is fundamentally mistaken. The most interesting claim is R2, which, I will argue, survives in a much modified form.

3.3 The Law of Causality

Following the time-honored technique used by impatient readers of mystery novels, let us turn to the last page of Russell's article, where Russell summarizes his major conclusions. He writes:

> We may now sum up our discussion of causality. We found first that the law of causality, as usually stated by philosophers, is false, and is not employed in science. We then considered the nature of scientific laws, and found that, instead of stating that one event A is always followed by another event B, they stated functional relations between certain events at certain times... We were unable to find any *a priori* category involved: the existence of scientific laws appeared as a purely empirical fact... Finally, we considered the problem of free will... The problem of free will *versus* determinism is... mainly illusory.... (26)

I will return briefly to Russell's discussion of free will (p. 49). It is clear that at the end of the day, Russell took his critique of the law of causality to be fundamental. What is that law? Russell cites two different formulations (6):

> The Law of Causation, the recognition of which is the main pillar of inductive science, is but the familiar truth, that invariability of succession is found by observation to obtain between every fact in nature and some other fact which has preceded it.[3] (Mill 1843, book 3, ch. v, §2.)

> this law means that every phenomenon is determined by its conditions, or, in other words, that the same causes always produce the same effect. (Bergson 1910, p. 199.)[4]

---

[3] Given the sort of regularity view of causation endorsed by Mill, this statement implies that every event has a cause.

[4] It is clear from context that Bergson does not endorse this principle, but is only presenting what he takes to be the normal understanding of it. Alan Hájek points out that Bergson's 'other words', repeated by Russell, do not express the same principle as that stated in the first clause. The principle 'same cause, same effect' would be trivially true if there were no causes; not so the principle that every phenomenon is determined by its conditions (at least if we count 'determining conditions' as causes).

Russell's central argument is succinct and powerful:

> The principle 'same cause, same effect'...is...utterly otiose. As soon as the antecedents have been given sufficiently fully to enable the consequent to be calculated with some exactitude, the antecedents have become so complicated that it is very unlikely they will ever recur. (8–9)

By the time you pack enough material into A to guarantee that B will follow, A is so complex that we can't reasonably expect it to occur more than once. (Indeed, it was Mill himself who taught us just how detailed any such A would have to be.) But *any* A that occurs only once will invariably be succeeded by any B that just happens to occur on that occasion. Although rough generalizations may play a role in the early stages in the development of a science, science is not constructed upon invariable regularities. Note that Russell's critique also serves as an effective critique of regularity accounts of causation, such as that put forward by Mill. This may go some way toward explaining why Russell did not clearly distinguish between claim R4 and the others.

At various points Russell takes pains to attack not merely the truth or utility of the law of causality, but also its *a priority*, as well as the *a priority* of any suitable replacement. This clearly shows that one of Russell's targets is the Kantian view that something like the law of causality is a necessary condition for us to gain knowledge of the empirical world. (See e.g. Kant 1781, Second Analogy, A 189–203, B 234–49; 1787, §§14–38.)

Note finally that Russell is not rejecting determinism per se. In fact he clearly embraces Laplacean determinism (pp. 13–14). So if anything, the situation is even worse for the law of causality than Russell claims, since there is strong evidence for indeterminism among quantum phenomena, and it is now known that even Russell's beloved classical gravitational astronomy does not entail determinism (Earman 1986, Norton, ch. 2 of this volume). The rejection of the law of causality is definitely something that Russell got right.

3.4 Misleading Associations

Claim R2 has two clauses, the first an observation about philosophy, the second a methodological recommendation. The first clause is phrased somewhat awkwardly in Russell's introduction: '...the word "cause" is...inextricably bound up with misleading associations...'. It might seem natural to paraphrase this by saying 'the word "cause" has caused a great deal of confusion', although there are obvious reasons why this phrasing is undesirable. Russell's discussion of free will seems aimed in part to establish this first clause: it illustrates one area in which causal talk has led to confusion. Ironically, Russell's own paper also serves to illustrate the point. Russell's failure to distinguish between claims R1, R3, and R4 is no doubt due, in part, to ambiguities in words like 'cause' and 'causality'. The former can be a noun or a verb, the latter can describe a relation between events, or a principle governing such relationships. Indeed, David Lewis (1986a, p. 175) makes a point of avoiding the word 'causality' (he calls it a 'naughty word') precisely because it has such a double connotation.

The second clause of R2, that the extrusion of the word 'cause' (and its cognates) from the philosophical vocabulary is desirable, is clearly supposed to follow from the first clause; it also seems to follow from R1. We will discuss this methodological dictum in greater detail later.

3.5 On the Notion of Cause

R1 is not stated explicitly, although the title certainly suggests that the coherence of the notion of cause is to be scrutinized, and much of the argumentation on pages 2 through to 8 of Russell's paper is naturally construed as support for R1.

Russell challenged the coherence of the notion of cause in part by criticising existing attempts to give a definition of this concept. In particular, he considers the following purported definition, one of three drawn from Baldwin's *Dictionary*:

Cause and effect...are correlative terms denoting any two distinguishable things, phases, or aspects of reality, which are so related that whenever the first ceases to exist the second comes into existence immediately after, and whenever the second comes into existence the first has ceased to exist immediately before. (Cited in Russell 1913, p. 2)

Russell then raises a Zenoesque paradox for this definition: If cause and effect are distinct and temporally contiguous, then they cannot both be point-like; at least one of them must be temporally extended. Assume, for the sake of argument, that the cause has some finite duration, say from

t_0 to t_1. (There are tricky topological issues about whether the temporal endpoints are included within the duration of the cause. Since the cause and effect are contiguous, one, but not both, must be occurring at t_1. But let us put this issue aside.) Since the effect does not begin until t_1, the first half of the cause (with duration from t_0 to $t_{0.5}$), is not really efficacious: the pattern of invariable association can be disrupted by contingencies occurring in the interval between $t_{0.5}$ and t_1. We can repeat this process *ad infinitum*, showing that at most one instantaneous time-slice of the cause can be efficacious, contradicting our assumption that the cause is extended. I will not examine this particular argument in detail, since the many difficulties with a simple regularity analysis of causation are widely known. Moreover, 90 years later we still lack a widely accepted analysis of causation.

In the twenty-first century, however, we have become sufficiently cynical to reject arguments of the form:

There is no adequate philosophical account of X, despite a rich history of attempts to provide one. Therefore, the concept X is incoherent or fundamentally confused.

This argument embraces an optimism about the power of philosophical analysis that is no longer warranted. With hindsight, then, we should reject Russell's argument for R1 on the basis of his critique of Baldwin's definition.

There is, however, an interesting general point that emerges from Russell's critique. The problems for Baldwin's definition resulted in part from its combination of two distinct elements: cause and effect are supposed to stand in a relation of temporal contiguity, and also of invariable association. Many of the problems that face more recent attempts arise for the same general reason. Our notion of cause seems to involve (at least) two different dimensions: cause and effect *covary* in some way; and they stand in certain kinds of spatiotemporal relations to one another. Regularity, probabilistic and counterfactual theories (such as Mackie 1974, ch. 3, Cartwright 1983, and Lewis 1986b, respectively) attempt to capture the covariance aspect of causation, while process theories (e.g. Salmon 1984) attempt to capture the broadly spatiotemporal aspects. There have been some interesting attempts to combine these (Menzies 1989, Dowe 2000, Schaffer 2001), but none has received widespread acceptance. Thus Russell raises the possibility that our notion of cause may be incoherent because it attempts to combine incompatible elements. But as we noted before, this does not follow simply from the fact that no attempt to combine these elements has met with perfect success, especially in light of the partial success that has been achieved, and the insights that have been gained by such attempts.

So far, however, we have not really touched on Russell's central argument for RI (and by extension, for the others as well). This argument involves as a central premise the following claim:

P. The word 'cause' does not appear in the advanced sciences.

This premise seems to play a dual role. First, there is something like an inference to the best explanation from P to RI: RI, if true, would explain why advanced sciences do not explicitly employ causal concepts. (This explanation would presumably appeal to the tacit premise that advanced sciences do not make use of notions that are incoherent.) Second, P seems to undermine a primary motivation for thinking that there *must be* a coherent notion of cause. Any attack on RI, then, will have to address the premise P, and its dual role in supporting RI. The critiques of Cartwright (1983) and Suppes (1970) do just that. Cartwright attempts to provide us with an alternative rationale for having a notion of causation, one that does not appeal to the supposed role of causal concepts in science. Suppes directly attacks premise P itself. More subtly, I think Suppes' critique serves to undermine the abductive inference from P to RI: to the extent that P is true, it can be explained without appeal to the incoherence of causal notions. I now turn to a more detailed discussion of these two critiques.

3.6 Cartwright's Criticism

Cartwright reports having received the following letter:

It simply wouldn't be true to say, 'Nancy L. D. Cartwright... if you own a TIAA life insurance policy, you'll live longer.'
But it is a fact, nonetheless, that persons insured by TIAA do enjoy longer lifetimes, on the average, than persons insured by commercial insurance companies that serve the general public. (Cartwright 1983, p. 22)

TIAA—the Teacher's Insurance and Annuity Association—offers life insurance for university professors and other individuals involved in education. Owning such a policy is *correlated* with longevity because only educators qualify to purchase such a policy. Educators tend to live longer than average for a number of reasons: they are comparatively well-compensated and receive good health benefits, they tend to be well-informed about health care issues, their careers are not especially dangerous, and so on. Nonetheless, purchasing a TIAA life insurance policy is not an *effective strategy* for achieving longevity.

By contrast, quitting smoking, reducing consumption of saturated fat, and exercising are (to the best of our current knowledge) effective strategies for achieving longevity. Cartwright then claims that

> ...causal laws cannot be done away with, for they are needed to ground the distinction between effective strategies and ineffective ones. (Cartwright 1983, p. 22).

It's not clear to me that *laws* are needed, insofar as laws are normally thought to be exceptionless, fully general, and so on. Perhaps we can get by with regularities, or patterns of functional dependence. But apart from that, I believe that Cartwright is fundamentally correct. There is a crucial distinction between the *causal* laws, regularities, and such like, and the *non-causal* ones. The former can, while the latter cannot, be exploited for the purposes of achieving desired outcomes. Ultimately, this distinction is what causation is all about.

This provides us with one explanation for the absence of overtly causal notions in gravitational astronomy: there is no real prospect for manipulating the positions or masses of the heavenly bodies. By contrast, the fields in which there is the greatest interest in causation today tend to be ones with a practical orientation, such as econometrics, epidemiology, agronomy, and education. But I also think that there is a deeper explanation for the absence of causal talk in gravitational astronomy.

Cartwright's pragmatic orientation provides us with a helpful way of thinking about causation. As our paradigmatic causal system, consider a collection of billiard balls bouncing around on a low friction billiard table. This is not literally a closed system, but for stretches of time it remains largely unaffected by influences other than those intrinsic to the system itself. We could write down 'laws' describing the motions of the billiard balls during these periods of isolation, although these will only be *ceteris paribus* generalizations that break down under certain conditions. These generalizations can be used to predict the motion of the balls so long as the relevant *ceteris paribus* conditions continue to hold.

Now suppose that I intervene to change the motion of a particular ball. From the point of view of the 'laws' describing the isolated system, this intervention is 'miraculous', it is something that is not even countenanced within those generalizations. Now some of the original generalizations will be violated, while others remain valid (at least as applied to the motions of certain of the balls). The latter, but not the former, can be used to predict the effects of the intervention, and hence to support effective strategies. These are the genuinely causal generalizations. Looked at in a different way, the genuinely causal counterfactuals (what Lewis (1979) calls the 'non-backtracking counterfactuals') are the ones whose antecedents are (or rather, would be) made true by external interventions. This accords with Lewis' idea that the antecedents of non-backtracking counterfactuals are (typically) made true by 'miracles', except that in our case the antecedents are only miraculous relative to the 'laws' of the isolated system. From an external perspective, there is nothing miraculous about intervening on an otherwise isolated system. This picture is elaborated in much greater detail in Woodward and Hitchcock (2003), Hitchcock and Woodward (2003), and Woodward (2003). It also informs important recent work in causal modeling, such as Pearl (2000) and Spirtes, Glymour and Scheines (2000).

This picture runs into deep problems, however, when we try to expand the laws of the system so as to include the source of the intervention within the system itself. In particular, it's going to be extraordinarily difficult to find causation within a theory that purports to be universal—about *everything*. Russell's example of gravitational astronomy has just this character: the laws of classical gravitational astronomy were taken to apply to the universe as a whole (and presumably the 'final theory', if there is one, will also be universal in this sense). When the system under study is the universe as a whole, there is no external vantage point from which we can interfere with the system, and from which we can sensibly talk about miracles within the system. How to find or even understand causation from within the framework of a *universal* theory is one of the very deep problems of philosophy. In that regard, this problem is in good company. Nietzsche wrote that 'As the circle of science grows larger, it touches paradox at more places' (*The Birth of Tragedy*, cited in Putnam 1990). This theme has been taken up in a paper by Hilary Putnam (1990). There are many places in science (broadly construed) where our theories work perfectly well so long as there is an external perspective which is not covered within the theory itself, but when the theory 'grows larger'—when it tries to encompass the external perspective as well—it becomes embroiled in paradox. Putnam discusses two central examples. The first is the measurement problem in quantum mechanics. Even in its classical (non-relativistic) form, that theory makes extraordinarily accurate predictions about the outcomes of various measurements that can be performed on a system. Once we include the measuring apparatus within the system, however, the normal rules of quantum mechanics (Schrödinger's equation and Born's rule) yield contradictory predictions for the behavior of the apparatus. The problem is sufficiently serious that otherwise sensible physicists have put forward outlandish metaphysical hypotheses (such as the existence of parallel universes, and a fundamental role for consciousness in physics) in hopes of resolving it.

Putnam's second example is the liar paradox. We can talk of sentences being true or false, and even construct powerful semantic systems, so long as the sentences in the target language do not contain the words 'true' and 'false'. These terms appear in the meta-language, the one used within the external perspective used to analyze the target language. But once we introduce the terms 'true' and 'false' into the target language, we are faced with the liar sentence: 'This sentence is false.' This sentence can be neither true nor false, on pain of contradiction, and our semantic theory 'touches paradox'.

The problem of free will and determinism also has something of this character. We are perfectly happy to apply our (hypothetical) deterministic theories to determine the consequences of our decisions, once made. Indeed, as compatibilists such as Hume (1748, §8) and Hobart (1934) have pointed out, freedom would be *undermined* if there were no systematic connection between our choices and their physical outcomes. Problems arise, however, when we try to bring the decision-making process itself within the scope of the deterministic theory. Some libertarians have postulated that *agents* can be causes that nonetheless lie outside of the causal order, thus guaranteeing the existence of an external perspective that cannot be brought under the scope of a deterministic theory, a constraint upon the diameter of the circle of science. This is yet another example of the extremes to which philosophers have been driven by the desire to avoid paradox.

3.7 Suppes' Criticism

The second main line of response that I want to pursue comes from Patrick Suppes' book *A Probabilistic Theory of Causality* (1970). In the introduction, Suppes launches an attack on Russell's key premise P. After citing the passage from Russell with which we began, Suppes writes:

> Contrary to the days when Russell wrote this essay, the words 'causality' and 'cause' are commonly and widely used by physicists in their most recent work. There is scarcely an issue of *Physical Review* that does not contain at least one article using either 'cause' or 'causality' in its title. A typical sort of title is that of a recent volume edited by the distinguished physicist, E. P. Wigner, 'Dispersion relations and their connection with causality' (1964). Another good example is the recent article by E. C. Zeeman (1964) 'Causality implies the Lorentz group.' The first point I want to establish, then, is that discussions of causality are now very much a part of contemporary physics. (Suppes 1970, pp. 5–6)

Since Suppes wrote this almost 40 years ago, I conducted a quick and unsystematic internet search of the *Physical Review* journals (a series of 9) from 2000 to 2003 and found 76 articles with 'cause', 'causes', 'causality', or some similar term in the title. Here are the first three examples listed:

> 'Tree Networks with Causal Structure' (Bialas et al. 2003)
> 'Specific-Heat Anomaly Caused by Ferroelectric Nanoregions in Pb(Mg[sub 1/3]Nb[sub 2/3])O[sub 3] and Pb(Mg[sub 1/3]Ta[sub 2/3])O[sub 3] Relaxors' (Moriya et al. 2003)
> "Observables" in causal set cosmology' (Brightwell et al. 2003)

So Suppes' observation remains true today.

Taking the above passage in isolation, it may be natural to construe Suppes as claiming that physics has advanced since Russell's time, and that we have now learned that causation *is*, after all, an essential notion in philosophical theorizing. But Suppes' point is really quite different. Continuing from the previous quote, Suppes writes:

> The reasons for this are, I think, very close to the reasons why notions of causality continue to be an important ingredient of ordinary talk... At the end of the passage quoted from Russell, there is an emphasis on replacing talk about causes by talk about functional relationships, or more exactly, by talk about differential equations. This remark is very much in the spirit of classical physics, when the physical phenomena in question were felt to be much better understood at a fundamental level than they are today. One has the feeling that in contemporary

physics the situation is very similar to that of ordinary experience, namely, it is not possible to apply simple fundamental laws to derive exact relationships such as those expressed in differential equations. What we are able to get a grip on is a variety of heterogeneous, partial relationships. In the rough and ready sense of ordinary experience, these partial relationships often express causal relations, and it is only natural to talk about causes in very much the way that we do in everyday conversation.   (Suppes 1970, 6)

Suppes' point, then, is not so much that physics has advanced, but that it has, in a certain sense, *regressed*. Physics, like other branches of science, seeks not only to provide a systematic treatment of the phenomena presently within its scope, but also to uncover new phenomena. At any given time, then, there are bound to be many phenomena that are deemed to be within the scope of a field such as physics, yet still await systematic treatment within that field. Russell was writing at a highly unusual time in the history of physics, or at least of gravitational astronomy, when existing physical theory was thought to have successfully encompassed (almost) all of the phenomena within its scope. This period was short-lived: Einstein turned gravitational astronomy on its head less than three years later.[5] In the meantime, physics has uncovered a slew of phenomena that have not yet succumbed to comprehensive mathematical analysis. Often, it is natural to describe these phenomena in causal terms.

Russell's premise P, then, is strictly false. Suppes' attack is, nonetheless consistent with a more modest formulation along the following lines:

P* There are advanced stages in the study of certain phenomena when it becomes appropriate to eliminate causal talk in favor of mathematical relationships (or other more precise characterizations).

We can explain the truth of P* without appealing to the incoherence of causal notions. To say that one thing *causes* another is to state something very important about the relationship, as we have learned from Cartwright. Nonetheless, it is to characterize the relationship only in a qualitative way. When it becomes possible to state the relationship in more precise terms, we do so. But it does not follow that the relationship has ceased to be causal. To the extent that Russell's premise P is true, it does not support his conclusion R1.

[5] There is no evidence that Russell was aware of developments that were already underway within (a very small part of) the German physics community. Indeed, by 1912–1913, general relativity was not even on the radar screen of the British physics community (Andrew Warwick, personal communication).

3.8 Conclusions

So let us take stock. Russell was right that science in general is not grounded in some 'law of causality'. Moreover, he was right that the word 'cause' does not appear in sciences 'like' classical gravitational astronomy. There are, however, very few fields like this; classical gravitational astronomy is unusual in a number of respects: it is remote from practical applications, it purports to be a universal theory, and its principles can be captured in simple mathematical formulas. But it does not follow from any of this that our notion of causation is incoherent. Rather, following Cartwright (1983), we should recognize that the notion of causation is indispensable, not because it is fundamental to science, but because the distinction between causal relationships and non-causal relationships grounds the distinction between effective and ineffective strategies. This distinction is very real, and one that we are all aware of (at least implicitly).

Cartwright's argument gives rise to an interesting grammatical point. 'Cause' can be a verb, as when we say 'lightning storms cause fires'. 'Cause' can also be a noun, as in 'the lightning storm was the cause of the fire'. In addition, 'causation' is a noun, referring either to an abstract concept, or to a specific relation that holds between certain events (or facts, or properties, or what have you). 'Causality' is also a noun, often used as a synonym for 'causation', but also having a more archaic meaning akin to 'determinism', and associated with the 'law of causality' attacked by Russell. Cartwright's argument, however, focuses our attention on a less-discussed cognate: the adjective 'causal'. This adjective applies to relationships, such as laws, regularities, correlations, or patterns of functional dependence. It is this adjective that picks out the most central causal notion. The noun 'causation' is best thought of as referring not to a specific relation, but rather to a property that is possessed by some relationships and not by others.

This analysis meshes nicely with Suppes' critique. Central to Suppes' argument is the idea that claims of the form 'A causes B' can be *precisified*. For example, it may be possible to characterize functional relationship between A and B in mathematical terms. The different possible precisifications correspond to different causal relationships that might hold between A and B.

I shall Follow Chalmers (1996), who distinguishes between the 'easy' and 'hard' problems of consciousness, and divide the problems of causation into the 'easy' ones and the 'hard' ones. The hard problems concern the origins of the distinction between the causal and the non-causal. These include finding the sources of causal asymmetry and causal modality (i.e. what makes causal relations non-accidental), and locating causation within a global scientific theory of the sort that physics aims to provide. We can of course expect to make some progress on these hard problems, just as we have made progress on the liar paradox and the measurement problem. Moreover, we may expect to learn a great deal about causation in the process of tackling these problems. But we should not expect these hard problems to go away any time soon.

The 'easy' problems, by contrast, concern distinctions among the different kinds of causal relationships. Like Chalmers' 'easy' problems of consciousness, these problems are actually pretty hard, but they should at least be tractable within existing philosophical methodology. Unfortunately, I think much of the energy here has been misdirected, by arguing over just which causal relation *is* causation per se. This is particularly prominent in the literature on counterfactual theories of causation, where there is a cottage industry of trying to define *causation* in terms of counterfactuals that are already assumed to be, in a broad sense, caus*al*.[6] Within this literature, authors have addressed such questions as: is *causation* transitive? Now, I cannot disparage this work in any general way, since I have engaged in precisely this debate myself. (The answer is 'no', by the way; see Hitchcock 2001a, 2001b). But there is a need to appreciate that the target of analysis here is not causation per se, but rather a specific kind of causal relation called 'token' causation (I find this terminology misleading, and prefer 'actual' causation, but I will not attempt to argue with entrenched usage). Many philosophers, especially metaphysicians, seem to think that token causation just *is* causation. This belief is belied by the fact that there is an entire field devoted to causal representation and causal inference that barely mentions token causation. For example, token causation is never discussed in Spirtes, Glymour and Scheines (2000), and appears only in the final chapter of Pearl (2000). Presumably, these authors are nonetheless talking about *causal* inference and *causal* modeling. Token causation is in fact a rather narrowly circumscribed causal notion that is involved when we make retrospective assessments of responsibility. (See Hithcock MS for

---

[6] More specifically, the accounts are formulated in terms of 'non-backtracking' counterfactuals, without broaching the controversial issue of whether these can be analyzed in non-causal terms.

further discussion). It applies, for example, when attributing moral or legal responsibility, or when analyzing system failure in engineering.

The real work that I think needs to be done is that of providing useful taxonomies for causal relationships. Suppes' critique of Russell suggests one way in which we might classify causal relationships: by functional form. When we are dealing with precise quantitative variables, it may be possible to capture a causal relationship between variables in a mathematical equation. But even when a causal relation does not lend itself to precise mathematical characterization, we can still ask questions about the qualitative structure of the relationship: Do increases in one variable lead to increases or decreases in another? Is this relationship monotonic? Do the details of the effect depend upon the details of the cause in a fine-grained way, or is it more of an on-off affair? Do the various causal variables act independently, or do they interact with one another?

In addition to questions about the functional form of a causal relationship, we can also ask questions about the paths of causal influence: Does one factor cause another directly (relative to some level of graining), or via some third factor? Does the cause work by direct physical contact, or by removing an obstacle impeding the effect? Does one factor influence another in more than one way?

Let me say a little more about this last question, since it is a topic that I have explored in some detail (Hitchcock 2001b). Consider Hesslow's well-known example involving birth control pills. Thrombosis is a worrisome potential side effect of birth control pills. And of course, birth control pills are very effective at preventing pregnancy. Now it turns out that pregnancy itself is a significant risk factor for thrombosis. (Since birth control pills essentially mimic the hormonal effects of pregnancy, it is not surprising that the two have similar side effects.) These facts are represented in Figure 3.1, where the thickness of the line represents the strength of the influence, and the solidity of the line represents the direction of the influence. For many women—such as young, fertile, sexually active, non-smokers—birth control pills *lower* the overall chance of thrombosis. How should we describe this situation? Do birth control pills *cause* thrombosis, or *prevent* it? I describe this situation by saying that there are two different routes whereby the consumption of birth control pills affects thrombosis: one in which birth control pills act 'directly' (relative to the coarse level of grain used here), and one in which pills affect thrombosis by affecting pregnancy. Birth

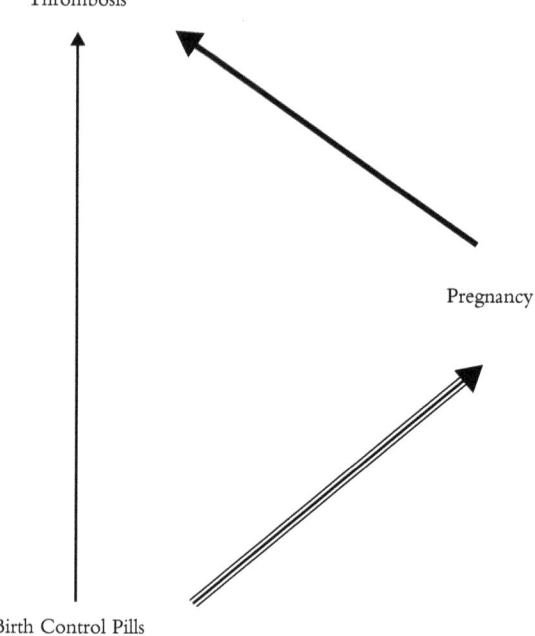

Figure 3.1

control pills have a *component effect* along each of these routes: along the direct route, the effect is positive, that is, the promoting of thrombosis; along the second route, the effect is preventative. The consumption of oral contraceptives also has a *net* effect on thrombosis, which is preventative. The terminology is intended to evoke an analogy with net and component forces in Newtonian Mechanics, although in the case of effects, there is nothing resembling the elegant additive rule for combining forces.

Net and component effects will be relevant in different circumstances. If a woman wants to know whether taking birth control pills is an effective way to prevent thrombosis, then we need to look at the net effect: if she wants to avoid thrombosis, she should take the pills.[7] On the other hand, if we want to know whether the manufacturer of the pills can be held liable for a particular woman's thrombosis, we look at the direct component effect, rather than the net effect.

[7] Of course, you might think that she could gain the benefits without paying the costs by using an alternative form of birth control. In fact, however, even the relatively small difference in efficacy between birth control pills and other forms of contraception may be enough to produce a negative net effect.

This discussion illustrates just one dimension along which causal relationships can be classified; there are many others. And despite the flaws of various extant theories of causation, these theories often serve well enough to make the kinds of distinction that are needed. For example, within a probabilistic framework (such as that of e.g. Cartwright 1983), it is relatively easy to account for the difference between net and component effects. In probabilistic theories of causation, one evaluates the causal relevance of C for E by comparing the probabilities of E, conditional on C and *not-C*, *while also conditioning on other relevant factors*. The difficulty, of course, lies in saying just what these other factors are; without providing an adequate answer to this question, we will not have an adequate theory of causation. Nonetheless, even this basic framework is helpful. Suppose that the set of background factors conditioned on does *not* include pregnancy; then, we would expect that conditioning on pill consumption would *lower* the probability of thrombosis. This captures the net effect of pregnancy on thrombosis. But now suppose that we do condition on pregnancy (or its absence). Then, we would expect that conditioning on oral contraception would increase the probability of thrombosis; that is, among women who do in fact become pregnant, those who take birth controls are more likely to be stricken by thrombosis, and likewise among women who do not in fact become pregnant. This probability comparison captures the component effect of pregnancy of thrombosis along the direct route. In effect, by conditioning on pregnancy, we eliminate any effect of birth control pills on thrombosis along the indirect route, leaving only the effect along the direct route.[8] Thus the probabilistic notion of conditionalization, while it may be inadequate for purposes of analyzing causation, can nonetheless be used to shed light on the distinction between net and component effects.

Let us turn, finally, to Russell's second claim:

the word 'cause' is so inextricably bound up with misleading associations as to make its complete extrusion from the philosophical vocabulary desirable

I agree with Russell that the use of the word 'cause' can be misleading. This is not because the notion of cause is fundamentally misguided, but rather because there are many types of causal relationship. When we say

---

[8] Capturing the component effect along the indirect route is trickier. It requires either graining more finely, so that there will be intermediate factors along the route that is direct in the coarse-grained picture; or using a special kind of chaining procedure. By far the best technical account of route-specific causal dependence is Pearl (2001).

that taking birth control pills causes thrombosis, we are saying something that is true in one sense, and false in another. Absent clarification, such usage has the potential to cause confusion. What is needed is not so much the elimination of the word 'cause', as the introduction of more precise terms for characterizing causal concepts.

There is in all of this some good news. Many philosophers are, of course, interested in the notion of cause for its own sake. But philosophers are also interested in causation because it seems to be an ingredient in many other philosophically interesting concepts: freedom, moral responsibility, perception, rational choice, explanation, and perhaps also knowledge and reference. There is an old-fashioned view, due in no small part to Russell's critique, that goes something like this:

> We cannot observe causal relations directly, therefore, unless causation can be analyzed in terms of empirically respectable notions, causation remains a part of spooky metaphysics. Until causal theorists are able to provide such an analysis, we should avoid the use of causal notions in our philosophical projects.

This view holds many philosophical projects hostage until causal theorists achieve the impossible. The alternate picture I am presenting offers a much more sanguine vision of how causal theorists may contribute to other philosophical projects. There are of course many hard philosophical problems that may well be held hostage to the hard problems of causation: the problem of free will and determinism within a physical world may well be of that sort. But there are many philosophical projects that stand to benefit from progress on the easy problems of causation.

Consider causal theories of reference, for example. The central idea here is (very roughly) that part of what it means for our word 'tiger' to refer to tigers, is for our current usage of that word to be caused by some appropriate baptismal event. How should I react to this proposal? My reaction (the right reaction, I hasten to add) is *not* the following:

> How can it help to explain reference in terms of something as mysterious as causation? Until someone can explain to me what causation is in terms of something more respectable, like empirical regularities or even the laws of physics, I just don't understand it.

Rather my reaction is along these lines:

> What, exactly is the causal relationship that's supposed to hold between the baptismal event, and my current usage? Will any old causal relationship do?

For example, suppose the relationship between a baptismal event and our current usage is as follows:

The warriors of a particularly bellicose tribe were on their way to massacre the members of a new upstart tribe, but were prevented from doing so when they were waylaid by tigers. They had never seen tigers before, and by sheer coincidence, named them 'lions' then and there. This tribe was soon wiped out completely by voracious tigers. As it happens, the upstart tribe that was saved by the tigers was the first to speak proto-indo-european. Thus if the baptismal event, the initial encounter with the tigers, had not occurred, then we would be speaking a very different language, and would not be using the word 'lion' today.

Is this the right sort of connection between a baptismal event and our current usage? It certainly seems to be causal. If this had really happened, would it follow that our word 'lion' refers to tigers? If not why not?

More generally, for just about any philosophical concept that has a causal dimension, not just any old causal relationship will be relevant. Analytic work thus stands to benefit a great deal from a better taxonomy of the ways in which things can be causally related to each other. It is in attacking these 'easy' taxonomical problems that causal theorists have the potential to provide the greatest benefit for philosophy at large.

References

Bergson, H. (1910). *Time and Free Will: An Essay on the Immediate Data of Consciousness* (F. L. Pogson trans.). London: George Allen and Unwin.

Bialas, P. Z., Burda, J., Jurkiewicz, Krzywicki, A. (2003). 'Tree Networks with Causal Structure'. *Physical Review, E* 67: 066106.

Brightwell, G., Dowker, F., García, R. S., Henson, J. and Sorkin R. D. (2003). '"Observables" in Causal Set Cosmology'. *Physical Review, D* 67: 084031.

Cartwright, N. (1983). 'Causal Laws and Effective Strategies' in *How the Laws of Physics Lie*. Oxford: Oxford University Press, 21–43.

Chalmers, D. (1996). *The Conscious Mind*. Oxford: Oxford University Press.

Dowe, P. (2000). *Physical Causation*. Cambridge: Cambridge University Press.

Earman, J. (1986). *A Primer on Determinism*. Dordrecht: Reidel.

Hitchcock, C. (2001a). 'The Intransitivity of Causation Revealed in Equations and Graphs'. *Journal of Philosophy*, 98: 273–99.

____ (2001b). 'A Tale of Two Effects'. *The Philosophical Review*, 110: 361–96.

____ (MS) 'Token Causal Structure'.

Hitchcock, C. and Woodward, J. (2003). 'Explanatory Generalizations, Part II: Plumbing Explanatory Depth'. *Noûs*, 37: 181–99.

Hobart, R. E. (1934). 'Free Will as Involving Determinism and Inconceivable without It'. *Mind*, 43: 1–27.

Hull, D. (1992). 'That Just Don't Sound Right: A Plea for Real Examples', in J. Earman and J. D. Norton (eds) *The Cosmos of Science*. Pittsburgh: University of Pittsburgh Press, 428–57.

Hume, D. (1748). *An Enquiry Concerning Human Understanding*.

Kant, I. (1781). *A Critique of Pure Reason*.

—— (1787). *A Prolegomenon To Any Future Metaphysics*.

Lewis, D. (1979). 'Counterfactual Dependence and Time's Arrow'. *Noûs*, 13: 455–76. Reprinted in Lewis (1986b), 32–52.

—— (1986a). 'Postscripts to "Causation",' in Lewis (1986b), 172–213.

—— (1986b). *Philosophical Papers, Volume II*. Oxford: Oxford University Press.

Mackie, J. L. (1974). *The Cement of the Universe*. Oxford: Clarendon Press.

Menzies, P. (1989). 'Probabilistic Causation and Causal Processes: A Critique of Lewis'. *Philosophy of Science*, 56: 642–63.

Mill, J. S. (1843). *A System of Logic*. London: Parker and Son.

Moriya, Y., Kawaji, H., Tojo, T. and Atake, T. (2003). 'Specific-Heat Anomaly Caused by Ferroelectric Nanoregions in $Pb(Mg_{1/3}Nb_{2/3})O_3$ and $Pb(Mg_{1/3}Ta_{2/3})O_3$ Relaxors'. *Physical Review Letters*, 90: 205901.

Norton, J. (this volume). 'Causation as Folk Science'.

Pearl, J. (2000). *Causality: Models, Reasoning, and Inference*. Cambridge: Cambridge University Press.

—— (2001). 'Direct and Indirect Effects', in *Proceedings of the Seventeenth Conference on Uncertainty in Artificial Intelligence*. San Mateo CA: Morgan Kauffman, 411–20.

Putnam, H. (1990). 'Realism with a Human Face' in J. Conant (ed.) *Realism with a Human Face*. Cambridge: Mass: Harvard University Press, 3–29.

Russell, B. (1913). 'On the Notion of Cause'. *Proceedings of the Aristotelian Society*, 13: 1–26.

Salmon, W. (1984). *Scientific Explanation and the Causal Structure of the World*. Princeton: Princeton University Press.

Schaffer, J. (2001). 'Causes as Probability-Raisers of Processes'. *The Journal of Philosophy*, 98: 75–92.

Spirtes, P., Glymour, C. and Scheines, R. (2000). *Causation, Prediction, and Search*, Second Edition. Cambridge, MA: MIT Press.

Suppes, P. (1970). *A Probabilistic Theory of Causality*. Amsterdam: North-Holland.
Woodward, J. (2003). *Making Things Happen: A Theory of Causal Explanation*. Oxford: Oxford University Press.
Woodward, J. and Hitchcock, C. (2003). 'Explanatory Generalizations, Part I: A Counterfactual Account'. *Noûs*, 37: 1–24.

4

Causation with a Human Face

JIM WOODWARD

The recent literature on causation presents us with a striking puzzle. On the one hand, (1) there has been an explosion of seemingly fruitful work in philosophy, statistics, computer science, and psychology on causal inference, causal learning, causal judgment, and related topics.[1] More than ever before, 'causation' is a topic that is being systematically explored in many different disciplines. This reflects the apparent usefulness of causal thinking in many of the special sciences and in common sense. On the other hand, (2) many[2] philosophers of physics, from Russell onwards, have claimed that causal notions are absent from, or at least play no foundational role in, fundamental physics, and that at least some aspects of ordinary causal thinking (e.g. the asymmetry of the cause–effect relation) lack any sort of grounding in fundamental physical laws. If we also think that (3) if causal notions are appropriate and legitimate in common sense and the special sciences, then these notions must somehow reflect or derive from features of causal thinking (or true causal claims) that can be found in fundamental physics, then (1) and (2) appear to be (at the very least) in considerable tension with one another.

I will not try to systematically evaluate all of the many different claims made by philosophers about the role of causation in physics—the topic strikes me as very complex and unsettled and is anyway beyond my

---

Thanks to Chris Hitchcock, Francis Longworth, John Norton, Elliott Sober, and an anonymous referee for very helpful comments on an earlier draft and to Clark Glymour for extremely helpful correspondence in connection with Section 4.5.

[1] See, for example, Spirtes, Glymour, and Scheines (2000), Pearl (2000), and Gopnik and Schultz (forthcoming).

[2] Field (2003) and Norton (this volume), are among those expressing skepticism of one sort or another about the role of causal notions in physics.

competence. While I find the unqualified claim that causal notions are entirely absent from fundamental physics unconvincing (for reasons described on pp. 68–9), I am also inclined to think that there is something right in the claim that there are important differences between, on the one hand, the way in which causal notions figure in common sense and the special sciences and the empirical assumptions that underlie their application and, on the other hand, the ways in which these notions figure in physics. The causal notions and assumptions that figure in common sense and the special sciences do not always transfer smoothly or in an unproblematic way to all of the contexts in which fundamental physical theories are applied and common sense causal claims often do not have simple, straightforward physical counterparts. My aim in this paper is to explore some of the features of the systems studied by the upper-level sciences and the epistemic problems that they present to us that make the application of certain causal notions and patterns of reasoning seem particularly natural and appropriate. I will suggest that these features are absent from some of the systems studied in fundamental physics and that when this is so, this explains why causal notions and patterns of reasoning seem less appropriate when applied to such systems. There is thus a (partial) mismatch or failure of fit between, on the one hand, the way we think about and apply causal notions in the upper-level sciences and common sense and, on the other, the content of fundamental physical theories. Russell was right about the existence of this mismatch even if (as I believe) he was wrong in other respects about the role of causation in physics.

My general stance is pragmatic: the legitimacy of causal notions in the upper level sciences is not undermined by the disappearance (or non-applicability) of some aspects of these notions in fundamental physics. Instead, causal notions are legitimate in any context in which we can explain why they are useful, what work they are doing, and how their application is controlled by evidence. I thus reject claim (3) above: even if the claims of philosophers of physics about the unimportance or disutility of causal thinking in physics are correct, it still would be true that causal thinking is highly useful—indeed indispensable in other contexts.

My discussion is organized as follows. Section 4.1 comments briefly on the role of causation in physics. Section 4.2 discusses the role of causal reasoning in common sense and the special sciences. Section 4.3 sketches an interventionist account of causation that I believe fits the upper-level sciences and common sense causal reasoning better than competing approaches. Sections 4.4–4.8 then describe some of the distinctive features of the systems investigated in such sciences and way in which we reason about them. These include the fact that the causal generalizations we are able to formulate regarding the behavior of such systems describe relationships that are invariant only under a limited range of changes (Section 4.4), that such systems are described in a coarse-grained way (Section 4.5), that the systems themselves are located in a larger environment which serves as a potential source of 'exogenous' interventions (Section 4.6), that they have certain other features (including the possibility of 'arrow-breaking') that make the notion of an intervention applicable to them in a natural and straightforward way (Section 4.7), and that the epistemic problem of distinguishing causes and correlations, which arises very frequently in the special sciences, seems much less salient and pressing in fundamental physics (Section 4.8). Section 4.9 then explores the issue of the reality of macro-causal relationships in the light of the previous sections.

4.1 Causation in Physics

As I have said, my primary focus in this essay will be on the assumptions that guide the application of causal reasoning in common sense and the special sciences and make for its utility and how these differ from some of the assumptions that characterize the underlying physics. However, to guard against misunderstanding, let me say unequivocally that it is *not* part of my argument that causal notions play no role in or are entirely absent from fundamental physics. I see no reason to deny, for example, that forces cause accelerations. Indeed, as Smith (forthcoming) observes, there are numerous cases in which physics tells us how a local disturbance or intervention will propagate across space and time to affect the values of other variables, thus providing information that is 'causal' in both the interventionist sense described below and also in the sense captured by causal process theories such as Salmon (1984) and Dowe (2000).

It also seems uncontroversial that as a matter of descriptive fact various causally motivated conditions and constraints play important roles in physical reasoning and the application of physical theories to concrete situations. Thus advanced solutions of Maxwell's equations are commonly discarded on the grounds that they are 'non-causal' or 'causally anomalous' (Jackson 1999, Frisch 2000), candidates for boundary conditions involving non-zero fields at infinity or accelerations that are not due to forces may be rejected on the grounds that they are unphysical or acausal (Jackson 1999), various 'locality' conditions may be motivated by causal considerations and so on. However, it is also unclear exactly what this shows. Suppose (what is itself a disputed matter)[3] that in every case such causal constraints could be replaced by a more mathematically precise statement that does not use the word 'cause'. Would this demonstrate that causal notions play no fundamental role in physics or should we instead think of the replacement as still causal in content and/or motivation but simply more clear? Should we think of the constraints as in every case holding as a matter of fundamental law, which would allow us to say that whatever causal content a physical theory has is fixed by its fundamental laws, or should we instead retain the usual distinction between laws and boundary conditions, and (as seems to me more plausible and natural) hold that some of the causal content of the theory is built into assumptions about initial and boundary conditions, and is not carried by the laws alone? I will not try to resolve these questions here.

While causal claims and considerations are not absent from physics, certain commonly held philosophical assumptions about the role of such notions in physics and their connection to 'upper level' causal claims seem much more dubious. For example, in contrast to the assumption that all fundamental physical laws are causal (e.g. Armstrong 1997), many do not have a particularly causal flavor if only because of their highly abstract and schematic quality. Thus applications of the Schrödinger equation to particular sorts of systems in which a particular Hamiltonian is specified and specific assumptions about initial and boundary conditions adopted seem more 'causal' than the bare Schrödinger equation itself and similarly for many other examples.[4] When a fundamental physical theory is applied globally, to the entire universe, it is arguable that some of the presuppositions for the application of the notion of an intervention are not satisfied (cf. Section 4.6). Hence it may be unclear how to interpret what is going on in causal terms when 'causal' is understood along interventionist lines. A

---

[3] See Frisch (2002).
[4] See Smith (forthcoming) for more extended discussion.

similar point holds for some quantum mechanical contexts.[5] If, alternatively (or in addition) we think of causal claims as having to do with unfolding of causal processes in time from some local point of origin, as a number of philosophers (e.g. Salmon 1984, Dowe 2000) do, then few if any fundamental laws are causal in the sense of directly describing such processes[6] (Smith forthcoming). In part for these reasons and in part for other reasons that will be discussed on p. 73, the widely accepted idea that all true causal claims in common sense and the special sciences 'instantiate' fundamental physical laws which are causal in character, with causal status of the former being 'grounded' in these fundamental laws alone is deeply problematic.[7] In short, while there is no convincing argument for the conclusion that causation 'plays no role' in fundamental physics, there is also good reason to be skeptical of a sort of causal foundationalism or fundamentalism according to which fundamental physical laws supply a causal foundation for all of the causal claims occurring in the special sciences, and according to which, every application of a fundamental physical theory must be interpretable in terms of a notion of 'cause' possessing all of the features of the notion that figures in common sense and the special sciences.

4.2 Causation in the Special Sciences

Whatever may be said about the role of causation in physics, it seems uncontroversial that causal claims play a central role in many areas of human life and inquiry. Causal notions are of course ubiquitous in common sense reasoning and 'ordinary' discourse. Numerous psychological studies detail

---

[5] For example, it is arguable, (cf. Hausman and Woodward 1999) that there is no well-defined notion of an intervention on the spin state of one of the separated particles pairs with respect to the other in EPR type experiments. This would represent a limitation on the application of an interventionist account of causation only if there was reason to suppose that there is a direct causal connection between these states. Hausman and Woodward argue that there is no such reason; hence that it is a virtue, rather than a limitation in the interventionist account that, in contrast to other accounts of causation, it does not commit us to such a connection.

[6] Nor, contrary to 'conserved quantity' accounts of causation of the sort championed by Salmon and Dowe, will it always be possible to characterize 'causal processes' in fundamental physical contexts in terms of the transference of energy and momentum in accordance with a conservation law. Typically, the spacetimes characterized in General Relativity lack the symmetries that permit the formulation of global (that is, integral versions of) conservation laws. In such cases, no spatially/temporally extended process (no matter, how intuitively 'causal') will possess well-defined conserved energy/momentum. See Rueger (1998) for more detailed discussion.

[7] Davidson (1967) is a classic source for this idea.

the early emergence of causal judgment and inference among small children and the central role that causal claims play in planning and categorization among adults.[8] There is also considerable evidence that our greatly enlarged capacities for causal learning and understanding and the enhanced capacity for manipulation of the physical world these make possible are among the most important factors separating humans from other primates (Tomasello and Call 1997). However, for a variety of reasons, it seems to me misleading to think of causation as merely a 'folk' concept that is absent from 'mature' science. For one thing, many disciplines that are commonly regarded as 'scientific', including such 'upper-level' or so-called 'special science' disciplines as the social and behavioral sciences, medicine and biology, as well as many varieties of engineering traffic extensively in explicit causal claims. In disciplines such as economics, one finds explicit discussion of various concepts of causation (e.g. 'Granger' causation, in the sense of Granger 1998, and the contrasting manipulationist conception associated with writers like Haavelmo 1944, and Strotz and Wold 1960) and various tests for causation. In portions of statistics, in literature on experimental design, and in econometrics, one finds a great deal of self-conscious thinking about the difference between causal and merely correlational claims and about the evidence that is relevant to each sort of claim. Similarly, in brain imaging experiments it is common to find neurobiologists worrying that such experiments provide (at best) correlational rather than causal knowledge. (The latter requiring information about what would happen under experimental interventions, such as trans-cranial magnetic stimulation or from 'natural experiments' such as lesions.) Although there are of course exceptions, many of the most perceptive practitioners of these disciplines do not seem to doubt the utility of thinking about their subjects in causal terms.

4.3 An Interventionist Account of Causation

When issues arise about the coherence or legitimacy of some notion, it is often a useful heuristic to ask what the notion is intended to contrast with—what is it meant to exclude or rule out, what difference are we trying to mark when we use it? I believe that when 'cause' and cognate notions are used in the special sciences and in common sense contexts, the

---

[8] For discussion of these studies within an interventionist framework, see Woodward (forthcoming b)

relevant contrast is very often with 'mere' correlations or associations. In particular, the underlying problematic is something like this: an investigator has observed some relationship of correlation or association among two or more variables, X and Y. What the investigator wants to know is whether this relationship is of such a character that it might be exploited for purposes of manipulation and control: if the investigator were to manipulate X (in the right way), would the correlation between X and Y continue to hold, so that the manipulations of X are associated with corresponding changes in Y? Or is it instead the case that under manipulation of X there would be no corresponding changes in Y, so that the result of manipulating X is to disrupt the previously existing correlation between X and Y, with manipulation of X being ineffective as way of changing Y? In the former case, we think of X as causing Y, in the latter the connection between X and Y is non-causal, a mere correlation that arises in some other way—for example, because of the influence of some third variable Z or because the sample in which the correlation exists is in some way unrepresentative of the population of interest. Many epistemic problems in the special sciences fit this pattern such as the following.

> (1) An investigator observes that students who attend private schools tend to score better on various measures of academic achievement than public school students. Is this because (a) attendance at private schools causes students to perform better? Or, alternatively, (b) is this a mere correlation arising because, for example, parents who send their children to private school care more about their children's academic achievement and this causes their children to have superior academic performance, independently of what sort of school the child goes to? For a parent or educational reformer the difference between (a) and (b) may be crucial—under (a) it may make sense to send the child to private school as a way of boosting performance, under (b) this would be pointless.[9]
>
> (2) It is observed that patients who receive a drug or a surgical procedure are more likely to recover from a certain disease in comparison with those who do not receive the drug or procedure. Is this because the drug or procedure causes recovery or is it rather because, for

---

[9] For discussion, see Coleman and Hoffer (1987).

example, the drug or procedure has been administered preferentially to those who were more likely to recover, even in their absence?

(3) There is undoubtedly a systematic correlation between increases in the money supply and increases in the general price level. Is this because money causes prices, as monetarist economists claim, or is it instead the case that the causation runs from prices to money or that the correlation is due to the operation of some third variable? The answer to this question obviously has important implications for monetary policy (cf. Hoover 1988).

In each of the above examples, investigators are working with a notion of causation that is closely associated with the contrast between *effective* strategies and *ineffective* strategies in the sense of Cartwright (1979): the causal connections are those that ground effective strategies.[10] Even if we are firmly convinced that causal notions play no legitimate role in fundamental physics, it is hard to believe that there is no difference between drugs that cure cancer and those that are merely correlated with recovery or that some procedures, such as randomized trials, are not better than others, such as consulting the entrails of sheep, for assessing claims about the causal efficacy of such drugs. It is this general line of thought that motivates interventionist or manipulationist accounts of causation. For the purposes of this paper, I will describe a relatively generic version of such a theory—details are provided in Woodward (2003) and broadly similar views are described in Spirtes, Glymour, and Scheines (2000), and in Pearl (2000).

The basic idea is that causal claims are understood as claims about what would happen to the value of one variable under interventions on (idealized experimental manipulations of) one or more other variables. A simple statement of such a theory that appears to capture a type level notion of causal relevance, called total causation in Woodward (2003), is:

(C) X is a total cause of Y if and only if under an intervention that changes the value of X (with no other intervention occurring) there is an associated change in the value of Y.

---

[10] My claim at this point is simply that *in the examples under discussion*, the contrast between those relationships that are causal and those that are merely correlational coincides with the contrast between those that will or will not support manipulations. It is of course a further and more difficult question whether it is plausible and illuminating project to explicate the notion of causation in general along 'interventionist' lines, even when there is no practical possibility of manipulation. See Woodward (2003, ch. 3) for additional discussion.

Providing an appropriate characterization of the notion of an intervention is a matter of some delicacy. To see what is at issue, consider the following familiar causal structure, in which atmospheric pressure A is a common cause of B, the reading of a barometer, and S, a variable representing the occurrence/non-occurrence of a storm, with no causal link from B to S or vice-versa.

Figure 4.1

Clearly, if I manipulate B by changing the value of A, then the value of S will also change, even though, *ex hypothesi*, B is not a cause of S. If we want the relationship between behavior under manipulation and causation embodied in (C) to hold, we need to characterize the notion of an intervention in such a way as to exclude this sort of possibility. Intuitively, the idea that we want to capture is this: an intervention I on X with respect to a second variable Y causes a change in the value of X that is of such a character that *if* any change occurs in the value of Y, it occurs only as a result of the change in the value of X caused by I and not in any other way. In the case of the *ABS* system such an intervention might be carried out by, for example, employing a randomizing device whose operation is independent of A and, depending just on the output of this device, setting the position of the barometer dial to high or low, in a way that does depend on the value of A. If, under this operation, an association between B and S persists, we may conclude that B causes S; if the association disappears B does not cause S.

One way of making the notion of an intervention more precise, suggested by the above example, is to proceed negatively, by formulating conditions on interventions on X that exclude all of the other ways (in addition to X's causing Y) in which changes in Y might be associated with a change in X. The following formulation, which is taken from Woodward (2003), attempts to do this.

Let X and Y be variables, with the different values of X and Y representing different and incompatible properties possessed by the unit u, the intent being to determine whether some intervention on X produces changes in Y. Then I is an intervention variable on X with respect to Y if, and only if, I meets the following conditions:

(IV)
- I1. I causes X.
- I2. I acts as a switch for all the other variables that cause X. That is, certain values of I are such that when I attains those values, X ceases to depend upon the values of other variables that cause X and instead only depends on the value taken by I.
- I3. Any directed path from I to Y goes through X. That is I does not directly cause Y and is not a cause of any causes of Y that are distinct from X except, of course, for those causes of Y, if any, that are built into the I–X–Y connection itself; that is, except for (a) any causes of Y that are effects of X (i.e. variables that are causally between X and Y) and (b) any causes of Y that are between I and X and have no effect on Y independently of X.
- I4. I is independent of any variable Z that causes Y and is on a directed path from I to Y that does not go through X.

I2 captures the idea that an intervention on X should place the value of X entirely under the control of the intervention variable I so that the causal links between X and any other cause Z of X are severed. This helps to ensure that any remaining association between X and Y will not be due to Z. For example, by making the value of B entirely dependent on the output of the randomizing device in the example above, we ensure that any remaining association between B and S will not be due to A. This idea lies behind the 'arrow breaking' conception of interventions described by Spirtes, Glymour and Scheines (2000), and Pearl (2000). I3 and I4 eliminate various other ways in addition to X's causing Y in which X and Y might be correlated under an intervention on X.

One may think of the characterization IV as attempting to strip away, insofar as this is possible, the anthropocentric elements in the notion of an experimental manipulation—thus there is no explicit reference in IV to human beings and to what they can or cannot do and the characterization is given entirely in causal and correlational language.[11] In this respect,

---

[11] Because IV characterizes interventions in causal terms (the intervention causes a change in X, must bear a certain relationship to other causes of Y etc.), it follows that any account of causation like

the resulting theory is different from the agency theory of causation developed by Menzies and Price (1993, see also Price 1991) which makes explicit reference to human agency and to the experience of free action in characterizing what it is for a relationship to be causal. This is not to deny, however, that other facts about the sorts of creatures we are and the way in which we are located in the world (e.g. the fact that we ourselves are macroscopic systems with a particular interest in the behavior of other macroscopic systems with certain features) have played an important role in shaping our concept of causation. I shall return to this topic later.

4.4 Limited Invariance and Incompleteness

The connection between intervention and causation described by (C) is very weak: for X to cause Y all that is required is that there be some (single) intervention on X which is associated with a change in Y. Typically, we want to know much more than this: we want to know *which* interventions on X will change Y, and *how* they will change Y, and under what background circumstances. On the account that I favor this sort of information is naturally expressed by means of two notions: invariance and stability. Consider a candidate generalization, such as Hooke's law (H) $F = -kX$, where $X =$ the extension of the spring, $F =$ the restoring force it exerts, and k is a constant that is characteristic of a particular sort S of spring. One sort of intervention that might be performed in connection with (H) will involve setting the independent variable in (H)—the extension of a spring of sort S—to some value in a way that meets the conditions (IV). Suppose that it is true that for some range of such interventions the restoring force will indeed conform to (H). Then (H) is *invariant* under this range of interventions. More generally, I will say that a generalization is invariant (*simpliciter*) if and only if it is invariant under at least some interventions. Such invariance is, I claim, both necessary and sufficient for

---

C will not be reductionist, in the sense that it translates causal into non-causal claims. (My inclination, for what it is worth, is to think that no such reduction of the causal to the non-causal is possible). At the same time, however, I would argue that vicious circularity is avoided, since an interventionist elucidation (along the lines of C) of X causes Y does not presuppose information about whether there is a causal relationship between X and Y but rather information about *other* causal relationships or their absence—e.g. information about the existence of a causal relationship between I and X. For more on the issue of reduction, see Woodward (2003).

a generalization to describe a relationship that is exploitable for purposes of manipulation and control and hence for the relationship to qualify as causal. Typically, though, causal generalizations will be such that they will continue to hold not just under some interventions but also under at least some changes in *background conditions*, where background conditions have to do with variables which do not explicitly figure in the generalization in question. When a generalization has this feature I shall say that it is *stable* under the background conditions in question.[12] For example, taken as a characterization of some particular spring, (H) is likely to be stable under some range of changes in the temperature of the spring, the humidity of the surrounding air, and changes in the spatio-temporal location of the spring. I will suggest later, on pp. 78–80, that it is a feature of the generalizations that we call laws that they are stable under a 'large' or particularly 'important' range of changes in background conditions.

When thus characterized, both invariance and stability are clearly *relative* notions: a generalization can be invariant under some range of interventions and not under others and similarly for stability under background conditions. This will be the case for (H)—if we intervene to stretch the spring too much or if we change the background circumstances too much (e.g. if we heat the spring to a high temperature), the restoring force will no longer be linear. It would be a mistake, however, to take this to show that (H) does not describe a genuine causal relationship, in the sense of causal that an interventionist theory tries to capture. (H) is causal in the sense that it does not express a mere correlation between X and F but rather tells us how F would change under some range of interventions on X. It thus has the feature that we are taking to be central to causation—it describes a relationship that can be used for manipulation and control. However, the range of invariance of (H) (the conditions under which it correctly describes how F would change under interventions on X) as well as its range of stability are, intuitively, rather limited, at least in comparison with generalizations that describe fundamental laws of nature. Some philosophers contend that it is a mark of a genuinely fundamental law that it holds under all physically possible circumstances. Even if

---

[12] In Woodward (2003), I did not distinguish between invariance and stability in this way, but rather used 'invariance' to cover both notions. Since it is invariance under interventions rather than stability that is crucial for causal status, it seems clearer to distinguish the two notions. For additional discussion of this notion of stability, see Woodward (forthcoming a).

we weaken this requirement to accommodate the fact that, as currently formulated, many generalizations that are commonly described as laws (like Schrödinger's equation and the field equations of General Relativity) do, or may, break down under certain conditions, it remains true that these generalizations hold, to a close approximation, under a large or extensive range of conditions and, furthermore, that the conditions under which they do break down can be given a relatively simple and perspicuous theoretical characterization. By contrast, the conditions under which (H) will break down are both extensive and also sufficiently disparate that they resist any simple summary.

This limited invariance and stability of (H) goes hand in hand with its *incompleteness*: the restoring force exerted by a spring is contingent on many additional conditions besides those specified in (H)—conditions having to do both with the internal structure of the spring and with its environment. If (but only if) we have a spring for which these conditions happen in fact to be satisfied, (H) will correctly describe how its restoring force will change in response to interventions on its extension. In this sense, (H) describes a causal relationship that is merely *locally* invariant and stable or a relationship of contingent or *conditional* dependency. In these respects, (H) is paradigmatic of the sorts of causal relationships that we typically are able to establish and operate with in the special sciences and in common sense.[13]

There are a variety of different ways in which such merely locally invariant/stable relationships can arise. One of the simplest possibilities, illustrated by the case of the spring, is that the value of one variable Y depends on the value of a second variable X according to some relationship $Y = G(X)$ when some third variable B assumes some value or range of values $B = [b_1, \ldots, b_n]$ but this dependence disappears or changes radically in form when B assumes other values outside this range. If, as a contingent matter of fact, (4.1) B usually or always takes values within the range of stability for $Y = G(X)$—at any event, around here and right now or for cases which are of particular interest for us—*and* if (4.2) as long as B is within this range, intervening to change the value of X will not change the value of B to some value outside this range or otherwise disrupt the relationship $Y = G(X)$, then this relationship will be at least locally invariant.

---

[13] See Woodward (2003) for a more extended defense of this claim.

Note that requirement (4.2) is crucial for invariance. Both Hooke's law (H) and the relationship between the barometer reading B and the occurrence of the storm S will break down under the right conditions. What distinguishes these relationships is (4.2): all interventions that change the value of B will disrupt the $B-S$ association but the corresponding claim is not true for (H). The distinction between the generalization describing the $B-S$ relationship, which is not invariant under *any* interventions (although it is stable under some changes in background conditions), and (H), which is invariant under *some* but not all interventions and thus has a limited but non-empty range of invariance, is a distinction *within* the class of generalizations that fail to hold universally, under all circumstances. This makes the notion of invariance particularly suited for distinguishing between causal and merely correlational relationships in the special sciences, since we cannot appeal to notions like universality and exceptionlessness to make this distinction.

I have been comparing generalizations like (H), which are invariant under only a rather limited range of interventions and stable under a limited range of background conditions, with other generalizations, like laws of nature, with a 'more extensive' range of invariance/stability. Can we be more precise about the basis for such comparisons? In the case of macroscopic causal generalizations we are particularly interested in invariance and stability under changes that are not too infrequent or unlikely to occur, around here, right now, and less interested in what would happen under changes that are extremely unlikely or which seem 'farfetched'. Consider the generalization,

(S) Releasing a 5kg rock held at a height of 2 meters directly above an ordinary champagne glass with no interposed barrier will cause the glass to shatter.

(S) is certainly invariant under some range of interventions which consist of rock releasings. It is also stable under many possible changes in background conditions of a sort that commonly occur around here—changes in wind conditions, humidity, ordinary variations in the structure of the glass and so on. Of course there are other physically possible interventions/background conditions under which (S) is not invariant/stable. For example, it is physically possible that the rock might be deflected by a meteor on its downward path or vaporized by a blast of high energy radiation from space in such a way that the glass is left intact. If we wished to add conditions to (S) that would ensure that its antecedent was genuinely nomologically sufficient for the shattering of the glass, we would need to exclude these and many other possibilities. However, occurrences of the sort just described are extremely rare at least at present in our vicinity. We may formulate a relatively invariant and stable generalization (albeit one that falls short of providing a nomologically sufficient condition) if we ignore them. This is the strategy that is typically followed in the special sciences. From the point of view of agents who are interested in manipulation and control, this strategy makes a great deal of sense: an agent who wishes to shatter the glass can 'almost guarantee' this outcome by dropping the rock. Similarly for a central bank that wishes to lower inflation by restricting growth in the money supply, assuming that the relevant generalization is invariant and stable under changes that are likely to occur.

Whether occurrences like meteor strikes are likely to occur around here is, of course, a contingent matter, dependent not just on the fundamental laws governing nature but on initial and boundary conditions that might have been otherwise. This illustrates one respect in which causal generalizations in the special sciences often rely on, or presuppose, various sorts of contingent facts that are not guaranteed by fundamental laws alone and why we should not expect that the 'truth-maker' for (or underlying physical explanation of) the claim that a generalization like (S) is relatively invariant and stable to be located just in facts about underlying physical laws. Note, though, that although contingent, it is an 'objective' matter (*not* a matter of idiosyncratic individual taste and opinion) whether the sorts of disrupters of (S) just described are likely to occur around here.[14]

4.5 Coarse-Graining

There is another, related feature of the causal generalizations of the upper level sciences that contributes to their distinctive character: the variables in upper level causal theories are extremely *coarse grained* from the point of view of fundamental physics. This encompasses several different ideas. First, typically a number of different microstates, distinguishable from the

---

[14] Whether all grounds for the assessment of relative invariance are similarly objective is a difficult question that I will not try to resolve here. For further discussion, see Woodward (2003) on 'serious possibility' and 'farfetchedness'.

point of view of fundamental physics, will realize the same value of the macro-variables of upper level causal theories. The relationship between thermodynamics and statistical mechanics (from which I take the notion of coarse-graining) is paradigmatic: if we think of a microstate as a complete specification of the position and momentum of each of the component molecules making up a sample of gas, then a very large number of such micro-states will realize a single value for the pressure and temperature of the gas. Similarly, a large number of different molecular states will constitute different ways of realizing the state or event which we describe as the shattering of a glass or the attainment of a certain number of years of schooling.

Second, while initial and boundary conditions in fundamental physics are often described by expressions specifying the exact values of variables at *each* space time point within some region, as when charge densities and field strengths are specified as a function of position in classical electromagnetism, causal generalizations in the special sciences often relate variables such that a single value of these characterizes an entire macroscopically spatially extended and perhaps temporally extended region, with boundaries that are not very precise from the point of view of fundamental physics. Typically, these regions and the objects associated with them, will be (from a macroscopic perspective) relatively connected and cohesive, or at least not too diffuse, discontinuous and gerry-mandered. As illustrations, consider generalizations relating the impact of a rock on a window and its subsequent shattering and position in a primate dominance hierarchy to level of serotonin expression. (high position in the hierarchy causes high levels of expression and vice-versa). The variables involved in these generalizations {*shattering, non-shattering*}, {*window breaks, window does not break*}, {*position in the dominance hierarchy is such and such*} and so on, take their values across extended spatio-temporal regions with imprecise boundaries. Thus the shattering occurs in the region of the window but beyond a certain level of discrimination, there may be no definite answer to exactly where this event is located or when it begins and ends. In addition, coarse grained variables may fail to completely partition the full possibility space as seen from the point of view of an underlying fine-grained theory.[15] Both episodes of shattering and of non-shattering are likely to be

---

[15] Thanks to Chris Hitchcock for emphasizing this point.

identified by means of their similarity to prototypical or paradigm cases. From a macroscopic perspective the possibility of an outcome intermediate between shattering and not shattering may not be recognized, even if this corresponds to some possible, albeit very unlikely microstate—e.g., the rock grazes the glass in such a way to crack it extensively and displace small portions of it while leaving most of the pieces in (almost) their previous position, so that the result is neither a prototypical shattering or non-shattering. To the extent that all this is so, it may also be unclear exactly which microstates are to be identified with or are realizers of the shattering.[16] However, to the extent that we are likely to be interested only in the contrast between the window's shattering and its not shattering at all, this indeterminacy will not matter much.

Finally, in common sense and the upper-level sciences, causal relata are often described as operating across spatiotemporal gaps or, alternatively, in a way that is non-specific about the spatiotemporal relationship between cause and effect. Recovery from a disease will typically occur some significant lapse of time after the administration of the drug that causes recovery. A slowdown in economic activity may be caused by the decision of the central bank to raise interest rates but it seems doubtful that there is any clear sense in which the latter event is spatiotemporally contiguous with the former. It is true that in many, but by no means all,[17] cases involving macro-causality, there will exist (from a more fine-grained perspective) a spatio-temporally continuous process linking the cause to its effect. However, even when such processes do exist, upper level causal generalizations often do not specify them and the correctness and utility of the upper level generalizations do not rest on our actually having information about such processes. This feature is captured nicely by interventionist accounts which take the distinctive feature of causal relationships to be exploitability for purposes of manipulation, regardless of whether there is a spatiotemporal gap between cause and effect.

---

[16] This is one reason, among many, why it is unsatisfactory to say, as Davidson does, that when rock impacts cause glass shatterings, both of these macroscopic events 'instantiate' an underlying law relating micrsostates.

[17] 'By no means all' because in some cases the underlying processes will involve causation by omission or by disconnection in the sense of Schaffer (2000) or some complex combination of these and spatio-temporally continuous processes. Presumably most cases of causation by omission and disconnection do not involve the direct instantiation of fundamental causal laws.

How do these features of causal generalizations in the upper level sciences compare with the laws of fundamental physics? In contrast to the incomplete relationships of limited invariance between coarse-grained factors that are characteristic of the upper level sciences, fundamental laws typically take the form of differential equations, deterministically relating quantities and their space and time derivatives at single spatiotemporal locations. In these equations, as Hartry Field has recently observed (2003), there are no spatial or temporal gaps of a sort that would allow for the possibility of outside influences intervening between the instantiation of the independent and dependent variables. In contrast to the imprecise spatiotemporal boundaries of the causal factors that are of interest in the upper-level sciences, the initial and boundary conditions required for solution of the equations of fundamental physics are described by specifying the exact values of the relevant variables at every spacetime point within some extended region. While the causal generalizations of the upper-level sciences are invariant and stable only under some limited range of changes, the ideal within fundamental physics is to find generalizations that are invariant and stable under all possible changes or, failing that, generalizations that break down only under well-specified extreme conditions and whose range of invariance and stability is thus far less limited than the generalizations of the special sciences. In other words, the ideal is to find generalizations that are complete (or nearly so) in the sense of incorporating independent variables describing, as nearly as possible, *all* quantities such that *any* variation in their values would lead to different values for the dependent variable in that generalization and which, taken together, are nomologically sufficient for (that is, physically guarantee, whatever else may happen) the value of the dependent variable. Again, this contrasts with the causal generalizations of the special sciences which are radically incomplete and fall well short of specifying such nomologically sufficient conditions.

As a number of writers have argued, to specify a set of conditions S that are genuinely nomologically sufficient for some event of interest E, we need (at least) a description of a cross section of the entire backward light cone for E—a description that specifies the values of the relevant variables at every point within this cross section. Anything less than this will leave open the possibility that the conditions S are satisfied and yet some influence compatible with S occurs which would exclude the occurrence of E. (A burst of high energy radiation from outer space that vaporizes the rock just before it strikes the glass etc.) If one mark of a fundamental law is that it describes such a nomologically sufficient condition, then it seems highly unlikely that there will be fundamental laws relating the localized, coarse-grained events in which the upper level sciences traffic.

In fact, if E is given a sufficiently precise microphysical description and the causes of E are also given a similar description and are taken to include every variable, some variations in the value of which would lead to the non-occurrence of E, then an even more disturbing consequence appears to follow—everything in the backward light cone of E will qualify as a cause of E. Suppose that E is the event of my headache disappearing at t and that in the absolute past of this event are (A) my ingestion of aspirin 30 minutes prior to t, and at about the same time, my next-door neighbor's sneezing (S) and my wishing (W) my headache would go away. Consider very fine-grained specifications ((E^*), (A^*), (S^*), and (W^*)) of the exact position and momentum of the fundamental particles realizing (E), (A), (S), and (W). The occurrence of A will alter the gravitational and electromagnetic forces incident on E (and hence E^*) but this will also be true of (S) and (W)—indeed any small variation in S^* and W^* will alter E^*, albeit in small ways. So will small changes in the gravitational influence of the distant stars. If we want a genuinely nomologically sufficient condition for E^* it looks as though we need an exact specification not just of A^*, but of S^*, W^* and much more besides.

As Field (2003) emphasizes, this is not just the unthreatening point that many other factors besides those that are salient to common sense are among the causes of E—that in ordinary discourse we pick out just a small part of the complete 'Millian' cause of E. (As when we say that the complete cause of the fire includes the presence of oxygen and the absence of a sprinkler system etc. as well as the more salient striking of the match.) This point is acknowledged by many theories of causation. Rather, the observation in the previous paragraph threatens to collapse the distinction between causal and temporal priority and with it the whole point of the former notion. Barring extraordinary circumstances we think that taking aspirin is (or at least may be) an effective strategy for making a headache go away and that sneezing and wishing are not. If we are forced to the conclusion that the latter are causes of E as well, the motivation (according to manipulationist accounts) for introducing the notion of causation in the first place—the intuitive connection between causation and manipulation and the contrast between causal and merely correlational relationships—appears lost.

Obviously the description of causes in ordinary life and in the special sciences does not take the form of a complete description of the values of relevant variables at every point on a surface intersecting the backwards light cones of effects of interest. The key to understanding how it is possible to provide genuine causal information without providing such a detailed micro-description can be found in the notions of coarse-graining and limited invariance/stability described above. One consequence of coarse-graining is that it makes it permissible to ignore certain causal factors that would be relevant at more fine-grained level of description. If the effect of interest is the exact position and momentum of some collection of particles, then all forces incident on these particles are causally relevant. Suppose, however, the effect variable is framed in a much more coarse-grained or chunkier way—for example, as recovery/non-recovery from a particular condition or disease. If our task is to find the variables, variations in the value of which account for the contrast between those experimental subjects who recover and others who do not, it will almost certainly no longer be true that everything in the backward light cone is relevant to this contrast. For example, it is extremely unlikely that any actual variations in the gravitational force exerted by the distant stars on the subjects will have anything to do with the contrast between recovery and non-recovery (either for a single subject or collection of subjects). Similarly, for whether the subjects or their neighbors sneeze, for what they wish for at the time they ingest medication, and for the color of the clothes they wear. Note that this is true even though, as already emphasized, each token event of sneezing, wishing and so forth will causally influence (in some respects) each of the micro-events that realize individual tokens of recovery or non-recovery. This is possible because, although the forces arising from, for example, the occurrence of a neighbor's sneeze will certainly influence the microstate of the subjects, they will not change these states sufficiently to turn a subject who otherwise would have recovered into a non-recoverer or vice-versa.[18]

---

[18] The choice of grain associated with the causal analysis of a situation is intimately related to the contrastive character of causal claims. As we alter the grain, we alter the potential contrastive foci that are available. If we employ a very fine-grained description of a shattered glass, we can ask why the shattered pieces are in exactly this particular configuration, rather than (or in contrast to) a slightly

To put the point more generally: not all causal relationships (or relationships of nomological dependency) among micro-events aggregate up to causal relationships among coarse-grained macro-events that are constituted by those micro-events. Instead, whether one gets causation at the macroscopic level will depend (among other things) on the particular coarse-graining that is chosen. Of course, we look for sets of coarse-grained variables which are such that not every variable in the set turns out to be causally related to every other, since under this possibility, the discovery of causal relations among macro-events would lose most of its interest and point. Although coarse-graining may look imprecise and arbitrarily selective from the point of view of the underlying physics, it makes the task of finding and describing causes much easier.

It may seem surprising, even counterintuitive, that causal and statistical dependence relationships involving fine grained microscopic variables do not automatically show up in causal and dependence relationships among the macroscopic variables that are realized by the fine grained variables. As we have seen, 'realization' is a fuzzy notion, but one simple framework for thinking about at least some cases of this sort is as follows:[19] Suppose that $Z_1, Z_2, \ldots, Z_n$ are fine-grained variables that stand in various statistical independence, dependence and conditional relationships to one another as represented by a joint probability distribution $P(Z_1, Z_2, \ldots, Z_n)$. Suppose that $F_1, F_2 \ldots F_n$ are functions that map, respectively, $Z_1, Z_2, \ldots, Z_n$, into more macroscopic variables, $F_1(Z_1), \ldots, F_n(Z_n)$. What can be said about

different configuration—indeed, the use of a fine-grained level of description naturally suggests this as an appropriate explanatory question and that the explanandum should be understood in terms of this contrastive focus. If, instead, we employ a much more coarse-grained description, according to which the only two possibilities are that the glass either shatters or does not, then only a very different contrastive focus is possible—we now ask why the glass shattered rather than not shattering at all. Obviously a differentiating factor that is relevant to the explanation of this second contrastive focus—e.g. that the rock struck the glass (rather than missing it entirely) explains the contrast between shattering and not shattering—may not be (in this case, is not) relevant to the explanation of the first, more fine-grained contrast. I would thus reject the anonymous referee's suggestion that coarse-graining and contrastivity are different, competing ways of understanding the cases under discussion. On my view, they are complementary and closely associated.

[19] Thanks to Chris Hitchcock for a helpful discussion and to Clark Glymour for some very helpful correspondence. The framework that follows corresponds to one very simple possibility, with each fine-grained variable being mapped directly into a macroscopic variable. Of course, there are many other possible relationships between the fine-grained and the macroscopic. For example, the macroscopic variable might be a sum or some other function of the values of some fine-grained variable taken by each of a large number of units. For some relevant results in this connection, see Chu, Glymour, Scheines, and Spirtes (2003).

how the (in)dependence relationships among these variables depend on the functions F_i, and the distribution P? It is not clear that there is any illuminating general answer to this question but here are some relevant observations that will help to motivate the claims made in the previous paragraphs. First, if Z_j and Z_k are unconditionally independent, then $F_j(Z_j)$ and $F_k(Z_k)$ will also be independent as long as F_j and F_k are measurable functions.[20] Second, if Z_j and Z_k are dependent, then $F_j(Z_j)$ and $F_k(Z_k)$ will be dependent if F_j and F_k are 1–1 functions. In addition, if the functions F_i are 1–1, they will preserve conditional independence (screening off) relationships among the Z_i (Glymour forthcoming). If F_j and F_k are not 1–1, then simple examples show that depending on the details of the case, it is possible to have Z_j and Z_k dependent but $F(Z_j)$ and $F(Z_k)$ independent.[21] Functions that are not 1–1 can also fail to preserve conditional independence relationships. If we think of coarse-graining as involving the use of functions from micro to macro-variables that are not 1–1 (different values of the micro-variable are mapped into the same value of the macro-variable) then coarse-graining can indeed lead from dependence at the micro-level to independence at the macro-level.

Observing that it is possible for macro-independence to emerge from micro-dependence is of course not at all the same thing as providing an interesting characterization of the conditions under which this will happen. In my view, one of the major puzzles about the relationship between causation at the macro-level and the underlying physics is why there is so *much* independence (or at least near or apparent independence) and conditional independence among macroscopic variables, given that at a microphysical level, everything seems so interconnected. Macroscopic independence is one of the conditions that allows us to discover and exploit macroscopic causal relations. Somehow this independence must result from coarse-graining and its interaction with facts about the underlying

---

[20] Glymour, personal correspondence.
[21] Two examples, the first my own and the second due to Hitchcock: First, suppose that X can take any of the values $\{1, 2, 3, 4\}$ and that the associated values of Y are respectively $\{1, 0, 1, 0\}$ so that Y is a deterministic function of X. Suppose that each value of X is equally probable and when $X = 1, X = 2, F(X) = 1$ and when $X = 3, X = 4, F(X) = 0$. Let $G(Y) = Y$ be identity. Then $F(X)$ will be independent of $G(Y)$. Second, suppose that X is vector valued: $X = (X_1, X_2)$ and that Y and X are dependent because Y and X_1 are dependent, while Y is independent of X_2. Let $F(X) = X_2$, and let G be identity. Then $F(X)$ will be independent of $G(Y)$.

dynamics, but I would be the first to acknowledge that I have said very little about how this happens.[22]

Somewhat surprisingly, the coarse-grained character of the variables figuring in the incomplete causal generalizations of the special sciences enables those generalizations to be more stable than they would be if they remained incomplete but were formulated in terms of more fine-grained variables. In fact, there are few incomplete dependency relationships relating fine-grained variables that are stable over some usefully extensive range of changes in background conditions; instead all or most incomplete dependency relationships that are stable over even a modest range of changes will relate coarse-grained variables. Finding generalizations describing dependency relationships involving fine-grained variables that are relatively stable will typically require finding generalizations that are complete or nearly so.

Suppose that the state F at time t of some macroscopically spatially extended region R of the world is specified by a conjunction of fine-grained properties $P_1 \ldots P_n = F$ and that the state of the world F' at some previous time t' which is nomologically sufficient for Rs being F at t is given by the fine-grained conjunction $P'_1 \ldots P'_n$. Then I claim that in most cases and for most specifications of $P'_1 \ldots P'_n$ and $P_1 \ldots P_n$, we are unlikely to find even relatively stable generalizations of form: 'If P'_i, then P_i'. That is, generalizations that relate just one of the properties P'_i to just one of the properties P_i are likely to be highly non-invariant and exception-ridden. A similar point will hold if the generalizations in question relate a conjunction of some (but not all) of the P'_i to some Pi. The reason for this is that most fine-grained properties P_i of R at t will generally depend not on some single P'_i but rather on the entire specification $F' \ldots$ —the entire conjunction $P'_1 \ldots P'_n$. Changing any one of the variables P'_k, $k \neq i$ (even slightly) will disrupt the relationship between P'_i and P_i. On the other hand, if C is some property holding in R at t that is sufficiently coarse-grained, it may well be true that we can find some coarse grained property C' holding at t' such that the generalization '*if* C', then C' is incomplete but relatively invariant and stable, over locally prevailing background conditions.

As an illustration, consider again an experiment in which an irregularly shaped, 5kg rock is dropped from a height of two meters directly onto an intact champagne glass, causing it to shatter. Suppose first that we specify

---

[22] See Strevens (2003) for interesting additional discussion of this issue.

this effect in a very fine-grained way: we take the effect to be the exact shape and location of all of the various fragments of the glass five seconds after its breaking. Let M_1 be such a specification for one of the fragments of the glass. Assume that the interaction is deterministic, and hence that there is some extremely complicated specification S of the state of the rock, the state of the glass, and the surrounding environment at the time of collision that is nomologically sufficient for M_1. Note that for S to be nomologically sufficient for M_1, S must include a specification not just of the momentum p of the rock, but also the shape of that portion of the rock that comes in contact with the glass, the composition and structure of the glass, and much more besides. If we try to formulate an incomplete generalization linking some incomplete specification of S (e.g. if we just specify the momentum p of the rock and nothing more) to M_1, this generalization will almost certainly be exception-ridden and relatively unstable. Given rocks that share the same momentum p, but differ in shape, orientation or larger environment when they strike the glass and so on, even slight changes in these variables will result in a set of fragments that are at least slightly different and similarly for slight changes in the composition, structure and so on of the glass.

Suppose instead that we adopt a much more coarse-grained description of this situation: we represent the effect by means of a variable Y that can take just two values: {1 = *glass shatters*, 0 = *does not shatter*} and the cause by means of a variable X that takes the following two values. {1 = *5kg rock is dropped from a height of 2 meters directly striking glass, no rock strikes glass*}. In most non-extraordinary background circumstances, an intervention that sets $X = 1$ will be reliably followed by $Y = 1$: that is, the generalization $Y = X$ describes a relatively stable dependency relationship. This illustrates how coarse-graining allows the formulation of incomplete generalizations that are relatively invariant, although at the cost of predictive precision regarding fine-grained details.

Although I lack the space for detailed discussion, it is worth noting that the use of coarse-grained variables affects many other aspects of causal reasoning in common sense and the special sciences. Consider the familiar screening-off conditions connecting causal claims and probabilities and the supposed fork asymmetry associated with these: we expect that common causes will screen-off their joint effects from one another but that conditioning on the joint effect of two causes will render them dependent except for certain very special parameter values. As Arntzenius (1990) notes, under determinism, if C, which occurs prior to the joint effects E_1 and E_2, screens them off from one another, there must also be an event C^* which occurs after E_1 and E_2 is causally affected by them, and that screens them off from one another. Why then are we tempted to suppose that there is a fork asymmetry? At least part of the reason is that typically the later screening-off event C^* will be very hard to see—it will be very diffuse, spread out and gerry-mandered, corresponding to no single macro-event in any coarse-graining we are likely to adopt. If, as we tacitly assume, the screening-off common effect must be a single event in a natural coarse-graining, then it becomes more plausible that there is a fork asymmetry. In this sense, the asymmetry is in part a product of the particular coarse-graining of the macroscopic world that we adopt. A similar story can be told, I believe, about how it is possible for the coarse-grained entropy of a system to increase despite the fact that fine-grained entropy is constant over time. In the paradigmatic case of a gas expanding into an evacuated chamber, we must coarse-grain in order to have an expansion of the phase volume with time, so that entropy increases. The expansion is irreversible relative to an appropriately chosen coarse-graining but not in relation to a fine-grained level of description.[23]

4.6 Interventions

I turn now to a different feature of the systems that are studied in the upper-level sciences and our relation to them that makes causal notions seem particularly useful or appropriate. This has to do with the fact that such systems are typically only a small part of a much larger world or environment which is outside the scope of the inquirer's interest but which can serve as source of interventions. Recall the basic idea of an interventionist account

---

[23] For reasons that are lucidly explained in Sklar (1993, pp. 346 ff), I believe that it does not follow from the fact that whether entropy increases is relative to the grain one chooses that whether or not entropy increase occurs is 'subjective'. Coarse-grained entropy is a different quantity than fine-grained entropy and these quantities behave differently. Relative to a specification of system and a level of description or graining for it, it is an objective matter whether there is entropy increase. A similar point holds for causation—once one fixes the variables one is talking about, it is 'objective' matter whether and how they are causally related. Causation is, one might say, 'variable relative' in the sense that, as illustrated, different choices of variables or grainings will lead to different conclusions about whether everything in the backward light cone is causally relevant to an episode of glass shattering but it is not 'description-relative' in the sense that whether or not one variable is causally relevant to another depends on how those variables are described. See Woodward (2003) for additional discussion.

of causation: causal claims are linked to counterfactual claims about what would happen under possible interventions. Clearly if such a view is to be remotely plausible, 'possible' must be understood in a liberal way. For example, there are many cases of causal relationships between X and Y in which an intervention on X is not practically or technologically possible for human beings. Woodward (2003) discusses this issue and concludes that for an interventionist account of what it is for X to cause Y to be workable, what is crucial is that counterfactuals describing what would happen to Y (or in the indeterministic case the probability distribution of Y) under an intervention on X 'make sense' and 'have determinate truth values', rather than whether human beings are able to carry out the interventions in question. But under what circumstances will interventionist counterfactuals have the quoted features?

While I will not try to provide a definitive answer to this question (and in fact doubt that there is an uncontroversially correct answer that covers all possibilities), I want to suggest that there are some circumstances and systems for which interventionist counterfactuals seem straightforward and unproblematic and other systems for which this is less obviously the case. Unsurprisingly, among the former systems are those investigated in the special sciences. The latter include global applications of fundamental physical theories to the whole universe or large portions of it. I emphasize again that I do *not* mean to claim that the notion of an intervention has no application in physics; the notion seems perfectly reasonable when applied to the right sort of small, non-global systems. I maintain, however, that the requirements for the sensible application of the notion of an intervention help to explain why forms of causal thinking that seem natural in the special sciences do not straightforwardly extend to more global physical contexts.

Consider a typical case in which interventionist counterfactuals seem unproblematic. A researcher wishes to know whether treatment with a drug D will cause an increase in the rate of the recovery from a certain disease. She envisions the following experiment. Subjects with the disease are randomly assigned on the basis of the outcome of the flip of a fair coin to a treatment group who receive D and a control group from whom D is withheld. Which subjects in the trial receive the drug is thus determined entirely by the random assignment process. Then the incidence of recovery in the treatment and control group is compared. In an experiment of this design it usually will be reasonable for the experimenter to assume that the random assignment process constitutes an intervention on who receives the drug with respect to the outcome of recovery. Of course this is a defensible empirical assumption that might in principle be mistaken—perhaps unbeknownst to the first researcher a second scientist controls the outcome of the coin toss with a magnet and arranges that all and only those with unusually strong immune systems are assigned to the treatment group. Usually, however, the assumption that this operation constitutes an intervention will be correct—whatever variables influence the outcome of the coin toss will not causally influence or be correlated with whether the subjects recover except via the route, if any, that goes from the outcome of the flip to ingestion of the drug to recovery. Even if the researcher does not in fact perform this experiment, it seems clear enough what would be involved in performing it and no reason to doubt that there is a determinate answer to the question of what would happen if it were to be performed.

One reason why interventionist counterfactuals seem unproblematic in this case is that we are dealing with what Judea Pearl (2000) has called a 'small world'—a system (the subjects in the experiment who are or are not given the drug) that is isolated enough from its environment that it can serve as a distinctive subject of causal inquiry but not so isolated (or 'closed') that the idea of outside influences in the form of interventions makes doubtful sense. Put slightly differently, the system of interest is located in a larger environment which serves as a potential source of 'exogenous' interventions. However, apart from this, the environment is of no direct interest to the researcher. In the example under discussion, the outcomes of the coin flip are exogenous and part of the environment in this sense: the researcher is not interested in, and does not need to be concerned with, modeling in detail the causal processes that produce these outcomes as long as it is true that these processes, whatever they may be, do not affect recovery, independently of treatment. In this sort of case, we can keep any contra-nomic miracles that the occurrence of interventions may seem to require safely offstage, in the environment.[24]

As several writers have remarked (Pearl 2000; Hitchcock in ch. 3 of this volume; Hausman 1998) a similar strategy is no longer possible when a fundamental theory is applied to the whole universe at once. Now

---

[24] See Woodward (2003, pp. 127 ff) for additional discussion.

there is no longer anything outside the system being modeled to serve as possible source of interventions and it may be quite unclear how one may legitimately model interventions as part of the system being studied.

As an illustration, consider the claim that (U) the state S_t of the entire universe at time t causes the state S_{t+d} of the entire universe at time $t + d$. On an interventionist construal, this claim would be unpacked as a claim to the effect that under some possible intervention that changes S_t, there would be an associated change in S_{t+d}. The obvious worry is that it is unclear what would be involved in such an intervention and unclear how to assess what would happen if it were to occur, given the stipulation that S_t is a specification of the entire state of the universe. Although I don't claim that it is *obvious* that the relevant interventionist counterfactuals make no sense or lack determinate truth values, it seems uncontroversial that a substantial amount of work would have to be done to explain what these counterfactuals mean.

Commenting on this point, Pearl writes: 'If you wish to include the whole universe in the model, causality disappears because interventions disappear—the manipulator and the manipulated lose their distinction.' (2000, p. 350). While I am less confident than Pearl that causality 'disappears' in these circumstances, I think that it is very plausible that causal ascription becomes less natural and straightforward—increasingly strained—when candidate causes expand to include the state of the entire universe.

4.7 Arrow Breaking

There are several other features of the systems that are studied in the upper-level sciences that make the application of the notion of an intervention seem particularly apt. As we noted in connection with the *ABS* (pressure, barometer, storm) system, a natural way to investigate the causal structure of complex systems is to take them apart. By breaking or disrupting certain causal relationships in a system one may create circumstances in which other causal relationships, if real, will reveal themselves in associations. Thus, if we disrupt the relationship between *A* and *B* by manipulating *B*, we expect any causal relationship between *B* and *S* to show itself in a correlation between *B* and *S* that persists under this manipulation. Similarly, in the drug experiment, the effect of randomization is to replace a situation in which, for example, subjects decide on their own whether or not to take a drug with a situation in which who does or doesn't get the drug is controlled by the randomization process. It seems unproblematic to suppose that the causal influence of the subject's decisions on whether they take the drug is 'turned off' when the randomization is instituted and this assumption is crucial to the inference we draw from the randomized trial.

These assumptions about the possibility of turning off or breaking certain causal influences in order to isolate and investigate others go hand in hand with the fact that causal generalizations on which common sense and the special sciences focus have only limited ranges of invariance and stability. Because these generalizations hold only for a certain range of conditions and break down outside of these, it is possible, either by actively creating situations in which these generalizations break down or finding naturally occurring situations in which this happens, to turn off the causal influences they describe and use this operation to investigate other generalizations or causal relationships that remain intact. Thus we can readily create situations which disrupt the causal connection between atmospheric pressure and the barometer reading or between the experimental subject's own decisions and whether he takes a drug and then determine whether the relationship between barometer reading and the occurrence of the storm, or between drug ingestion and recovery remain intact under this operation. Again, however, it seems less clear how to carry over this idea of breaking some causal influences in order to investigate others into all of the contexts in which theories of fundamental physics apply. As noted above, we think that it is a mark of fundamental laws that they either do not break down at all or break down only in very special and unusual situations—neither experimenters nor nature can create such situations in anything like the range of circumstances in which typical macroscopic causal relationships can be disrupted.[25] Although, as Woodward (2003, s. 3.5) argues, we can sometimes appeal to our theories themselves to tell us what would happen under interventions that are counter-nomic, the fact remains that in many physics contexts there may be no physically realistic operation corresponding to placing some variable of interest entirely under the control

---

[25] This is not to say that there is nothing that looks like arrow-breaking in experimentation in physics—for example, one may shield an apparatus from electromagnetic forces that would otherwise be operative.

of an intervention variable, and breaking all other causal arrows directed into it.

There is yet another feature of the notion of an intervention that influences its application to the sorts of systems that are studied in the special sciences. Consider again the use of a randomized experiment to test the claim that a drug produces recovery from a certain disease. I said above that (in the absence of improbable coincidences) such an experiment will approximate the conditions for an intervention on treatment with the drug with respect to recovery. Recall that one condition for a successful intervention is that the intervention I on X with respect to Y should not cause Y via a route that does not go through X, and that I should be independent of any variable Z that causes Y but not via a route that goes through I and X. However, there is an apparently natural line of thought, echoing the argument in Section 5, that questions whether these conditions are ever likely to be satisfied—either in the randomized experiment under consideration or any other realistic case. Consider the microstates s_i, characterized in terms of quantities provided by fundamental physics, that realize the values of the variable 'recovery/non-recovery' for each subject i. Surely, it might be argued, these states s_i will themselves be causally influenced (via a route that does not go through the putative cause variable, ingestion (or not) of the drug) by the states of the micro-variables that realize the values of the intervention variable and by the microstates of causes of the intervention variable. For example, each occurrence of the coin flip that implements the randomization will alter the position of various elementary particles and will have the consequence that various forces on the variables s_i will be different to what they otherwise would be. For all we know, this fact will show up in some correlation between these sets of variables on repeated flips. Moreover, some of these forces will operate independently of whether the drug does or does not cause recovery. A similar point will hold for whatever micro-variables influence the outcome of the coin flip. In short (it might be argued), at the level of fundamental physics, events will be causally interconnected in a way that precludes the satisfaction of the conditions for an intervention.[26]

---

[26] Michael Friedman, among others, has suggested in conversation that something like this is true. Another (perhaps better) way of putting the worry is that the contrast between those variables that influence the effect only through the cause and those that instead influence the effect via a route that does not go through the cause—a contrast that is at the heart of the notion of an intervention—becomes

I believe that this objection will only seem plausible if we fail to take seriously the observations about coarse-graining made earlier. The claims in the previous paragraph about the independent causal influence of the micro-variables realizing the intervention on the micro-variables realizing recovery are perfectly correct but it does *not* follow that the intervention itself or the causes of it have a causal influence on recovery that is independent of treatment with the drug. If $s_1 \ldots s_n$ are micro-variables that realize the macroscopic, coarse-grained variable X and $s_1^* \ldots s_n^*$ are micro-variables realizing the macroscopic coarse-grained variable Y, it is perfectly possible for some instantiations of some of the s_i to causally influence and be correlated with some instantiations of some of the s_i^* and yet for it to be false that X causally influences or is correlated with Y. Suppose that we have a population of patients with a disease, some of whom will recover and some of whom will not. For each patient we flip a coin and record the results but these results are not used to determine what treatment, if any, the patients will receive. Instead we simply observe whether or not each patient recovers. For just the reasons described in the previous paragraph, some of the micro-level variables the values of which realize the coin flip will causally influence the micro-level variables realizing instances of the recovery variable but of course it does not follow (and we do not expect that) there will be a correlation between the outcome of the coin flip for individual patients and whether they recover—we don't think one can use the outcome of the coin flip to predict who will recover. In just the same way we don't think that we can use the outcome of coin flips to predict the future of the stock market or who will win the next US presidential race despite the presence of causal influences among the micro-realizations of these variables. As already noted, it is possible for causal and statistical independence among groups of coarse grained variables to emerge from a web of complicated causal dependencies in which everything is influenced by everything else (in its backward light

---

unclear at the level of fundamental physics. That is, the whole notion that one variable might affect another via multiple distinct routes is itself a consequence of our adoption of a coarse-grained perspective and the distinctness of different routes itself disappears at a fine-grained level, where the correct causal representation (if there is one) is just a chain structure in which a succession of single arrows connects one total state of the universe to another. This last representation eliminates the worry about independent influences on the effect that do not go through the cause, but has the result that everything in the absolute past of an event is now causally relevant to it. Thanks to Chris Hitchcock for helpful conversation on this point.

cone and to some degree or other) at a more fine-grained level. This in turn makes it possible to actually carry out interventions, which require such independence.

4.8 Distinguishing Causes and Correlations

I noted in Section 4.1 that in upper-level sciences in which causal talk plays a major role we often face is a generic inference problem that looks like this: We know that two variables X and Y are correlated but we don't yet know what their causal relationship is—whether X causes Y, Y causes X, whether there is a third variable or set of variables which are common causes of both X and Y, whether the correlation is present only in a sample that is not representative of the larger population from which it has been taken, and so on. To an important extent, the role played by our notion of causation is tied up with elucidating the differences among these possibilities. It is interesting that the same inference problem seems to arise much less often in fundamental physics. While one can think of possible exceptions (e.g. perhaps the status of thermodynamic asymmetries or the status of various boundary conditions on the whole universe imposed in cosmology), physicists are not usually in the epistemic position of knowing that, say, a regularity holds globally among fundamental physical quantities but not knowing whether this regularity represents a causal or nomological relationship or whether the regularity is a mere correlation is produced by some third variable. Instead, the more usual situation is this: although it may not be clear whether some proposed generalization holds globally, there is little doubt about whether if it is true, it will fall into the category of a 'law' or (less commonly) that of an 'accidental' generalization. Consider the equations specifying the coupling of gravity to matter associated with the various gravitational theories (e.g. the Brans-Dicke theory) that once were regarded as alternatives to General Relativity. No one appears to have doubted that if one of these equations had turned out to be true, it would have been a law of nature—that is, no one took seriously the possibility that the equation might be true but only accidentally so, in the manner in which the relationship between parental income and child's scholastic achievement might be merely accidentally true or a mere correlation. So the inference problem that looms so large in the special sciences and which helps to give causal talk its point seems somehow much less pressing in fundamental physics.

What accounts for this difference? One factor that seems to be at work in fundamental physics contexts is the availability of a great deal of detailed background knowledge/expectations that guides our decisions about whether a generalization is appropriately regarded as a candidate for a law or merely accidentally true. This includes information to the effect that laws should satisfy various symmetry requirements—if a generalization fails to satisfy these requirements it is unlikely to be a law, regardless of whatever other descriptive virtues it may possess. It may also include information to the effect that the generalization of interest is derivable from known laws and initial conditions that are identified by theory as holding pervasively but merely contingently as may be the case for the second law of thermodynamics and certain cosmological regularities.

The situation in the special sciences is quite different. First, causal generalizations as well as generalizations that describe mere correlations do not even purport to hold globally. True causal generalizations in the special sciences are typically restricted to various more or less specialized systems and break down under a variety of conditions, not all of which are well understood. In the special sciences, the causal/accidental distinction is a distinction *within* the category of non-universal generalizations with many exceptions. Second, in many contexts in the special sciences, we lack the kind of background theoretical knowledge that would provide a basis for sorting generalizations into the categories of causal vs. merely accidental or correlational. At least in part for these reasons, the epistemic problem of deciding into which of these categories a candidate generalization falls has a kind of salience in the special sciences that it does not have in fundamental physics. With this comes a greater use of concepts and patterns of reasoning designed to mark this distinction.

4.9 Are Macroscopic Causal Relationships Real?

My strategy so far has been to draw attention to features of the systems typically studied in the special sciences that make characterizations that appeal to causal notions particularly useful and illuminating. I'm fully aware, however, that this strategy will seem unsatisfying to the metaphysically minded.

Even if it is granted that causal description is sometimes useful, this leaves untouched (it will be said) the question of whether causal relationships are 'real' or 'objective'. What if anything do the arguments in Sections 4.1–4.8 suggest about the literal truth of causal claims about macroscopic systems? Should such claims be construed purely instrumentally, as nothing but helpful fictions?

Let me approach this issue by means of an analogy. Consider the role of chance in systems that are governed by deterministic laws. In particular, consider a gambling device like a roulette wheel, the operation of which (let us suppose) is governed at the relevant level of analysis by deterministic laws and yet which generates outcomes that are (or look as though they are) independent and occur with stable probabilities strictly between zero and 1. How is this possible?

The broad outlines of an answer to this question go back at least to Poincaré and have been set out (and generalized) by Michael Strevens in a recent interesting book (2003). Loosely described, Strevens' treatment appeals to the following ideas. First, consider the 'evolution functions' that map the initial conditions in the system of interest that are relevant to the final outcome onto the various particular values of the macroscopic outcome variable. In the case of the roulette wheel, the initial conditions will include, for example, the initial position of the wheel and the angular momentum imparted to it by the croupier, and the outcomes will be a particular number or color. Assume that the dynamics of the system are such that it exhibits 'sensitive dependence to initial conditions'—more specifically, assume that nearby regions in the phase space of initial conditions map onto different outcomes, and that for an appropriately chosen partition of the phase space into small contiguous regions, the volume of the regions that are mapped into each of the outcomes is constant, or approximately so, within each cell of the partition. Then in repeated trials that impose any distribution of initial conditions that is appropriately 'smooth' (as Strevens calls it—this means that the probability density of the initial conditions is approximately constant within each cell of the partition), one will get outcomes that are independent and have stable probabilities. But why the restriction to smooth distributions? Although Strevens does not quite put it this way, one natural motivation is this: It is plausible to assume that in the case of real life gambling devices like roulette wheels any macroscopic (hence spatially and temporally extended) system (like the croupier) can only intervene on the wheel in a coarse-grained or macroscopic manner: that is, it cannot impose a non-smooth distribution on the relevant initial conditions of the system. It cannot, for example, fix (via a sort of Dirac delta function operation) a particular point value for the initial conditions of the wheel with probability 1, so that all other initial conditions receive zero probability.

Assume for the sake of argument that this claim is correct. What follows about the objective reality of chances (that is, the reality of non-trivial chances strictly between zero and 1) for such devices? If our criterion for the objective reality of such chances is whether they appear in fundamental physical laws then the answer is clear: the chances we associate with such devices are not objectively real, however useful they may be for summarizing the behavior of the devices and guiding betting behavior. Because the underlying laws are deterministic, the objective chance of, say, 'red' on any given spin of the wheel is always either zero or 1. Any other value for the probability of this outcome must be understood as a 'subjective' or 'merely epistemic' probability, reflecting our ignorance of exact initial conditions.

Looked at another way, however, this assessment is misleading. The non-trivial chances we ascribe to outcomes are a reflection of an interaction between perfectly objective facts about the dynamics of gambling devices and other facts (arguably also equally objective) about the kinds of distributions of initial conditions that a macroscopic agent or process is able to impose. Moreover, given these facts, it may well be true that there is nothing that any macroscopic process (the hand of any croupier or any similar sized intervention) can do in the way of imparting a spin to the wheel which will produce deviations from the ascribed chances, as well as no observations or measurements that might be made by a macroscopic observer that will allow for the prediction of anything more about outcomes than is already recorded by the chances. So if we mean by objective probabilities, probabilities that reflect patterns that will obtain under any distribution over initial conditions that a macroscopic agent is able to impose or learn about—that is, probabilities that are invariant/stable under changes in *these* distributions, even if they are not invariant/stable under other, conceivable distributions—then there are non-trivial objective probabilities associated with the roulette wheel. It would be quite mistaken to suppose that these probabilities are 'subjective' in the sense that, say, they are matters of individual taste, limited only by considerations of coherence, or that, given the facts about the dynamics and the initial conditions it is possible to impose, there is no fact of the matter about which probability assignments are correct. Instead the probabilities are 'real' in the straightforward sense that they reflect constraints not just on what macroscopic agents are able to learn about but also what they are able to do.[27]

My suggestion is that at least in some respects the status of causal relationships in macroscopic systems is similar. If our criterion for 'objectively real' is 'found (or grounded in a direct way) in the laws of fundamental physics alone', then, as we have seen, it is dubious that most macroscopic causal relationships are 'real'. Or, to put the point more cautiously, it seems doubtful that we will find, for each true causal claim in common sense and the special sciences, counterpart relations in fundamental physical laws, with all of the features that we ascribe to macroscopic causes. This is because macroscopic causal relationships do not depend just on facts about fundamental laws but also reflect other considerations as well—for example, the coarse graining operations associated with our status as macroscopic agents and the frequency with which various initial conditions happen to occur in our spatiotemporal vicinity. Among other things, coarse-grained factors and events, with ill-defined boundaries and spatiotemporal relations, are not plausibly regarded as instances of fundamental laws.

Just as with the chances we ascribe to deterministic gambling devices, nothing prevents us from adopting the grounded-in-physical-laws-alone criterion for what is 'real' as a stipulative definition. However, we need to take care that we do not read more into this stipulation than is warranted. In particular, as nearly as I can see, it is consistent with the 'unreality' of macroscopic causation, in the sense associated with the criterion just

---

[27] The anonymous referee worries that the roulette wheel exhibits merely a 'counterfeit' notion of non-trivial objective chance. But contrast the claim that the probability of red equals 0.5 on the next spin of a roulette wheel satisfying the conditions described above with the claim that, say, the probability of war between India and Pakistan in the next ten years is 0.5. In the latter case, we have no corresponding story about what would constitute repeated trials with the same chance set-up, and no story about how the interaction between an underlying dynamics and the initial conditions that a macroscopic agent (or other process) is able to impose yields stable frequencies etc. There are real differences between the processes that underlie the generation of outcomes in the case of the roulette wheel and the processes that will generate war (or not) between India and Pakistan. It is these differences that make us think that ascription of chances in the latter case is far more 'subjective' than in the former case. So while one can certainly stipulate a meaning for 'objective chance' according to which non-trivial objective chances require indeterminism, this has the unfortunate consequence of collapsing important distinctions among the behavior of deterministic systems.

described, that there are, in a straightforward sense, facts about whether manipulating macroscopic variable X will be associated with changes in macroscopic variable Y or whether instead the observed association between X and Y is a mere correlation. Similarly, there are facts about whether a relationship between X and Y that is exploitable for purposes of manipulation would continue to hold across various sorts of changes in background conditions. Moreover, such facts can be discovered by ordinary empirical investigation. If the line taken in this paper is correct, macroscopic causal claims (like 'chances' in a deterministic world) reflect complicated truths about (i) an underlying microphysical reality and (ii) the relationship of macroscopic agents and objects to this world. This second ingredient (ii) gives macroscopic causal talk a number of its characteristic features—coarse-grainedness, a focus on small worlds where this is a possibility of outside intervention, reliance on contingent facts about initial and boundary conditions and so on—but it does not make such claims 'subjective' in the sense of not being controlled by evidence, dependent on the idiosyncratic tastes or interests of individual investigators and so on. Just as the notion of chance seems pragmatically unavoidable when dealing with systems like roulette wheels (it is not as though we have some alternative available which works better), so also for 'causation' when dealing with many macroscopic systems.

4.10 Conclusion

I conclude with a puzzle/concession and a restatement of a moral. The former has to do with the role of explanation in physics. Suppose that causation (at least when construed along the interventionist lines that have been described) does not play a foundational role in fundamental physics. What follows for how we should think about explanation in such contexts? Are all explanations causal with the apparent result that at least in some respects and contexts, fundamental physics does not provide explanations? Does physics instead provide non-causal explanations and if so, how should we understand the structure of these?[28] I am not sure what to think about these questions, but also do not think that this uncertainty is in itself a

---

[28] I am grateful to the anonymous referee for raising these questions, which are undeniably relevant and important.

good reason to reject the conclusions about the role of causation in physics reached above.

Next, the moral. This has been implicit in earlier portions of my discussion, but merits explicit underscoring. It is a popular idea that true causal (and counterfactual claims) from everyday life and the special sciences cannot be 'barely true' but instead require grounding (or 'truth-makers') in fundamental physical laws. This idea strikes me as arguably correct if it is interpreted in the following way: given a true garden variety causal claim, there will be some associated in—principle physical explanation (or story or account, to use more neutral words) for its holding, and this will include, among other factors, appeal to fundamental laws. However, the idea in question often seems to be understood in a different and more restrictive way: as the claim that reference to fundamental laws alone (together with the claim that the events or factors related in the causal claim 'instantiate' the law or bear some other appropriate relationship to it) gives us all that is needed to state the grounds or truth-makers for causal claims. Here the idea is that the causal part of the content of all causal claims is somehow grounded in the fundamental laws themselves with nothing else required. If the argument of this paper is correct, this second interpretation of the grounding idea is mistaken. Typically, the grounds or truth-makers for upper-level causal claims like 'Cs cause Es' or 'particular event c caused particular event e' will involve many additional factors besides laws (and besides facts about whether C, E, c and e instantiate laws or are part of conditions that instantiate laws etc.). These additional factors will include very diffuse, messy, and non-local facts about initial and boundary conditions that do not obtain just as a matter of law and have little to do with whatever underlies or realizes C, E, c or e themselves (recall the discussion in s. 4.4). If so, attempts to provide sufficient conditions for Cs cause Es along the lines of 'Cs are (or are part of conditions that are) linked by fundamental laws to Es' are unlikely to be successful.

References

Armstrong, D. (1997). *A World of States of Affairs*. Cambridge: Cambridge University Press.

Arntzenius, F. (1990). 'Common causes and Physics', *Synthese*, 82: 77–96.

Cartwright (1979). 'Causal Laws and Effective Strategies', *Noûs*, 13: 419–37.

Chu, T., Glymour, C., Scheines, R. and Spirtes, P (2003). 'A Statistical Problem for Inference to Regulatory Structure from Associations of Gene Expression Measurement with Microarrays', *Bioinformatics 19*, pp. 1147–52.

Coleman, J. and Hoffer, H. (1987). *Public and Private High Schools.* New York: Basic Books.

Davidson, D. (1967). 'Causal Relations'. *Journal of Philosophy*, 64: 691–703.

Dowe, P. (2000). *Physical Causation.* Cambridge, UK: Cambridge University Press.

Field, H. (2003). 'Causation in a Physical World', in M. Loux and D. Zimmerman (eds). *Oxford Handbook of Metaphysics.* Oxford: Oxford University Press.

Frisch, M. (2000). '(Dis-)Solving the Puzzle of the Arrow of Radiation', *British Journal for the Philosophy of Science,* 51: 381–410.

Frisch, M. (2002). 'Non-Locality in Classical Electrodynamics', *British Journal for the Philosophy of Science*, 53: 1–19.

Glymour (forthcoming). 'Mental Causation and Supervenient Science'.

Gopnik, A. and Schulz, L. (forthcoming). *Causal Learning: Psychology, Philosophy and Computation.* Oxford: Oxford University Press.

Granger, C. W. J. (1998). 'Granger Causality', in J. Davis, D. Hands, U. Maki (eds), *Handbook of Economic Methodology.* Aldershot: Edward Elgar, 214–16.

Haavelmo, (1944). 'The Probability Approach in Econometrics', *Econometrica*, 12 (Supplement): 1–118.

Hausman, D. (1998) *Causal Asymmetries.* Cambridge, Cambridge University Press.

Hausman, D. and Woodward, J. (1999). 'Independence, Invariance and the Causal Markov Condition', *The British Journal for the Philosophy of Science,* 50: 521–83.

Hitchcock, C. (this volume). 'What Russell got Right'.

Hoover, K. (1988). *The New Classical Macroeconomics.* Oxford: Basil Blackwell.

Jackson, J. D. (1999). *Classical Electrodynamics.* New York: Wiley.

Menzies, P. and Price, H. (1993). 'Causation as a Secondary Quality'. *British Journal for the Philosophy of Science*, 44: 187–203.

Norton, J. (this volume). 'Causation as Folk Science'.

Pearl, J. (2000). *Causality: Models, Reasoning and Inference.* Cambridge, Cambridge University Press.

Price, H. (1991). 'Agency and Probabilistic Causality', *British Journal for the Philosophy of Science*, 42: 157–76.

Rueger, A. (1998). 'Local Theories of Causation and the Aposteriori Identification of the Causal Relation', *Erkenntnis*, 48: 25–38.

Salmon, W. (1984). *Scientific Explanation and the Causal Structure of the World.* Princeton: Princeton University Press.

Schaffer, J. (2000). 'Causation by Disconnection', *Philosophy of Science*, 67: 285–300.

Sklar, L. (1993). *Physics and Chance: Philosophical Issues in the Foundations of Statistical Mechanics.* Cambridge: Cambridge University Press.

Smith, S. (forthcoming). 'Armstrong on the Relationship between Causation and Laws'.

Spirtes, P., Glymour, C. and Scheines, R. (2000). *Causation, Prediction, and Search,* 2nd edn. New York, NY: MIT Press.

Strevens, M. *Bigger Than Chaos.* (2003). Cambridge: Harvard University Press.

Strotz, R. H. and Wold, H. O. S. (1960). 'Recursive vs. Non-recursive Systems: An Attempt at a Synthesis', *Econometrica,* 28: 417–27.

Tomasello, M. & Call, J. (1997). *Primate Cognition.* New York: Oxford University Press.

Woodward, J. (2003). *Making Things Happen: A Theory of Causal Explanation.* Oxford: Oxford University Press.

Woodward, J. (forthcoming a) 'Sensitive and Insensitive Causation', *Philosophical Review* 115 (1).

Woodward, J. (forthcoming b) 'Interventionist Theories of Causation in Psychological Perspective', in A. Gopnik and L. Schulz (eds), *Causal Learning: Psychology, Philosophy and Computation.* New York: Oxford University Press.

5
Isolation and Folk Physics

ADAM ELGA

There is a huge chasm between the sort of lawful determination that figures in fundamental physics, and the sort of causal determination that figures in the 'folk physics' of everyday objects. For example, consider a rock sitting on a desk. In everyday life, we think of the rock as having a fixed stock of dispositions—the disposition to slide on the desk when pushed, to shatter when struck by a sledgehammer, and so on. When a strong interaction comes the rock's way, the rock's dispositions determine how it will respond. More generally, we think of the behavior of an ordinary object as being determined by a small set of conditions. The conditions typically specify the object's dispositions to respond to various sorts of interference, and describe the sorts of interference that the object in fact encounters. Call this the 'folk model' (Norton ch. 2 in this volume).

In fundamental physics, no small set of conditions suffices to determine an ordinary object's behavior. Instead, differential equations describe how the exact physical state of the world at one time[1] lawfully constrains its state at other times (Russell 1913). In the worst case–the case of non-local laws—one would have to specify the entire state of the world at one time, in order to determine the state of even a small region at some future time.[2] And even if locality holds in the sense of relativity theory (so that no influences travel faster than light), one would still have to specify the state of a huge region of the world (Field 2003, p. 439). For example, in order

---

For helpful discussion and correspondence, thanks to the Corridor Group, Sheldon Goldstein, Chris Hitchcock, and Jim Woodward

[1] Here for convenience I speak as if rates of changes of physical quantities at a time count as part of the state of the world at that time.

[2] Here I have assumed determinism, but a similar point holds under indeterministic laws.

to determine what a rock will do in the next 0.05 seconds, one would have to specify the exact present state of the entire earth.

What to make of this chasm between the two sorts of determination? One reaction would be to utterly renounce the folk model. The most extreme portions of Russell (1913) can be seen as advocating that reaction. That reaction would be too extreme. The folk model is useful. It seems to capture important features of the world. So there must be *something* right about it. What?

Norton (ch. 2 in this volume) answers: the folk model is approximately correct, in certain limited domains. Here is the idea. When an old scientific theory is superseded by a new one, sometimes the new theory allows us to derive that, in a certain limited domain, the old theory is approximately correct. For example, one can derive from General Relativity that the classical theory of planetary orbits is approximately correct, provided that spacetime isn't too curved. Norton argues that the folk model can be recovered as approximately correct, in an analogous way.

On this picture, our fundamental laws have a very special feature. They are such as to make the folk model approximately true in certain domains (including the domain of the mundane comportment of medium-sized dry goods).[3] Not all laws have that feature. Some fundamental laws, represented by perfectly respectable differential equations, do not make the folk model even approximately true in any domains at all.

So there is a story to tell about how *our* laws yield the approximate truth of the folk model in certain domains. My aim is to tell part of that story.

5.1 Extreme Locality

Why do ordinary objects behave in ways that fit the folk model? The first step in answering is to explain why any objects exist in the first place. Why isn't there just an amorphous soup, for example, or no stable matter at all? Answering such questions is beyond the scope of this paper.[4] But the questions are still worth posing, in order to emphasize that their answers are

---

[3] Compare Norton (ch. 2: section 4.4): 'Our deeper sciences must have quite particular properties so that [entities figuring in a superseded theory] are generated in the reduction relations.'

[4] For a particularly accessible introduction to quantum-mechanical derivations of the stability of matter, see Lieb (1990). For more technical treatments, see Lieb (1976) and the references therein.

not obvious. It is not automatic that a system of fundamental laws should allow for stable objects—huge quantities of particles that tend to move as a unit. Thankfully, our laws do.

Now: given that there are relatively stable middle-sized objects, why does their behavior roughly accord with the folk model? In answering, it is best to start by considering the *extreme locality* of the folk model.

Suppose that our fundamental physical laws are local, in the sense that the speed of light is the maximum speed of signal propagation.[5] That sort of locality guarantees that the rock on your desk is isolated from very distant goings-on. For example, when it comes to what the rock will do in the next ten seconds, it absolutely does not matter what is going on now at the surface of the sun.[6] That is because not even light could get from the sun to your rock in ten seconds.

Locality, in the above technical sense, gives a certain guarantee that objects are isolated from distant goings-on. But the guarantee only concerns *very* distant goings-on, since the speed of light is so high. Example: for all the guarantee says, the observable behavior of your rock in the next second depends on whether someone blinks right now at the opposite end of the Earth. (Remember that light can cover the diameter of the Earth in a twentieth of a second.) Indeed, for all the guarantee says, everything on the Earth could be so massively interconnected that any change anywhere on the Earth would make huge, unpredictable differences everywhere else within a twentieth of a second.

Thankfully, things aren't nearly so interconnected. When it comes to the behavior of the rock sitting on your desk in the next second, you can pretty much ignore what's going on right now at the other end of the Earth. It's not that you have a guarantee that your rock's imminent behavior is *utterly independent* of what's going on at the opposite end of the Earth. For if there were a supernova there right now, that would certainly make a big difference to your rock in less than a second (Field 2003, p. 439). You don't have a guarantee of *absolute* isolation. Instead, you have an assurance that, subject to some very weak background conditions (for example, the condition that there will be no gigantic explosions, and that the mass of the Earth will remain roughly unchanged), the rough behavior of your

---

[5] More carefully: suppose that the physical state at any point of spacetime is nomically determined by the state on time-slice of the back light cone of that point.

[6] Here the relevant 'now' is, say, the one determined by your rest frame.

rock is independent of what is going on at the opposite end of the Earth. Indeed, you have a similar assurance that the rough behavior of your rock is independent of what is going on down the block.[7] That is what it means for the generalizations that figure in the folk theory to be 'extremely local'.

In short, when it comes to the rough behavior of your rock, you can often treat it as if it were isolated from distant influences. Note the qualifications, though. Your rock isn't *really* isolated from distant influences. For example, the exact microscopic trajectories of the rock's molecules are sensitive to goings-on in the next room, due to gravitational effects. But who cares about the exact trajectories of rock molecules? When it comes to getting around in the world, the rough macroscopic behavior of rocks is much more important.

Now: *why* is it that you can treat the rough macroscopic behavior of your rock as if it were independent of distant influences? There are two factors.[8]

The first factor is that the forces acting on your rock from afar are either negligibly tiny or nearly constant. There are four forces to consider: strong, weak, electromagnetic, and gravitational. The strong and weak nuclear forces fall off in strength dramatically at greater than atomic-scale distances. Electromagnetic forces are stronger and longer-ranged, but—around here anyway—there are few strongly charged macroscopic objects. So when it comes to the strong, weak, and electromagnetic forces, objects are only subject to tiny distant influences.

The one remaining force, gravitation, operates with significant strength even at long distances. But in our neighborhood of the universe, mass distributions do not rapidly and massively fluctuate. So gravitational forces on Earth don't change much over short distance or time scales. As a result, we can treat the gravitational force as a fixed background. Relative to that background, changes in distant matters make only negligible gravitational differences to medium-sized objects on Earth.[9] The bottom line is that differences in distant matters of fact only make for tiny differences in the forces acting on ordinary objects.

---

[7] You have that same assurance even if the laws turn out to violate relativistic locality (due to quantum-mechanical entanglement, for example).

[8] I am indebted in the following two paragraphs to helpful correspondence with Jim Woodward.

[9] The same cannot be said of larger objects. For example the tides depend on differing gravitational forces (due to the moon) at different places in the Earth. So when modeling the tides, we are not free to think of the Earth as being in a near-constant gravitational field. Thanks here to Frank Arntzenius.

Here enters the second factor: statistical-mechanical considerations show that tiny differences in the forces acting on the rock are very unlikely to affect its rough macroscopic behavior. How does that go? On the assumption of determinism, one standard story is that the fundamental laws supply a probability distribution over initial conditions of the universe. That probability distribution induces a probability distribution over the states of typical rocks sitting on tables—a distribution that counts it as unlikely that small differences in forces would affect the rough behavior of those rocks.

Putting the two factors together, we can conclude that differences in distant matters of fact are unlikely to make a difference to the macroscopic behavior of your rock. So when the folk model says that the behavior of your rock depends only on the nature of the rock, and on the strong interactions that come the rock's way (e.g. shaking of the table), it does not go too far wrong. Furthermore, what goes for your rock goes for many ordinary objects.

5.2 Default Behavior, and the Importance of Isolation

The folk model ascribes default—or 'inertial'—behaviors to many systems. And the model says how such systems are disposed to deviate from their default behaviors, if they encounter interference (Maudlin 2004). For example, the default behavior of the rock on your desk is to just sit there and do nothing. And your rock is disposed to slide along the desk if pushed.

Given this framework, one can partially represent the causal structure of a situation with a graph. Each node in the graph represents a system, and arrows represent interactions in which one system perturbs another from its default behavior.[10]

However, the framework is useless unless systems can to a large degree be treated as isolated from their surroundings. For consider a system whose rough behavior is sensitive to a wide range of variations all over the place. That system will not have a single behavior that is stable enough to be usefully treated as a default. Furthermore, no manageable list of deterministic dispositions will capture interesting regularities about the behavior of the system. And if one insisted on representing such a system

---

[10] Such graphs are similar to the 'interaction diagrams' from Maudlin (2004: 439).

in the sort of causal graph just described, the system would require so many incoming arrows that the graph would be useless.

In contrast, the folk model is useful to us because so many systems can be treated as isolated from so much of their environments. As a result, the generalizations that figure in the folk model are fairly simple, and the associated causal graphs (of ordinary situations) are fairly sparse (see Woodward in ch. 4 of this volume).

So it really is crucial to the applicability and success of the folk model that many systems can be treated as isolated from much of their environments. The same is true of many special sciences. For example, consider the second law of thermodynamics, according to which closed (isolated) systems never decrease in entropy. Strictly speaking, the law *never* applies to reasonable-sized systems, since long-range gravitational effects ensure that such systems are never completely isolated. The law has practical applications only because many systems (e.g. gases in sealed cannisters) can be treated for many purposes as if they were isolated.

5.3 Sensitive Systems

The rock sitting on your desk can be treated as isolated from much of its environment. But not all ordinary systems can be treated as isolated in this way. For example, some rocks may be precariously balanced; others may be a hair's breadth away from cracking in half. In other words, some systems are *sensitive*: their rough behavior depends sensitively on small differences in forces—and hence on distant matters, even when the dependence is not mediated by strong nearby interactions. How well does the folk model fit such systems?

In general, of course, the folk model needn't fit such systems very well at all. But here on Earth, many sensitive systems are either *detector-like* or *quasi-chancy*. And the folk model accommodates detector-like and quasi-chancy systems quite well. I will explain each category in turn.

5.3.1 Detector-Like Systems

A system is *detector-like* if it is sensitive to distant influences, but only those of a very particular kind (or of a small number of kinds). Examples: devices that measure tiny seismic vibrations, fancy light detectors, cosmic ray detectors, and spy devices that eavesdrop on distant rooms by bouncing lasers off window panes. The rough behavior of such systems is sensitive to distant influences. And such influences needn't be mediated by a strong interaction. For example, a good light detector can register the presence of a single photon. But such systems are not sensitive to just *any* old distant influence. Seismic detectors only are sensitive to vibrations in the ground, light detectors to light (and usually just light coming from a particular direction), and so on.

Detector-like systems can easily be incorporated into the folk model. For though they are sensitive to distant influences, they are sensitive to only very particular distant influences. So they can be treated as isolated, *excepting* the particular influences that they detect. Think of it this way. The folk model is useful because so many objects can be treated as isolated from so much. Insensitive systems (such as rocks on tables) are insensitive to distant influences, and so are well described by the folk model. But detector-like systems are *almost completely* insensitive to distant influences, since they are sensitive to such a narrow range of distant influences. As a result, they too are well described by the folk model.

So much for detector-like systems. Before turning to quasi-chancy systems, however, we will need some background on the nature of statistical explanation.

5.3.2 *Statistical Explanation*

Some physical processes are downright ruled out by fundamental dynamical laws. For example, classical mechanics downright rules out a process in which a motionless, isolated particle suddenly accelerates though it is subject to no force.

That is one way for fundamental laws to explain why a particular sort of process does not occur. But it is not the only way. For example, consider a process in which an unsuspended boulder hovers in mid-air. Such a process is perfectly compatible with the fundamental dynamical laws (all that is needed is a sufficient imbalance between the number of air molecules hitting the bottom of the rock, and the number hitting the top). But nevertheless our laws make such a process exceedingly *unlikely* (Price 1996). Perhaps the laws are indeterministic, and they ascribe a very low chance to such a process. Or perhaps the laws are deterministic, and only a very small range of lawful initial conditions lead to the occurrence of such a process. In the former case, the explanation appeals to the objective chances that figure in the indeterministic laws. In the latter case, the explanation appeals to an objective probability distribution over initial conditions of the universe (see Loewer 2004 and Albert 2001). But either way, the explanation depends on an objective probability distribution over lawful histories. That distribution determines which sorts of histories the laws count as likely or typical, and which sorts the laws count as unlikely or anomalous.[11] The following discussion will appeal to such objective distributions.

5.3.3 Extreme Quasi-Chancy Systems

Some systems are sensitive to a great range of distant influences. An extreme example of such a system is the *Brownian amplifier*: a device that includes a tiny speck of dust haphazardly floating in a sealed glass container. The amplifier makes a sound every ten minutes: a whistle if the speck's last fluctuation was to the left, and a beep if it was to the right.[12] Even under the assumption of determinism, the Brownian amplifier is *incredibly* sensitive to distant influences. For example, consider a tiny change in the amplifier's distant environment: displacing a 1 lb moon rock by one foot. Very shortly, the pattern of sounds produced by the amplifier will be completely different in the original and the displaced-rock scenarios.

It seems therefore that the amplifier *cannot* be treated as isolated (or even as almost isolated). For its behavior—even its rough macroscopic behavior—depends on the state of pretty much every chunk of matter for miles around. But there is a trick. Think of the amplifier as having a chancy disposition: the disposition to beep-with-chance-50%-and-whistle-with-chance-50%. That *chancy* disposition is *stable* with respect to distant influences. In other words, if you treat the amplifier as if it were a chancy device, faraway goings-on will not affect the *chances* you should ascribe to it.

The fundamental laws license your treating the amplifier in this way. Here is why. Restrict attention to creatures with powers of observation and control rather like ours. The fundamental laws make it extremely unlikely

---

[11] Of course, such explanations are worthwhile only when the histories in question are grouped into relatively natural categories. For example, the laws will count *any* single history—specified in microscopic detail—as extremely unlikely. But that fact doesn't make every history anomalous. Thanks to Roger White for raising this objection.

[12] The Brownian amplifier is a variant of a device described in Albert (2001).

that such creatures can do better in predicting or controlling the amplifier's behavior than to treat it as a device that produces a random sequence of beeps and whistles.

To evaluate this claim, consider a gambler who repeatedly places bets (at fair odds) on what sound the amplifier will make next.[13] If the gambler can do better than to treat the amplifier as a 50/50 chance device, then she will be likely to win money over a long sequence of bets. She can do that only by following an appropriate rule. For example, suppose that the gambler thinks that the machine is very likely to produce the same sound as it last produced. Then she will follow the rule: 'Bet on beep whenever the last sound was a beep.'[14] Or suppose that she thinks that the machine is somehow coupled to the value of the Euro. Then she might follow the rule 'Bet on beep whenever the Euro just increased in value with respect to the Dollar.'

The fundamental physical laws make it extremely unlikely that any such rule would allow the gambler to cash in. That is the sense in which the laws make it unlikely that we can do better than treat the amplifier as a 50/50 chance device, in predicting its behavior.[15]

Here is why the laws make it unlikely that the gambler cashes in. For the purposes of this discussion we may assume that the laws are deterministic.[16] Now consider rules of the form: 'Bet on beep whenever condition C holds', where condition C is one that the gambler is capable of detecting before placing her bet. The laws make it very likely that, of the sounds that Brownian amplifiers make after condition C holds, about half are beeps. For example, the laws make it likely that, of the sounds that Brownian amplifiers make immediately after the Euro has risen with respect to the dollar, close to half are beeps.

And why is *that*? Why is it that creatures like us are not capable of sensing conditions that distinguish upcoming beeps from whistles? After all, we have assumed determinism, and so we have assumed that such conditions exist. (One such condition is a gigantic particle-by-particle specification of every state of the world that leads to the machine beeping next.) The

---

[13] Assume that the gambler must bet several minutes in advance of the sound in question.

[14] To 'bet on beep' is to bet that the next sound the machine will make will be a beep.

[15] The connection between randomness and 'invariant frequencies under admissible place selections' is inspired by Von Mises' definition of an infinite random sequence (see van Lambalgen 1987).

[16] The arguments carry over in a straightforward way under indeterminism.

ISOLATION AND FOLK PHYSICS   115

reason is that creatures like us cannot detect such complicated conditions. We can detect relatively simple macroscopic conditions, such as 'the rock is on the left side of the desk'. With the help of special apparatuses, we can even detect a certain very limited range of conditions concerning microscopic matters. But remember that the Brownian detector is sensitive to the position of nearly every chunk of matter for miles around. It is also sensitive to the detailed trajectories of the air molecules in the chamber that holds the dust speck. A condition that managed to pick out upcoming beeps would have to put detailed, horrendously complicated constraints on all of these matters, and more. Creatures like us have no hope of detecting such conditions.

There is one loose end. Where did the 50 percent come from? Why is it that for any condition we can detect, that condition is followed by beeps approximately *50 percent* of the time? Why not 10 percent? Or no stable percentage at all? Strevens (1998) has offered a beautiful answer in the tradition of Poincaré's (1905) 'method of arbitrary functions'.[17] Here is the basic idea.

Suppose that you must choose a color scheme for a black-and-white dart board.[18] Your goal is to guarantee that close to half of the darts thrown at the board land in a black region. The thing for you to do is to choose a scheme that (1) alternates very rapidly between black and white, and (2) is such that in any smallish square region, about half of the region is colored black. One such scheme is an exceedingly fine checkerboard pattern.

Such a scheme is a good idea because the distribution of landing-places on the board is likely to be relatively smooth.[19] For example, it is unlikely that any player will be accurate enough to cluster all of her throws in

---

[17] See also Keller (1986), Diaconis and Engel (1986), and Engel (1992).
[18] This example is adapted from Diaconis and Engel (1986).
[19] There are three subtleties here. *First*: the notion of smoothness employed here is *not* the technical notion of being continuous and infinitely differentiable. Instead, in the present case it is that the probability density for different dart locations varies slowly on the length scale set by the fineness of the checkerboard pattern. (An appropriately generalized notion of smoothness applies to other cases. Throughout this chapter, it is this notion of smoothness that I employ.) *Second*: the smoothness of a distribution depends on how the space of outcomes is parameterized. In the present case, the relevant parameterization is the natural one, in terms of distances on the board as measured in standard units (Strevens 1998, p. 241). *Third*: since the conclusion concerns a whole sequence of throws, the relevant distribution is over *sequences* of landing places. That way, the resulting notion of smoothness rules out bizarre dependencies between the throws. For example, it rules out a distribution according to which the first throw is uniformly distributed, but subsequent throws are guaranteed to land in the same spot as the first one.

a single square millimeter of the board. As a result, conditions (1) and (2) make it extremely likely that about half of the darts will land on black. In other words, the board's color scheme allows a weak *qualitative condition* on the distribution of landing-places (that the distribution is smooth) to provide a near-guarantee that about half of the darts land on black (Strevens 1998, p. 240).

The dynamics of the Brownian amplifier accomplish an analogous trick. Any smooth probability distribution over the state of the amplifier's environment makes it very likely that the amplifier will beep about as often as it whistles. Indeed, something stronger is true. Think back to the gambler who suspects that an increase in the strength of the Euro indicates an upcoming beep. Any smooth probability distribution over the state of the amplifier's environment makes the following very likely: of the sounds that the amplifier makes immediately after the Euro has gotten stronger, close to half are beeps. In other words, a gambler who guides her betting by the condition of the Euro is unlikely to cash in. And the same is true for any other condition simple enough for creatures like us to detect. That is why, no matter what (relatively simple) condition a gambler uses to select her bets, the laws count it as very unlikely that she will cash in by betting on the amplifier. In other words, the laws count it as very likely that creatures like us (who are trying to predict the sounds that the amplifier will make) can do no better than to treat the amplifier as if it were a 50/50 chance device.[20]

So the folk model *can* apply to the amplifier. The amplifier is *quasi-chancy*: it has stable dispositions to *act as a certain sort of chance device*. Understood in this way, it is reasonable to treat the amplifier as isolated from distant influences.

The same is true of many other systems that are sensitive to a wide variety of distant influences. By the time a system is sensitive to a wide enough range of distant influences that it no longer counts as detector-like, it very often ends up *so* sensitive that we can do no better than treat it as if it were genuinely chancy. The folk model can accommodate such

---

[20] Strevens (2003, s. 2.5) contains a detailed discussion of where smooth distributions come from. I do not understand the situation well enough to know the relationship between that explanation, and the one given above. But I acknowledge a debt to Strevens (1998) and Strevens (2003) in my thinking on these matters.

quasi-chancy systems by ascribing to them stable dispositions to produce particular chance distributions.

5.3.4 Less Extreme Quasi-Chancy Systems

The Brownian amplifier is the most extreme sort of quasi-chancy system, since creatures like us absolutely cannot do better than treat it as a 50/50 chance device. It is worth considering systems that are less extreme in this respect. Consider, for example, roulette wheels in casinos in the 1970s. One might think that humans cannot do better than treat such wheels as roughly uniform chance devices. But it turns out that measurable conditions (the initial velocities of the wheel and the ball) yield significant information about what quadrant the ball will land in. Indeed, gamblers have attempted to use such information to exploit Las Vegas casinos (see Bass (1985), as cited in Engel (1992, p. 96)).

So, unlike the Brownian amplifier, people *can* do better than to treat 1970s roulette wheels as uniform chance devices. Nevertheless, doing better requires the sort of information and knowledge of detailed dynamics that few people possess. So for most purposes, it is still a good approximation to think of the wheels as if they were uniform chance devices. The same is true for many ordinary systems. Such systems are not so sensitive as to *utterly rule out* that creatures like us could do better than treat them as chance devices. But they are sensitive enough that they are indistinguishable from chance devices by people with the sort of information ordinarily available. Such systems include precariously balanced rocks, leaves fluttering to the ground, light bulbs that are poised to burn out, crash-prone computers, and perhaps even the ping-pong-ball devices used to choose lottery numbers.

There is a tradeoff between simplicity and generality in whether to treat such systems as chancy. On the one hand, one can treat them as deterministic, in which case they will count as sensitive to a wide range of factors, and as having quite complicated dispositions. The nodes that represent the systems in causal graphs will have many incoming arrows. In representing the systems this way, one gains generality at the cost of complication. On the other hand, treating such systems as if they were chancy will simplify matters greatly, since they will count as isolated from much more of their environments. Such representations gain simplicity at the cost of accuracy and generality.

5.4 Conclusion

Folk models of everyday situations are enormously useful. What makes them useful is that so many ordinary objects can be treated as isolated from so much of their environments. As a result, we can often ascribe to objects salient default behaviors, from which they may be perturbed by interactions of only very particular kinds.

An important reason we can treat so many systems as isolated in this way is a combination of *in*sensitivity, and *super*sensitivity. Some systems are insensitive to small differences in forces and initial conditions. In combination with an appropriate statistical assumption, this licenses us to treat such systems as isolated. Other systems are very sensitive to differences in forces and initial conditions—but many such systems are *so* sensitive that we do better to treat them as chancy devices. By ascribing chancy dispositions to such systems, we can again treat them as mostly isolated. Again, this is licensed by an appropriate statistical assumption.

In attempting to reconcile the folk model of causation with fundamental physical laws, Russell focused on dynamical laws. Little wonder, then, that he thought a reconciliation was impossible. For as we have seen, the dynamical laws do *not* on their own underwrite the usefulness or approximate correctness of the folk model. They do so only in conjunction with statistical assumptions: either probabilistic laws, or laws that supply a probability distribution over initial conditions.[21]

References

Albert, D. Z. (2001). *Time and Chance*. Cambridge, MA: Harvard University Press.
Bass, T. A. (1985). *The Eudaemonic Pie*. Boston: Houghton Mifflin.
Diaconis, P. and Engel, E. (1986). 'Some Statistical Applications of Poisson's Work', *Statistical Science*, 1(2): 171–4.
Engel, E. (1992). *A Road to Randomness in Physical Systems*. Berlin: Springer.
Field, H. (2003). 'Causation in a Physical World', in M. Loux and D. Zimmerman (eds) *Oxford Handbook of Metaphysics*. Oxford: Oxford University Press.

---

[21] Compare to Field (2003): 'the notion of causation, like the notions of temperature and entropy, derives its value from contexts where statistical regularities not necessitated by the underlying [dynamical] physical laws are important'.

Keller, J. B. (1986). 'The Probability of Heads', *American Mathematical Monthly*, 93(3): 191–7.
Lieb, E. (1976). 'The Stability of Matter', *Review of Modern Physics*, 48: 553–69.
─── (1990). 'The Stability of Matter: From Atoms to Stars', *Bulletin of the American Mathematical Society*, 22: 1–49.
Loewer, B. (2004). 'David Lewis' Humean Theory of Objective Chance', *Philosophy of Science*, 71 (5): 1115–25.
Maudlin, T. W. (2004). 'Causes, Counterfactuals and the Third Factor', in J. Collins, N. Hall, and L. Paul (eds). *Causes and Counterfactuals*. Oxford: Oxford University Press.
Norton, J. D. (ch. 2, this volume). 'Causation as Folk Science'.
Poincaré, H. (1905). *Science and Hypothesis*. New York: Dover.
Price, H. (1996). *Time's Arrow and Archimedes' Point*. New York: Oxford University Press.
Russell, B. (1913). 'On the Notion of Cause', *Proceedings of the Aristotelian Society*, 13: 1–26.
Strevens, M. (1998). 'Inferring Probabilities from Symmetries'. *Noûs*, 32: 231–46.
─── (2003). *Bigger than Chaos: Understanding Complexity through Probability*. Cambridge, MA: Harvard University Press.
van Lambalgen, M. (1987). 'Von Mises' Definition of Random Sequences Reconsidered', *Journal of Symbolic Logic*, 52(3): 725–55.
Woodward, J. (ch. 4, this volume). 'Causation with a Human Face'.

6

Agency and Causation

ARIF AHMED

Can causation be explained in terms of human agency? The view that it can I call the 'agency theory'. The most common objection to it is that it is circular. To be an agent means to bring something about. 'Bringing about' is a causal notion. So we need to understand 'cause' before we understand 'agency': the correct explanation must go the other way. Agency is to be explained as a kind of causation (e.g. Hausmann 1997, p. 17). I shall call this the 'circularity objection'.

Whatever other problems the agency theory faces, I claim that this one can be answered. My aim here is not to give a full defence of the agency theory. It is only to show that the appeal to agency need not involve circularity.

I begin (Section 6.1) by distinguishing various commitments of the agency theory and attempt to settle which ones are prima facie circular. Then (Section 6.2) I describe the most recent and plausible version of the agency theory. This theory *is* vulnerable to the objection. The authors squarely face the objection and I describe and criticize their reply. In Sections 6.3 and 6.4, I consider what kind of argument *would* rebut the circularity objection. In Section 6.5, I try to provide such an argument. In Section 6.6, I consider objections.

6.1 Varieties of the Agency Theory

A number of views fall under my description of an 'agency theory'. Here are some of them:

I am most grateful to E. Craig, H. Mellor, S. Olsaretti, M. Potter, and also to an anonymous reader of this volume.

(A) Constitutive: the concept of agency features in a necessary, sufficient and elucidatory condition for 'X is a cause of Y'.
(B) Constitutive: wherever there is causation there is agency.
(C) Phylogenetic: the concept of causation arose in the Western world out of the concept of agency.
(D) Ontogenetic: you and I have a concept of causation that arose from our concept of agency.
(E) Ontogenetic: you and I have a concept of causation that arose from our being agents.
(F) Counterfactual: nobody would have had the concept of causation if they hadn't had the concept of agency.
(G) Counterfactual: nobody would have had a concept of agency if they hadn't been agents themselves.

Somebody has held each of these doctrines. Berkeley (1710, ss. 25–9) seems to have been committed to all of them except (A) and (C). Reid (2001) held (D). Collingwood (1940, p. 291, 322) held (C) and (D). Von Wright (1975, pp. 108–13) held (G). Menzies and Price (1993) hold (A), (D) and (F) and perhaps (E) and (G). Which of these claims is vulnerable to the circularity objection?

The circularity objection doesn't apply to (B), (E), (G) or (F). This is clear enough for (B) which says nothing about direction of explanation. So any reliance of one's *grasp* of agency on a prior grasp of causation doesn't touch it. Similarly for (E) which only says that the concept of causation arises from our *being agents*, not from our having the concept of agency. And the same goes for (G). As for (F)—any holistic or 'locally holistic' model of concept possession allows for mutual counterfactual dependence. It is therefore compatible with (F) that the concepts of agency and causation might be mutually counterfactually dependent like those of Newtonian mass and force: I could not have possessed either without the other. So (F) is arguably compatible with its converse and therefore arguably immune to the circularity objection.

But the objection does apply to (A), (C), and (D). It applies to (A) because of the asymmetry of 'elucidatory' (which means: an account that can be understood by somebody who has no prior grasp of what is being accounted for). It applies to (C) and (D) because 'arose from' implies temporal precedence. In this paper I will examine a recent and sophisticated version of the agency theory. It is committed to (A) and (D) and is therefore vulnerable to the circularity objection. In the next section I shall outline the theory itself and describe and evaluate their defence of it.

6.2 Menzies and Price

6.2.1 Their Version of the Agency Theory

Menzies and Price (henceforth 'MP') say that X is a cause of a distinct event Y just in case bringing about X would be a means by which a free agent could bring about Y (1993, p. 191). And their account at least involves (A).

Two things apparently need clarification: the notion of 'bringing about' and that of 'means'. MP's position (1993, p. 194) is that the first notion doesn't need clarification because we directly observe it. The second notion is explained probabilistically. Bringing about X is a means for bringing about Y just in case, from the point of view of a free agent in a position to bring about X, $Pr(Y/X) > Pr(Y)$. That is: the probability of Y's occurring conditional on X's occurring is greater than the unconditional probability of Y. In brief: X causes Y iff from the point of view of an agent able to bring about X, doing so makes Y more likely.

It will clarify matters further if we compare this theory of causation with neighbouring ones.

The appeal to agency distinguishes the account from naïve as well as standard probabilistic accounts of causation (an example of a 'standard' account being Suppes 1970, ch. 2). The naive theory says that A causes B iff $Pr(B/A) > Pr(B)$ without restriction to any particular viewpoint. The trouble with this is that the right-hand relation is symmetric[1] whereas the causal relation is not. However when probability is calculated *from an agent's point of view* the matter changes. For an agent able to bring about or prevent A 'the conditional probability of B on A' ceases to mean the probability you would give to B on *learning* that A. It becomes the probability you would give to B on *bringing it about* that A. This solves the difficulty, for then '$Pr(B/A) > Pr(B)$' need not be interpreted as a symmetric relation. MP say that in regarding an event as a free action we

---

[1] Proof: By Bayes' Theorem we have $Pr(B/A) = Pr(B.A)/Pr(A) = Pr(A/B).Pr(B)/Pr(A)$. So $Pr(B/A)/Pr(B) = Pr(A/B)/Pr(A)$. So the quantity on the left of this equation is greater than one iff the right hand quantity is too.

in effect create for it an independent causal history. For example, in enquiring whether one's manipulation of an effect Y would affect the probability of its normal cause X, one imagines a new history for Y, a history that would ultimately originate in one's own decision, freely made. (1993, p. 191)

The effect of this restriction is that one doesn't regard the effect Y as making its normal cause X more likely. For the probability of X's happening is independent of that of Y once the latter has an independent causal history. The effect of adding an 'agency' story to the naive probabilistic account of causation is therefore to render it insensitive to *spurious evidential correlations*. A spurious evidential correlation is a case where there is a correlation between two types of event X and Y such that X is evidence for but not a cause of Y—e.g. between its raining (X) and its having been cloudy five minutes before (Y). The agency theory does what the naive theory cannot: it motivates a distinction between 'genuine' and spurious correlations and does not predict that the latter are causal.[2]

The account also differs from other agency-based theories of causation such as those of Berkeley and Collingwood. For Berkeley the paradigmatic causal relation is that between willing to do something and doing it. For Collingwood it is that between getting somebody to do something and his doing it. But for MP the paradigmatic relationship is that between means and end. The end is the aim in view and the means is the deliberate action that you perform in order to get it. They are saying that X causes Y iff a suitably placed agent—one who had X in his power—would regard X as a means of bringing about Y.

6.2.2 The Circularity Objection

The Price/Menzies account is the offspring of an agency and a probability account of causation and faces many objections that arise against one of its parents. I want to examine *one* of the difficulties inherited from the 'agency' side of things, namely the circularity objection. MP express that objection as follows:

The second and perhaps most frequently cited objection to the agency theory is that it necessarily involves a vicious circularity. The apparent circularity is plain

---

[2] The correlation between an effect and its cause is of course not the only kind of spurious correlation. Another kind is the correlation between common effects. But the Menzies/Price account can deal with this in the same way.

enough in our statement of the agency approach, according to which A is a cause of a distinct event B just in case *bringing about A* would be an effective means by which a free agent could *bring about B*. This statement contains two references to 'bringing about', which seems on the face of it to be a causal notion: doesn't an agent bring about some event just in case she causes it to occur? It would appear, then, that agency accounts are vitiated by the fact that they employ as part of their analyses the very concept which they are trying to analyse. (1993, p. 193)

Clearly this is at least an objection to (A).

As they have stated it the circularity arises twice: once for each occurrence of 'bring about'. But there are reasons for thinking that it arises in two *other* ways as well. First, they appeal to the notion of a 'means'—and isn't saying that A is a means to B just saying that A causes B? And second, they appeal to the idea of a free agent—and isn't a 'free agent' in their sense one whose actions are either uncaused or satisfy certain causal conditions? Indeed isn't it essential to their case that that is what they say (recall the talk of an 'independent causal history')?

So the MP analysis of causation makes *four* appeals to causation on the right hand side. If these are ineliminable then the theory is heavily circular and (A) is false. If it is impossible to grasp what is meant by 'bring about', 'means', and 'free agent' without already grasping what is meant by 'cause', then MP's theory cannot be elucidatory.

6.2.3 Their Reply

MP claim to have a systematic means of defending their theory against this and other objections. Noting the similarity between their account of causation and dispositionalist accounts of secondary qualities, they observe that each objection to the former is analogous to some objection to the latter and can be met in essentially the same way. The 'secondary quality' analogue of the circularity objection is that 'in stating that to be red is to look red to a normal observer in normal conditions, the theory employs the concept "red" on the right-hand side of the analysis' (1993, p. 194).

In the case of secondary qualities the reply is this:

The key to seeing that this theory is not circular is to recall that colour terms, like the terms for other secondary qualities, can be introduced by ostension. Thus a novice can be introduced to the concept 'looks red' by being shown samples of red: the salience of the redness in the samples and the novice's innate quality space should suffice for him to grasp the fact that the samples look alike in a certain respect. Thus, the dispositionalist explanation of the concept 'red' need not fall into the trap of circularity. The dispositionalist can explain the concept 'looks red' without having to rely on any colour concept. (1993, p. 194)

The analogous reply for agency is that:

we all have direct experience of acting as agents. That is, we have direct experience not merely of the Humean succession of events in the external world, but of a very special class of such successions: those in which the earlier event is an action of our own, performed in circumstances in which we both desire the later event, and believe it is more probable given the act in question than it would be otherwise. To put it more simply, we all have direct personal experience of doing one thing and thence achieving another. It is this common and commonplace experience that licenses what amounts to an ostensive definition of the notion of 'bringing about'. In other words, these cases provide direct non-linguistic acquaintance with the concept of bringing about an event; acquaintance which does not depend on prior acquisition of any causal notion. (1993, p. 194)

Before evaluating that reply I will make one clarification. The passage might give the impression that MP are trying to defend (D) and not (A) against the circularity objection. Their very next sentence reinforces this: 'It is true that Hume himself considers the possibility that "the idea of power or necessary connexion" *derives from* the fact that we are conscious of "the internal power... of our will."' (my emphasis). In fact the passage will if true sustain *both* (A) *and* (D) against the circularity objection (though obviously it *entails* either). For the passage argues against the concept of agency being essentially an offshoot of that of cause. It says that you *can* grasp agency without grasping causation.

6.2.4 Problems With Their View

We noted that the problem of circularity arose in four distinct places in the MP theory: the two occurrences of 'bringing about', the occurrence of 'means', and the occurrence of 'free agent'. Let us consider these and ask whether the reply just considered is satisfactory; and let us follow the secondary quality analogy by assuming that their term 'direct experience' means something like 'sensory acquaintance' (I consider other possibilities at objection (iii)).

First consider the relation of means to end. MP say: 'we all have direct personal experience of doing one thing and thence achieving another'. I might agree with this sentence if you take away two letters. What we all have direct experience of is doing one thing and *then* achieving another. I cannot see that we have direct experience of anything that distinguishes 'thence' with its causal implications from 'then' which lacks them. I cannot see that the sequences in which ends are brought about by means look any different from the sequences in which the former merely *succeed* the latter. At least that distinction is no more visible among certain means–ends sequences than it is among sequences to which I stand only as a spectator.

What about 'bringing about'? The same remark of MP's is relevant to this part of the circularity objection: 'We all have experience of doing one thing and thence achieving another.' But it seems to me false to say that we have direct experience of *doing* something. Of course there are senses of 'experience' in which this is true; but if 'experience' means sensory acquaintance then it is not true. That was the point of Wittgenstein's question: if you subtract the fact that my arm goes up from the fact that I raise my arm, what is left? If the question was about what *sensory* difference there is between these cases then the answer is: often enough none at all.[3]

Finally, consider the notion of a free agent. It is rather hard to see how MP's reply engages with the problem. The problem was supposed to be that 'free agent' means one whose actions have a special kind of *causal* history. Suppose that we *do* have sensory acquaintance with free agency. So we have sensory acquaintance with the fact that our actions have a special kind of causal history. But then we can distil the concept of causation *directly* from experience without having to go via the MP definition in terms of means, ends, and agent probability. If I know from experience what it is to be a free agent then I must *already* have the concept of cause. Thus this part of the reply seems to be a defence of (E), *not* a defence of (A). It implies only that *when* I act freely I come to have causal concepts. But this is independent of whether you can grasp causation in terms of independently graspable *concepts* of means, ends, and agent probability.

Now MP might respond as follows. They might say that the description of 'free agency' as a causation-involving notion was inessential to its strict characterization and only appropriate *ex post facto*. The correct account is

---

[3] You might say that the idea that an agent brings about something involves the notion of agent causation, a notion that can be understood independently of the event-causation that MP are trying to analyse. So the appearance of circularity is an illusion, at least for this bit of the analysis. I'd be more sympathetic to this reply if I had the first idea of what agent causation was.

this: you have certain experiences of free agency, experiences that give you the concept thereof. From these you distil a certain means of calculating agent-probabilities, that is, ones evaluated under the assumption that you are acting freely in some respect. *We* describe these means in causation-involving way, for example, 'you assume that your acts have a special causal history'. But *you* do not yet think of them like this. For you the calculation of agent-probability is primitive: you don't think of it as involving any causal judgements at all. It is only *once* you have got this ability that you are ready to be introduced to the notion of cause: you are to judge of causes by employing the already acquired method of evaluating agent probabilities, one that involves, for example, conditionalizing on propositions about the past as well as the contemplated act and so forth. You might *then* think of agent probabilities as conditional probabilities calculated on the assumption that the event conditioned upon has no causes, or whatever. But describing them as such in no way reflects the conceptual priority of causation. An analogy: you might learn how to distinguish such geometrical figures as ellipses, circles and hyperbolae by their algebraic equations. You might later learn that and how these are all conic sections. And you might later describe this episode of your youth by saying that you then learnt how to distinguish the conic sections. But nothing in this implies that you need to grasp the concept of a conic section in order to grasp a technique describable *ex post facto* as 'a technique for distinguishing conic sections'.

I think that this is a reasonable enough reply apart from one point. Note that if it works then it will rebut the charge of circularity at *all four* points. For MP can now say that our judgement of agent-probability is *primitive*, and it is only *ex post facto* that we can introduce *any* of the four causation-involving expressions by which we may characterize it. The agent has certain experiences of free agency; these somehow provoke him to evaluate conditional probabilities in a certain way: but he is then quite innocent of the idea that these reflect *means* by which *a free agent* may *bring about* certain ends. He just forms a judgement of probability and acts accordingly. At this point he is ready for MP to tell him what a cause is. When they do, he will perhaps describe agent-probabilities in causal terms. But until then he can be said to make judgements of agent-probability on an experiential basis without going *via* the causal judgements that they ultimately support.

However there is one point on which I find this analysis implausible. Much of the rest of this paper is an attempt to subtract this implausibility.

MP are asking experience to play a genetic role that it cannot fill. Suppose I had a certain experience when and only when I performed an action. The trouble is that nothing about a blank 'experience of action' (construed on the model of 'experience of redness') will give me the concept of free action needed to underpin the theory. Remember, they eliminated spurious correlations by invoking the agent's point of view. From an agent's point of view, probabilities conditional on events in my power have to be calculated in a certain way. This strategy relies on our having a concept of agency rich enough to support the inference from 'X is in my power' to 'I need to conditionalize on X under certain special assumptions e.g. holding fixed the probabilities I attach to propositions about the past'. But how can a 'direct experience', that is a mere sensory episode, endow me with that kind of concept? Here the analogy with secondary qualities falters. The reason 'looks red' lacks the conceptual connections enjoyed, for example, by 'Newtonian force' is precisely that ostension is all you need to grasp the former concept. But the requisite concept of agency or 'bringing about' is more like a primary than a secondary quality. Its relative conceptual richness makes it more like 'hardness' than like 'looking red':

It is not possible to distil the concept of hardness solely out of the experiences produced by deformation of the skin which is brought into contact with the hard object, for it is not possible to distil out of such an experience the theory into which the concept fits.   (Evans 1985, p. 270)

and the same could be said of agency or 'bringing about', at least of such a concept of agency as MP need.

For example: suppose we know that punching people in adulthood is normally correlated with playground violence. So I might think that this man is more likely to have had a violent childhood if he goes ahead and punches that other man. But suppose that I have forgotten about my own childhood. Then when *I* am wondering whether to punch this man I will *not* think that if I do then it is more likely that I was violent in the playground. According to MP this reflects a difference in the way we evaluate the conditional probabilities. And my question is: how can the occurrence of a distinctive accompanying *experience* in the first-person case tell me that I should evaluate it in a different way?

I conclude that MP have not given a successful defence of their theory against the circularity objection. But they have pointed us in the right direction. And I think we can begin to go further in that direction if we attend to the distinction between genetic and analytic questions.

6.3 Concept-Introduction and Concept-Possession

Return to the analogy with secondary qualities. The analogue of the circularity objection was this: you cannot grasp what it is for something to satisfy '*x* looks red' without already grasping what it is for something to satisfy '*x is* red'. And the reply was that you *can* grasp the former without already grasping the latter:

> The key to seeing that this theory is not circular is to recall that colour terms, like the terms for other secondary qualities, can be introduced by ostension. Thus a novice can be introduced to the concept 'looks red' by being shown samples of red: the salience of the redness in the samples and the novice's innate quality space should suffice for him to grasp the fact that the samples look alike in a certain respect. Thus, the dispositionalist explanation of the concept 'red' need not fall into the trap of circularity. The dispositionalist can explain the concept 'looks red' without having to rely on any colour concept.   (Menzies and Price 1993, p. 194)

In Section 6.2.4, I criticized MP's application of this analogy on the grounds that the concept of agency cannot be introduced by simple ostension; or if it can then it will not be suited to playing the role required by their analysis of causation.

But there are actually *two* points in this passage that need to be distinguished; and even though the first of these does not carry over to the case of agency, the second may. These are as follows:

(i) It is possible to introduce the concept 'looks red' *by ostension* to one who does not yet grasp the concept 'is red'.
(ii) It is possible to grasp the concept 'looks red' without yet grasping the concept 'is red'.

Now consider the analogues of (i) and (ii) for agency and causation; these are respectively (iii) and (iv):

(iii) It is possible to introduce the concept of agency *by ostension* to one who does not yet grasp the concept of cause.

(iv) It is possible to grasp the concept of agency without yet grasping the concept of cause.

If (i) is true then (ii) is plausible; if (iii) is true then (iv) is plausible. The role of (i) in rebutting the circularity charge in the case of secondary qualities was that it gave grounds for (ii). But once we have (ii), (i) has done its work: a dispositionalist about colour who denied (i) but accepted (ii) on some other grounds would be just as secure against charges of circularity. Similarly the role of (iii) in rebutting the circularity charge in the case of causation was that it gave grounds for (iv). But once we have (iv), (iii) has done its work: an agency theorist who denied (iii) but accepted (iv) on some other grounds would be secure against charges of circularity.

After all, the difficulty was that no such agency theory as MP's *could* be elucidatory. For it is said, anybody who grasped the terms of their *analysans* must already have grasped their *analysandum*. But if (iv) is true then that is wrong. If (iv) is true then the account *could* be elucidatory. So if it is to be rejected then it must be on some grounds other than circularity, for example, that it gets the extension wrong. That sort of objection is beyond the scope of this paper.

The (i)/(ii) and (iii)/(iv) distinctions are both instances of a more general one between the processes whereby we can or normally do acquire a concept, and what possession of that concept actually amounts to. Constraints on the former can have consequences for the latter or at any rate consequences for dependences amongst the latter. If ostension is all it takes for somebody to acquire the concept C (e.g. 'looks red') then one might be able to show that possession of C will not depend on possession of some distinct concept D (e.g. 'is red'). But the questions are in principle separable.

MP's approach was to establish (iv) on the basis of (iii). In Section 6.2.4 I criticized this on the grounds that (iii) is false. But the falsity of (iii) doesn't imply the falsity of (iv). And it is (iv) that really matters. If we can find some grounds for (iv) that are independent of (iii) then we can rebut the circularity objection. That is what I aim to do.

Before doing this it is necessary to emphasize what *kind* of thing must be meant by 'grasp of agency' if it is to serve the role required by the agency theory. That role is evident enough: the whole point of introducing agency into a probabilistic theory was to introduce an asymmetry into one's judgements of probability. What they require therefore is for at least the following to be true: that whether or not one grasps the concept of causation, one can form a judgement of the probability of q conditional upon one's *bringing about p*, a conditional probability that differs from the standard conditional probability of p on q *sans phrase*.

This introduces an asymmetry because the agent-probability of q conditional upon p so understood may be different from the probability of q even though the agent-probability of p conditional upon q is the same as the probability of p. For the agent-probability of q conditional upon p is the standard conditional probability of q conditional upon *my bringing it about that p*. And while it is a theorem of the probability calculus that $\Pr(p/q) = \Pr(p)$ iff $\Pr(q/p) = \Pr(q)$, it does *not* follow, nor is it in general true, that $\Pr(p/I \text{ bring it about that } q) = (<)\Pr(p)$ iff $\Pr(q/I \text{ bring it about that } p) = (<)\Pr(q)$.

The requisite 'grasp of agency' must therefore be this: one has a grasp of agency to the extent that one is able to form judgements of agent-probabilities, that is, to the extent that one can be said to have an opinion about the probability of q conditional on *my bringing it about that p*. And the dispute over the Price/Menzies account may therefore be settled in the latters' favour if we can establish the following: that one might make judgements of agent-probabilities *without* being able to form causal judgements. Our endeavour will then be directed towards that question.

6.4 Circularity and Concept-Possession

Before I begin this enterprise, however, I need to say what would count as success in it. There has been a lot of loose talk so far about 'concepts' and their 'possession' or 'grasp'. What the circularity objection amounts to and how it should be answered depends on what 'grasp of a concept' amounts to.

To see this, consider a rather crude account of concept-possession: to possess the concept F is just to be able to sort any group of objects into two classes: the Fs and the non-Fs. The crudity of this model can be seen in two consequences. It entails that to possess a concept F is to possess all concepts co-extensional with F. It also entails that you can, for example, grasp the concept *female fox* without grasping the concept *female*. For it is logically possible that you should be able to sort any group of objects into female foxes and non-(female foxes) even though you cannot sort every group of objects into females and non-females.

On this view the circularity objection says that you need to be able to distinguish instances from non-instances of causation in order to be able to distinguish instances from non-instances of agency. But this is wrong. It is logically possible that someone could say correctly of any given event (under a given description or perceptual presentation) whether it was an instance of his agency, even though he could not say correctly, of every pair of events, whether they were causally related.

In fact, on this crude model of concept-possession very little would be circular. Suppose I analysed 'fox' as 'male fox or vixen'. That would not be circular. You *could* on this model possess the concept *male fox* and the concept *vixen* without possessing the concept *fox*. You might be somebody who could identify (distinguish instances from non-instances of) male foxes and vixens without yet being able to identify foxes. The analysis is therefore *not* circular because the *analysans* is a Boolean combination of concepts that you can grasp without grasping the *analysandum*.[4] But in some important sense it *is* circular. We therefore need a different account of concept-possession for the question of circularity to bite.

The reason this is so is that my crude model of concept-possession makes every concept inarticulate. It doesn't show how the possession of any concept C might involve the possession of any other concept G. Charges of circularity trade on the fact that the exercise of certain concepts relies on the exercise of other ones. Whereas my crude model is only plausible for 'secondary qualities': those concepts whose exercise is so to speak intellectually primitive.

For all other concepts—and as I argued in Section 6.2, agency had better be one of these—we need to proceed in some other way. We need some criterion for settling whether the exercise of a concept or the making of a certain judgement is reliant upon the exercise of another concept or the making of another type of judgement. In particular we need a criterion for settling whether the making of judgements of agent-probability relies upon the making of causal judgements. Could it be true for example to say that

---

[4] For the right-hand side is a Boolean combination of the concepts *male fox*, and *vixen*. Of course it is *also* a Boolean combination of the concepts *male*, *fox* and *vixen*. But on the account I am considering, your inability to grasp *those* three concepts without grasping the concept *fox* is a necessary but not yet sufficient condition for the analysis to be circular.

somebody thinks that $\Pr(q/I\ bring\ it\ about\ that\ p) = \Pr(q)$ even though he does *not* have any opinion as to the causal relevance or irrelevance of p to q? How are we to settle this?

Well, I can't give a straightforward account of what it *is* for somebody to possess one concept rather than another, and even if I could it would take us too far afield. But we can approach the question by proceeding indirectly. Instead of asking: what *is* it to possess a concept let us ask instead: on what *grounds*—usually partially but rarely exclusively *behavioural* grounds—can we say that somebody possesses or lacks a concept? Although answering that question in full generality would be just as hard as answering its predecessor, still we can suggest answers for one or two concepts: and these may be enough to settle the question of circularity. For we can settle it in MP's favour by showing the following: that we might have grounds in somebody's behaviour for saying that they possessed the concept of agency or agent-probability while also having grounds for withholding causal concepts from them. If we *could* have grounds of that sort then we *would* have reason to say that you can grasp the concept of agency without grasping that of causation.

Arguing in this way would in fact retain the parallel with secondary qualities. MP say that we can explain 'looks red' by ostension of appropriate samples to somebody who doesn't grasp 'is red', on the grounds that 'the salience of redness in the samples and the novice's innate quality space should suffice for him to grasp the fact that the samples look alike in a certain respect' (1993, p. 194). But talk of the 'innate quality space' is elliptical for talk about something that isn't really space at all. There is no suggestion that the subject's phenomenal concepts stand in literally spatial relations to one another or to anything else. Nor are we being told anything about his intracranial topography. Talk of 'similarity space' is just a way of representing relations of greater or less *perceptual* similarity, where we can understand perceptual similarity in some such behavioural way as this: subjects are shown to find a perceptually more similar to b than to c if, when they are so conditioned as to respond in one way to b and in another way to c, they will respond to a as they do to b. If that *is* what is meant by 'similarity space' (whether it is innate or conditioned doesn't really seem to matter) then the passage just quoted from MP is saying this: it is certain patterns of response on the part of the subject that give us reason to suppose that he grasps what it is for something to look red. And for this claim to be enough to defuse the charge of circularity it must further be being supposed (quite plausibly) that just that pattern of response does *not* give grounds for supposing that the subject grasps what it is for something to *be* red. Thus there could be somebody of whom it was reasonable to assert on the basis of his behaviour that he grasped what it is for something to look red and to deny on the same basis that he grasped what it is for something to *be* red.

I want to pursue a similar strategy with regards to agency and causation. I want to suggest that behaviour could be imagined that provides grounds for attributing a notion of 'agent probability' but not one of causation. But we are to understand 'behaviour' in the last sentence rather more broadly than in the context of chromatic dispositionalism. It is no longer enough to imagine patterns of behaviour as brutish as those that might be displayed by someone who has just learned 'looks red' if only for the reasons stated at the beginning of this section. The behaviour that grounds attributions of agent-probability will include the exercise of *some* other judgements. It is only that those other judgements will not include causal ones. If such forms of behaviour really are possible, then it seems that we will have good reason to believe (iv) (or as much of it as MP need). Now I argued that (iv) provides as much of a defense against circularity for the agency theory of causation as does (ii) for the dispositionalist theory of colour. Hence my argument will, if successful, establish MP's conclusion—that the circularity objection is unreasonable—while dispensing with their disputable premise (iii).

6.5 Agent-Probability

How might an agent manifest a grasp of agent-probability? What does he have to do or say to make a reasonable observer think it likely that he is making judgements of the form $\Pr(p/I \text{ bring about } q) = n$? I will give three interlocking arguments. The first one runs like this. (1) It is reasonable to attribute judgements of *non*-agent probability on the basis of the agent's betting behaviour. (2) It is reasonable to attribute judgements of *non*-agent probability on the basis of the agent's betting behaviour only if it is *also* reasonable to attribute judgements of *agent*-probability on the same basis.

Therefore (3) it *is* reasonable to attribute judgements of agent-probability on the basis of the agent's betting behaviour.

The second argument proceeds as follows. (4) It is reasonable to attribute causal judgements to an agent on the basis of his betting behaviour only if it somehow explains that behaviour. (5) Causal judgements do *not* explain the agent's betting behaviour. Therefore (6) It is *not* reasonable to attribute judgements of causation on the basis of the agent's betting behaviour.

The third argument uses the conclusions of the first two arguments as premises: (3) It is reasonable to attribute judgements of agent-probability on the basis of the agent's betting behaviour; (6) It is *not* reasonable to attribute judgements of causation on the basis of the agent's betting behaviour. And it has a third premise: (7) It is possible for an agent's behaviour to include the relevant betting behaviour and nothing else that makes it plausible to attribute causal concepts to him. The conclusion is therefore (8). It is possible to manifest grasp of agent-probability without manifesting grasp of causation. The conclusion of this third argument lends plausibility to—though indeed it does not entail—(iv): the claim that you can *grasp* agent-probabilities without grasping causation. I conclude that MP were right to reject the circularity objection.

6.5.1 *The First Argument*

Since the argument is valid I need only give grounds for believing its premises.

Premise (1) might be motivated as follows. Suppose that an agent is offered a choice between two bets. One of the bets is risky but the potential payoffs are high; the other bet is a sure thing but the payoff is low. To fix things, let us imagine that he is offered a choice between a free bet that pays out $1 if it rains tomorrow and nothing if it doesn't, and a free 'bet' that pays out $k in either case. Then the payoff matrix looks as in Table 6.1.

The suggestion is that the probability with which the agent judges that it will rain tomorrow may reasonably be taken as that value of k for which

Table 6.1

	Rain	No Rain
Bet 1	$1	$0
Bet 2	$k	$k

the agent is indifferent between Bet 1 and Bet 2 (we know that $0 \leq k \leq 1$ when k takes that value since if $k < 0$ then the agent will definitely prefer Bet 1 and if $1 < k$ then the agent will definitely prefer Bet 2).

This is a well-known proposal that obviously needs refinement. Rather than give a full-blown defence of (1) (which would involve extravagant lengthening of this essay) I want to suggest that at least some of the more obvious refinements are unlikely to affect the first argument, since its use of (1) is independent of those refinements.

It may be said that (1) is true only on the assumption that the subject's marginal utility for money in dollars is constant. But in fact this assumption is false, as follows from the fact that you would miss $1 less if you were a millionaire than if you only had $2 to your name. We could get around this by saying instead that we can use betting behaviour, not to assign absolute values to subjective probabilities, but simply to *compare* them. Let us say that a 'bet on p' is a free bet that pays out $1 if p and $0 if $\sim p$. Then we can weaken our suggestion to the following. If the value of k for which an agent is indifferent between a bet on p and Bet 2 is greater than the value of k for which the agent is indifferent between a bet on q and Bet 2, then the agent regards p as more likely than q. The plausibility of this proposal is independent of whether the agent attaches constant marginal utility to dollars, as is evident from the fact that it can plausibly be rephrased in such a way as to eliminate reference to k altogether. An agent may be said to think p more likely than q just in case he prefers a bet on p to a bet on q, in the sense of 'bet on p' just stipulated. This is a genuine weakening of the original suggestion and therefore weakens any conclusions that it supports. But since our whole concern is with *relative* probabilities, or more precisely *relative conditional* probabilities, it will become apparent that this weakening is harmless.

It may also be said that the original suggestion for attributing judgements of subjective probabilities implies an implausible degree of determinacy in the latter, for it implies that there is just *one* value of k for which the agent is indifferent between Bet 1 and Bet 2. But is it really plausible that an ordinary human being would prefer Bet 1 to Bet 2 when $k = 0.895421$ but not when $k = 0.895422$? However this objection does not affect the argument at all, for it does not affect the proposal of the last paragraph, that we may attribute *comparative* subjective probabilities on the basis of preference behaviour that involves no such fine discrimination.

I turn to premise (2) which states that it is reasonable to attribute judgements of *non*-agent probability on the basis of the agent's betting preferences only if it is *also* reasonable to attribute judgements of *agent-probability* on the same basis. The premise will seem plausible if we consider cases where premise (1) seems to break down in a certain way. These are cases where (1), when interpreted as a method for attributing relative probabilities, seems to imply the absurdity that there are propositions p and q such that the agent regards p as more likely than q *and* regards $\sim p$ as more likely than $\sim q$.

To see this let p be the proposition that the subject will have a cup of tea in the next five minutes and let q be the proposition that on its next toss this fair coin will land heads. Now it is evident that any rational subject prefers a bet on p to a bet on q; for one can accept the bet on p, have the tea and collect the money. At least one will prefer the bet on p if one has access to a kettle and if having a cup of tea carries no positive costs. (If it is *not* costless then one will prefer a bet on p if it costs less than 50 cents on the assumption that the marginal utility of US currency is constant for amounts $\leq \$1$. In what follows I shall ignore this complication.) So the subject thinks p more likely than q. But it is *also* evident that any rational subject prefers a bet on $\sim p$ to a bet on $\sim q$; for one can accept the bet on $\sim p$, not have a cup of tea, and collect the money. At least one will prefer the bet on $\sim p$ if having a cup of tea carries no *negative* costs. So the subject thinks $\sim p$ more likely than $\sim q$. But this is absurd: it follows from the standard axioms that $\Pr(X) + \Pr(\sim X) = 1$ for any X and hence $\Pr(q) < \Pr(p)$ if and only if $\Pr(\sim p) < \Pr(\sim q)$.

Perhaps this anomaly owes something to the first-personal character of the proposition p: the agent is presented with a proposition that he would express as '*I* will have a cup of tea'. But this is not so: if he was the only person in a room then we could equally have taken p to be the proposition that *somebody* in the room has a cup of tea: he will still prefer a bet on p and a bet on $\sim p$ to *either* a bet on q *or* a bet on $\sim q$.

Should we then abandon the thought that you can attribute judgements of relative probability to somebody on the basis of his betting behaviour? I think we should not; instead we should state more precisely just what *kinds* of relative probability judgement are warranted by the behaviour. The reason that the 'cup of tea' example causes problems is because the agent does *not* regard the probability of p to be independent of his *betting* on p. Betting on p will raise the probability that p is true; betting on $\sim p$ will raise the probability that p is false. Instead of saying that the betting behaviour of the agent reflects the relative probability he attaches to the propositions that he is offered a bet on, we should say that the betting behaviour of the agent reflects the relative probability that he attaches to the propositions that he bets on *conditional on his betting* upon them. This allows us to deal straightforwardly with the 'cup of tea' example. The fact that the agent prefers a bet on p to a bet on q does not show that he thinks $\Pr(q) < \Pr(p)$; it shows that he thinks that $\Pr(q/\text{I bet on } q) < \Pr(p/\text{I bet on } p)$. And this, in conjunction with the fact that the agent prefers betting on p and betting on $\sim p$ to a bet on either of q or $\sim q$, does *not* imply that he believes an absurdity. It shows that he believes that $\Pr(q/\text{I bet on } q) < \Pr(p/\text{I bet on } p)$ and that $\Pr(\sim q/\text{I bet on } \sim q) < \Pr(\sim p/\text{I bet on } \sim p)$. And these two propositions are compatible with the probability axioms.

But if that is what we say in the 'cup of tea' case then why not *also* say it in other cases too? I mean: why not say it when the subject does *not* think that the probability of his betting one way or another makes any difference to the probability of the proposition that is bet upon? Suppose you prefer betting on the proposition that it rains tomorrow to betting on the proposition that it snows tomorrow. Then on the present proposal what that shows by *itself* is *not* that you think rain more likely than snow but only that you think that rain *given that you bet on rain* is more likely than snow *given that you bet on snow*. If in addition we assume (as indeed we normally do) that the subject believes that $\Pr(\text{snow/bet on snow}) = \Pr(\text{snow/bet on rain})$ and that $\Pr(\text{rain/bet on rain}) = \Pr(\text{rain/bet on snow})$ then we *can* infer that the subject thinks snow more likely than rain. This account of how to extract evidence for probabilities from the subject's choices of bet will make the same attributions as the earlier suggestion in cases where the subject regards his choice of bet as probabilistically inert. In cases where it is *not* inert, as in the cup of tea case, it will make more plausible attributions (since they are compatible with the probability axioms). So assuming this second account to be true in *both* the cases where the bet is inert *and* in cases where it is not yields an account that is just as plausible, at least in its probability attributions, as the more piecemeal story which applies the first account to the former case and the second account to the latter case. Moreover, assuming the second account to be correct in both cases will allow us to provide a uniform account of both cases. I suggest therefore that we *should* apply it to both cases. We should say that somebody who prefers a bet on *p* to a bet on *q* thereby evinces the judgement that Pr(*q*/I bet on *q*) < Pr(*p*/I bet on *p*).

If that argument is correct then it shows the following: that *if* it is reasonable to attribute comparative judgements of probability to an agent on the basis of his betting behaviour then they are not (or not just) judgements on the relative probability of the propositions that he is offered a bet on. They are judgements of the probability of those propositions *conditional upon* the proposition *that he makes that bet*. But the latter are exactly what was meant by judgements of agent-probability: they are probabilities conditional upon the agent's performing a certain *action*. Thus my 'cup of tea' example has shown this: that *if* we can estimate an agent's probability judgements on the basis of his betting choices then we must be able to estimate his *agent-probability* judgements on that basis. This was premise (2). If my arguments for premises (1) and (2) make their conjunction plausible then conclusion (3) is plausible too, because (3) follows from (1) and (2).

It might be objected that (2) is false on the grounds that we *can* attribute probability judgements to an agent without attributing *agent*-probabilities to him. After all, the example relied on the assumption that the agent was aware that betting was something that was *up to him* rather than a reflex or tic. If he had thought (perhaps mistakenly) that it was a reflex or tic then his 'choice' would *not* reveal anything about his judgements of agent-probabilities, that is, probabilities conditional on a proposition that describes him as *doing* something. A reflex or tic cannot be regarded as something that you *do*. If an agent thought that 'his' bets were made by movements of his *hand* or *vocal cords* but not by *him* then his taking of a bet on *p* when a bet on *q* was available would show something about his probability judgements. It would show that he thought Pr(*q*/My hand bets on *q*) < Pr(*p*/My hand bets on *p*). But it wouldn't show anything about his agent-probabilities, for he doesn't think that the event of his hand making one or another bet could be regarded as an action.

I agree that an agent who was thus deluded would not by his behaviour evince a judgement of relative agent-probabilities. But nor would he evince a judgement of relative probabilities at all, for he would not evince a *preference* for a bet on *p* to a bet on *q*. It is only if you think that you have a *choice* between a bet on *p* and a bet on *q* that your betting on *p* evinces a preference for betting on *p* over betting on *q*. The mere unexploited availability of a bet on *q* does not show that you think *p* more likely than *q* unless you are *aware* of that availability. So while the consequent of (2) is indeed false in this case, so is its antecedent. Hence the case does not constitute a counterexample to (2).

One might object that (2) is false on the grounds that it is not judgements of *agent-probability* that we should attribute to the agent but judgements of *causal relevance*. Thus the right thing to say in the 'cup of tea' case is *not* that the betting-behaviour is explained by judgements of probability conditional on propositions describing my potential actions. We should rather explain it by attributing to the agent the belief that his betting on taking a cup of tea is *causally* relevant to whether or not he has a cup of tea. I shall discuss this point in the context of premise (5) in the second argument.

That—except for the unfinished business of the last paragraph—concludes the case for (3), the proposition that it *is* reasonable to attribute judgements of agent-probability on the basis of the agent's betting behaviour.

6.5.2 *The Second Argument*

The second argument proceeds as follows. (4) It is reasonable to attribute causal judgements to an agent on the basis of his betting behaviour only if it somehow explains that behaviour. (5) Causal judgements do *not* explain the agent's betting behaviour. Therefore (6) It is *not* reasonable to attribute judgements of causation on the basis of the agent's betting behaviour. The second argument is valid so in order to defend its conclusion I need only defend its premises.

In defence of (4): suppose that nothing in the agent's behaviour needs explanation by reference to causal beliefs on the part of that agent. Then it would be gratuitous to suppose that that agent's behaviour evinces causal beliefs. I am not saying here anything special about *causation*. Suppose, for example, that nothing is added to our explanation of the agent's behaviour by saying that he makes judgements about horses as opposed to other horse-like mammals (e.g. donkeys or mules). This might be true if he responds in exactly the same way towards horses as he does towards mules or donkeys (e.g. he might try to sit on them or feed them sugar). In this case nothing in his behaviour is explained by saying that he has beliefs specifically about *horses* as opposed to equine mammals in general. And so it is plausible that we have no reason to *attribute* horse-beliefs to him in particular. The intuition that makes this conclusion plausible is the one that makes (4) plausible.

It is harder to motivate (5). That premise states that we do *not* need to advert to causal judgements to explain an agent's betting behaviour. I shall motivate it in two steps. First I describe and then cast doubt upon the supposed route by which the agent's causal beliefs *are* supposed to explain his betting behaviour. Then I shall suggest an alternative explanation that is compatible with the undisputed facts that suggested such a route.

The argument for thinking that there is such an explanatory route might be put like this. It seems that if the argument for premise (2) of the first argument is correct then premise (5) must be false. The argument for premise (2) implied that an agent's betting behaviour typically evinces judgements of agent-probability. The fact that I prefer a bet on p to a bet on q, if it evinces a judgement of relative probability at all, evinces the judgement of agent probability that $\Pr(q/\text{I bet on } q) < \Pr(p/\text{I bet on } p)$. But it is plausible that *this* judgement is *itself* either identical to or motivated by a judgement of *causal* relevance.

To see why, consider the 'cup of tea' example again. I said that we can attribute judgements of relative probability to an agent *only* if we attribute relative judgements involving *agent*-probability, for example, the probability of having a cup of tea conditional on his betting that he does as against the probability of the coins landing heads conditional on his betting that *it* does. But what explains these judgements other than judgements of causal relevance? Isn't the best explanation of the agent's betting preferences the fact that he thinks that his betting on having a cup of tea has a positive *causal* relevance to whether or not he has a cup of tea, whereas he thinks that betting on heads is causally *ir*relevant to heads? If so then we ought to conclude one of two things. *Either* the 'cup of tea' example does evince agent-probabilities but these are themselves explained by judgements of causal relevance. *Or* the notion of agent-probability is irrelevant and its attribution uncalled for: the agent's behaviour is explained solely by his causal beliefs by themselves. (This second way of putting the point is equivalent to the objection to premise (2) of the first argument that was stated but not rebutted in the penultimate paragraph of 6.5.1.) In either case premise (5) is false.

But I can't see *how* causal beliefs are supposed to do the explaining. There are two ways this might happen corresponding to the two disjuncts of the objection. First: it might be because the causal belief generates the action by itself without going via anything that deserves to be called a judgement of agent probability. But how does it do that? How can my belief that a certain asymmetric binary relation holds between two event-types, or possible event-tokens, or variables, motivate me to act in a certain way? We could dramatize the question by defining 'effecthood' to be the converse of causation. Then why is it that the belief that A-type events stand in a *causal* relation to B-type ones motivate me to bring about this A-type event whereas the belief that A stands in an *effectual* relation to B does *not* motivate me to bring about A? Unless we say something more about what distinguishes causal beliefs from effectual beliefs, we are left with an explanatory gap at the crucial point. Of course you might say this: what distinguishes causal beliefs from effectual beliefs is that causal beliefs but not effectual ones just *are* beliefs about what will happen if you *act* to bring about the first relatum. But that gives the game away to MP (though see more on this in my reply to objection (i)).

Second: one might say that causal beliefs explain actions via the judgements of agent-probability that they produce. So while the *content* of causal beliefs can be specified independently of beliefs about one's actions, nevertheless causal beliefs can explain the latter because of ancillary beliefs that we have about actions, which together with those causal beliefs imply something about agent-probabilities. One might do this by (i) attributing to our agent the belief that free actions are correlated with *no* causes but only with their effects, and hence knowledge of what an event of type A typically *causes*, but not knowledge of what typically causes *it*, will be relevant to my decision to bring about A. Or one might do this by (ii) saying, for example, that one regards causes as raising the probability of their effects conditional on their non-effects, and that when we judge of agent-probabilities we conditionalize not just on the contemplated action itself but also on those states of affairs or events to which we think the action is causally irrelevant. But *ad* (i): only a few philosophers think that one's free acts have no causes. Most people act freely, or think they do, while also thinking that their acts *do* have causes: 'I'm punching you because of what you said to me'. But if the attribution of causal beliefs is a compulsory part of the explanation of action then it should apply to more than just a few philosophers. And *ad* (ii): this just pushes the problem back a step, because now I want to know *why* when p describes a contemplated action we suddenly start to conditionalize upon the non-effects of p rather than just—as we normally do—upon those propositions whose truth or probability we regard as *epistemically* independent of p.[5]

That concludes the first part of my argument for (5): now for the second part. It is not in dispute that the agent-probabilities that we attribute to somebody are *correlated* with his specifically causal beliefs if he has any. This is clearly illustrated by Newcomb's Problem, which is as follows. Suppose that a rich and generous psychologist (henceforth 'The Predictor') offers you two boxes, one opaque and one transparent. You have a choice between taking both boxes and taking just the opaque one. You get to keep whatever is in the boxes; you can see that the transparent box contains some positive amount k (less than a million dollars). The Predictor tells you the following: that yesterday he made a prediction concerning which boxes you would take. If he predicted that you would take the opaque box only, he put a million dollars ($1M) in it. Otherwise he put nothing in it. He has almost always made correct predictions with a variety of subjects including many of your own past time-slices. The payoff matrix looks as in Table 6.2. What ought you to do?

I am not concerned here with what the right answer is. It seems to me that there are powerful arguments for taking both boxes and also for taking just the opaque box. What concerns me is what the *intuitive* answer is, and I am in no doubt that the intuitive answer is that you should take both boxes.[6] After

Table 6.2

	Pred (Both Boxes)	Pred (One Box)
Both Boxes	$k	$(k + 1M)
One Box	$0	$1M

---

[5] Dummett (1978, pp. 349–50) implies that when judging of agent probabilities we *do* conditionalize or partition just on those propositions that we regard as epistemically independent of p. But then the arbitrariness pops up somewhere else: why is it that if p describes a contemplated action, the agent regards propositions describing the *non-effects* of p (e.g. those that describe events in the past) as being epistemically independent of p? The point, which is compatible with everything Dummett says, is that we are still left with an explanatory gap between specifically *causal* beliefs and the actions that those beliefs are supposed causally to explain.

[6] It is sometimes argued that if you are 100 percent certain that the psychologist has got it right then the intuitive answer is that you should take just one box. The way I am describing it, you are not quite 100 percent certain that this is so, if only because the psychologist has on occasion got it wrong. In this case I think intuition pretty unequivocally tells us to take both boxes. This is so even if the psychologist has got it right very nearly all of the time, say 99.9 percent of the time. Cf. Nozick (1990, p. 232).

all the $1M is either already there or it isn't; in either case you may as well take both boxes and get the extra $k. And this will remain true however small k becomes (as long as it remains positive). And this illustrates the fact that your agent-probability goes in step with your causal beliefs. The near-universal preference for taking both boxes, however small k is, evinces the judgement that Pr(Pred Both/I take both boxes) = Pr(Pred Both/I take one box); and this aligns with the equally extensive judgement that whether you take one box or two is causally irrelevant to The Predictor's prediction.

But the undisputed fact just illustrated, that one's agent-probabilities and hence one's actions are *correlated* with one's causal judgements, does not imply that the latter in any way *explain* the former. It might instead be that neither explains the other but that there is still a *common* explanation for both (compare: the correlation between a barometer's prediction and subsequent weather). What might this common explanation be?

Here is a suggestion. It is clear that correlations between variables in known cases cause one to make certain probability judgements concerning unknown cases. If I know that 85 percent of observed Alzheimer's sufferers were found to have been non-smokers then I will in the absence of other information consider *this* hitherto unobserved Alzheimer's sufferer more likely to have been a non-smoker than a smoker. But the relative probabilities that one assigns to propositions about unobserved cases cannot simply be read off from the relative strengths of correlations in observed cases. Among emeralds so far observed, grueness has been just as frequent as hardness (where something is *grue* if green if first observed before t, otherwise blue). Still one thinks it more likely that *this* hitherto unobserved emerald is hard than that it is grue. This suggests that we do *not* simply extrapolate from observed frequencies to arbitrary new cases; whether we do or not might depend on other beliefs that we have about those new cases. My belief that it is now after t affects my present judgements of likelihood of grueness but not my present judgement of likelihood of hardness.

I do not know any general explanation or justification for the fact that some but not other background beliefs inhibit extrapolation from observed to unobserved cases with regard to some but not other correlated properties. But there must *be* some explanation, and so there must *be* some additional variables whose values in any particular case, together with facts about our experiences and our background beliefs concerning that case, jointly explain the probability judgements that we make concerning it. The background belief that it is now later than *t* will inhibit my extrapolation of the correlation between emeraldhood and grueness; it will not inhibit my extrapolation of the correlation between emeraldhood and hardness. Similarly my belief that we are *here* at 300m above sea level will inhibit my extrapolation of the correlation between something's being water and its boiling at 100 degrees; it will *not* inhibit my extrapolation of the correlation between emeraldhood and either grueness or hardness. Now it may be true that these variations themselves depend on the extrapolation of *other* correlations, or lack thereof, to the present case, for example, a positive correlation between the boiling point of water and its height above sea level, or zero correlation between observed colour of an emerald and time. But that just raises again another question of the same type: what explains why I extrapolate from (lack of) correlation between *colour* and time but not from that equally evident correlation (or lack thereof) between *schmolour* and time (where two things have the same schmolour if they are both grue)? These are variations in what Carnap called the rate at which we learn from experience. We learn at different rates in different directions, and Hume's point was that experience by itself neither explains nor justifies the rate at which we learn from it in any given direction. Whatever the explanation is, whether it be doxastic or only neurophysiological, there seems no evident reason to suppose that it involves causal beliefs.

Now I suggest that both agent-probabilities, where they occur, and the causal beliefs that go with them, might both be explained in just this way. Consider the Newcomb scenario. We have observed a strong correlation between one's choice (one box or two) and The Predictor's prediction. But that doesn't mean that we will extrapolate this correlation to all future cases. In particular it may be that when one regards the proposition that one will take both boxes as something within one's present control, one will *not* extrapolate from that correlation, though one may extrapolate from some others. And it may be that this apparent anomaly is really not anomalous, but an instance of just the same variation in rates of learning from experience that we remarked in other cases. Therefore it may be that the fact that we ignore certain correlations when forming agent-probabilities is no more mysterious than, because explicable on the same kind of basis as, the fact that we ignore the correlation between emeraldhood and grueness when forming a judgement about *this* emerald. Moreover it may be that the very same processes that cause us to generate agent-probabilities on the basis of evidence are the ones that cause suitably educated agents to form judgements of *causal* relations. In this way the correlation between causal judgements and judgements of agent-probability might be compatible with the idea that neither judgement need explain the other.

The last three paragraphs have been extremely speculative and are not strictly necessary for such a case as I can make for premise (5). That case rested on two points that I should like to conclude by emphasizing. First, that there is no obvious explanatory route from causal judgements to the kinds of behaviour that evince judgements of agent-probability. And second, that the fact that there is a *correlation* between causal judgements and agent-probabilities does not mandate us to say that there must be *some* such explanatory route. These two points at least place the onus on the circularity objector: if he wants to press his point against MP then he must defeat our prima facie case for saying that causal judgements do *not* explain the betting behaviour that evinces judgements of agent-probability.

6.5.3 *The Third Argument*

The third argument has two premises for which I have already argued indirectly: (3) that it is reasonable to attribute judgements of agent-probability on the basis of the agent's betting behaviour; and (6) that it is *not* reasonable to attribute judgements of causation on the basis of the agent's betting behaviour. And it has a third premise: (7) that it is possible for an agent's behaviour to include the relevant betting behaviour and nothing else that makes it plausible to attribute causal concepts to him. Its conclusion is therefore (8) that it is possible to manifest grasp of agent-probability without manifesting grasp of causation.

I have said all that I can in favor of (3) and (6); it remains for me to argue for (7). That argument proceeds as follows: I describe a case of the sort that (7) says is possible; my description contains no evident contradiction; therefore a case of that sort *is* possible.

Imagine somebody who displays betting behaviour of the sort that we have described. Thus he might prefer a bet on rain to a bet on snow, and he may prefer taking two boxes in the Newcomb case to taking one box, and so on. If the first and second arguments are successful then this is grounds for attributing agent-probabilities to him but not yet causal beliefs. But imagine further that he is blind to causation in all other respects. Thus he is perfectly aware of just the same regularities in nature as the rest of us; and

AGENCY AND CAUSATION   147

he uses them for prediction and retrodiction in just the same way as you and me. Only he seems to have no use for any expression for anything that deserves the name of causation. Thus nothing in his *explanatory* strategies, for example, requires special use of anything like causation. This might happen if for example his explanations of events do not advert to any temporally directed and hence asymmetric binary relation. Insofar as he is able to give explanations of events it is by subsuming them under regularities; but he sees no reason in principle why the *explanans* should always be temporally prior to the *explanandum*: a generalization of the form 'Whenever A happens at t B happens at $t-1$' is for him in no way specially ill-suited as an explanation of B. He may indeed remark upon the fact that the more succinct explanations *do* generally run in one temporal direction; he may even be able to explain *this* fact in terms of some very high-level (non-causal) generalization. But beyond this he feels no need to go.

Now it seems to me: (i) that somebody like this evinces causal beliefs only in his betting behaviour if anywhere; and (ii) that such a person is possible. My grounds for (i) are really that this is my intuitive response to the example. I have two grounds for (ii). First, that such persons are possible because my description of them contains no evident contradiction; and second, that such persons are possible because such persons have been actual. Berkeley is one example, Russell another.

It follows from (i) and (ii) that (7) is true. And it follows from (3), (6) and (7) that (8) is true: It is possible to manifest grasp of agent-probability without manifesting grasp of causation. But (8) gives us grounds to believe MP's claim (iv): that it is possible to grasp the concept of agency (by which I mean agent-probability) without yet grasping the concept of cause.

What does this mean? I think it means that (A) is true: one *could* give an explanation of causation to somebody in terms of agency. Or at least it means that (A) survives the circularity objection (though it may be open to others). Suppose that we had somebody like our imaginary character described in my defence of (7). We could say to him: look, take all the event-types that you think are correlated with events of type A in certain circumstances. Now ask yourself *which* of those correlations would inform your judgements of agent-probabilities. That is: if B is correlated with A, would you be prepared to extrapolate that correlation to cases where you think that *this* A-type event is under your present control? If the answer is yes then the A–B correlation is one that supports a judgement of *causal* relevance; if no then it isn't. We are in this explanation exploiting the agent's capacity to conduct counterfactual thought-experiments (perhaps describable as 'off-line' reasoning). And we are exploiting his ability to form judgements of agent-probability. But if my arguments have been right then we are *not* exploiting anything that could be described as a prior grasp of causation itself. Thus (A) is true, or at least true for all that the circularity objector has said: the concept of agency features in a necessary, sufficient and elucidatory condition for 'X is a cause of Y'.

This completes my defence of MP against the circularity objection. I want to close by looking at a few objections. Before doing that it is worth summarizing the strategy of Section 6.5 and stressing in consequence how little I have shown

The implicit strategy of this section was that the whole issue of circularity should be interpreted as one of conceptual dependence: it is the question whether grasp of one concept, c_1, requires grasp of another, c_2. And I assumed that we can answer *that* question by adopting an 'interpretative' stance: can there be interpretative grounds for attributing c_1 to an agent while withholding c_2? When c_1 = agency and c_2 = causation, I have suggested why I think the answer is yes and hence why I find the circularity objection unappealing.

My concentration upon issues of conceptual dependence has resulted in a complete extrusion of questions about how we do in fact learn about causation and agency. It may indeed be true that we learn about them both together, with bits of our knowledge about the one relying historically upon bits of our knowledge about the other. It may also be true that man's awareness of himself as an agent itself went hand in hand with the flowering of causal concepts. Nothing I have said, therefore, is in any way a defence of (C) or (D). But I do not think that MP are committed to (C), and I think that much that is interesting in what they say survives repudiation of (D).[7] As I have interpreted the question of circularity, it is no more dependent on (C) or (D) than is any alleged circularity in the JTB theory of knowledge dependent on the actual interrelations between the onto- or phylo-genesis of the concept of knowledge and the concept of belief. The question I have addressed is whether the concept of agency makes a notionally separable

---

[7] And so much of what they say would survive the criticism of Woodward (2003, p. 126).

AGENCY AND CAUSATION   149

contribution to our intellectual life. This is independent of whether the actual story of its actual assimilation is in fact cleanly separable.

Nor do I have anything to say about whether the notion of agency contributes to the *meaning* of causation. It may be that the explanation just suggested is a reference-fixing rather than a sense-determining device; it may be that there are other more direct routes to a grasp of causation. But even if that is true it is compatible with (A); at any rate it is compatible with saying that (A) is not threatened by circularity.

6.6 Objections

(i) *Objection to 6.2.3*. You conclude the section by saying that according to MP you *can* grasp agency without grasping causation, and that is the view that you try to defend. But can't they be saying that there is no gap here—that grasp of agency just *is* grasp of causation? Indeed there is evidence that that *is* what Price thinks (Price 1991, p. 169).
Reply. We need to distinguish between 'grasp of agency' in the sense of actually being able to form agent-probabilistic judgements in decision-theoretic contexts, and 'grasp of agency' in the sense of being able to form judgements about how one *would* assess probabilities if one *were* counterfactually in control of some actual event. The former is what I said you can have without having a grasp of causation. But the latter is what we need in order to grasp causation, if grasp of causation requires the ability to make causal judgements about events that were not in fact in one's control. It may be that MP would wish to say that the latter just *is* grasp of causation; my point was only that the *former* does not suffice for it.

(ii) *Objection to 6.2.4.* The causal primitivist might welcome the sort of account that MP provide. For the causal primitivist might say that we grasp the causal primitive by having direct observation of at least one of its instances. Here the question becomes, *if* one thinks that one directly observes agential bringing-abouts, why does one need any further explication of the notion of causation. So why not just go full-blown primitivist here?
Reply. This amounts to saying that when we act we have direct experience of *causation*, not just direct experience of agency; it is true that MP say

things that suggest this line (1993, p. 195, n. 14). My first difficulty with this is that at best it is a defence of (E), not (A). It is saying that when you act you have experience of causation itself, and that is quite compatible with the view that you need to grasp causation in order to grasp agency, being a specification merely of the conditions under which you learn about the former and not of its conceptual ingredients. So causal primitivism is not a defence of (A) against the circularity objection.

My second difficulty is that Hume seems to me right to say that we no more have experience of causation in cases where we are agents than in cases where we are not. MP criticize him for concentrating on the experience of the will–action sequence rather than on the experience of the means–ends sequence. But what he says (that if all we experience elsewhere is constant conjunction then all we experience *here* is constant conjunction) applies to both cases. Then causal primitivism gives no special place to agency at all: it is just one of many kinds of experience that afford a grasp of causation. The causal primitivist theory may be true. But I am not concerned with whether the causally primitivist theory is true. I am concerned with whether the agency theory is circular.

(iii) *Objection to 6.2.4.* You considered the case where 'experience' is understood to mean 'sensory acquaintance', whatever *that* is. But it might mean a number of other things. (a) Sometimes 'You have experience of being an F' just means that you *are* an F. And (b) there is *an* everyday sense of experience—however we ought to cash it out—on which as Michotte (1963) showed, we *do* directly experience causation. Why then can we not experience agency in this sense too?

Reply. Ad (a) The trouble with this is that *being an F* is not sufficient for a grasp of F-ness. We are all mammals as well as agents. In this sense therefore we all have 'experience of being mammals'. But you can't infer that we have 'direct non-linguistic acquaintance' with the concept *mammal*. For many (human) mammals do *not* grasp the concept. So again the circularity objection stands. It is easy to see that we all have 'experience' of agency in the bare sense that we *are* agents. But it is hard to see what follows about grasping the concept.

Ad (b): It might be that a conceptually sophisticated person *can* directly observe agency even though it isn't given in mere sensation. I have said nothing against agency's being directly observable in this sense. But MP were saying how we *acquire* the concept of agency. They were claiming that we could acquire it without already possessing that of causation. So they need to exhibit a perceptual process that does *not* rely on its subject's already possessing the concept of agency *or* that of causation. And it must be one from which the subject can *gain* the former concept. To say that a conceptually well-equipped person can see agency is not enough. We need to show that they could do this while *lacking* the concepts of agency and of causation. Otherwise we don't have an account of how they *get* the concept of agency in the first place. That is why the analogy with secondary qualities is important to MP. If it worked, it would show how one could, through the operation of perception *unencumbered by theory*, achieve grasp of the concept of agency without already possessing that of causation.

(iv) *Objection to 6.2.4.* You quoted Wittgenstein's question about what is left over if I subtract the fact that my arm rises from the fact that I raise my arm. Isn't it the fact that I *try* to raise my arm?

Reply. That may be true. But I don't think it will help MP, for we are normally not *aware* of trying to do something when we successfully do it. In everyday activity the mind seems to be focused wholly on the thing that one is doing, or at any rate focused wholly outwards.[8]

(v) *Objection to 6.4.* You talk there about concept-possession etc. But isn't it a vague and in fact immaterial question what counts as grasp of a concept? So when you argue that somebody might possess the concept of agency without possessing that of causation, couldn't we reply: you might as well say that he *does* possess a grasp of causation but is ignorant about its extension. What hangs on what we say here?

Reply. I agree that no sharp line can be drawn between lacking a concept and grasping it but being ignorant of commonly known facts. But that is compatible with distinguishing between a firm and a weak grasp of a concept. And the circularity issue then becomes this: is it true that somebody who already had a grasp of agency could *improve* whatever grasp of causation he might be said to have by appeal to the Menzies/Price account? I have argued in effect that the answer is yes, on the grounds that somebody might have a firm grip of agent probabilities while lacking as

---

[8] Cf. O'Shaughnessy (2004, p. 198).

yet the kinds of belief and explanatory strategies that we should regard as relatively central to a grasp of causation.

(vi) *Objection to 6.5.2*. You argued there that we can explain agent-probabilities on the basis of the very same observed correlations that supposedly explain causal beliefs. But you had said at 2.4 that agent-probabilities *cannot* be derived from experience since no experience can place the right constraints on the way that we evaluate probabilities! Isn't this inconsistent?

Reply. No. At 6.5.2 I said that we evaluate agent-probabilities by *bringing* to experience certain propensities to respond thereto; and it is the hypothetical application of these propensities to events beyond those that normally trigger them that explains our causal judgements. More specifically the picture was this. Somehow, though not through a characteristic experience, we come to regard some situations as ones in which we are as we say free to act in one way or another (though this way of putting it probably over-intellectualizes it). In those situations we form conditional probabilities that are *based* on experience and knowledge of correlations, but in a different way from the way we form conditional probabilities when we are spectators. The difference arises because in circumstances where we regard ourselves as agents, we extrapolate differently from observed correlations than in cases where we are spectators.

(vii) *Objection to 6.5.3*. It is essential to the MP account, as you there imply, that it involves *counterfactuals*. To know whether Brutus' stabbing caused Caesar's death, it is not enough to conditionalize on Brutus' stabbing Caesar under the supposition that I am *now* in control of it. For now I know for sure that Caesar died then, and so conditionalizing on the stabbing under that supposition cannot raise its probability. What must be meant is this: *if* I *had* been in control of the stabbing, and if I *had* been ignorant of certain things that I now know, and if I *had* known certain things of which I am now ignorant (e.g. Caesar's state of health just prior to the stabbing), then I *would* have regarded the probability of death conditional on the stabbing as having been higher? But (a) this is a bizarre counterfactual: surely you are not claiming that such outré questions enter into the down-to-earth question whether the stabbing caused the death; (b) it may be that we cannot understand counterfactuals without causal concepts; (c) how are we to characterize the things that you are supposed not to know other than this: as the propositions whose truth was *causally dependent* on the stabbing?

Reply. I confess that this objection raises difficulties for the MP line that I cannot at once see how to answer. *Ad* (a) Perhaps one might retreat to making a claim of this sort: our judgements of what an event of *type* A in circumstances of *type* C causes or caused exploit the same statistical correlations as those that we exploit when judging whether to *perform* A in circumstances of type C, and this not because the latter depends on the former but because they have a common source. Thus we are explaining beliefs about type-causation rather than analysing propositions about token-causation. We are saying that the former are generated by knowledge of regularities in the same way as one's agent-probabilities are so generated. And my argument could then be understood as saying that this *explanation* involves no circularity. *Ad* (b) That may be true in many cases. But first, I was concerned only with whether the MP appeal to *agency* was a source of circularity, not with whether other ingredients of their account introduce it. And second, it may be that counterfactuals about what one would have *believed* in certain circumstances can be evaluated by some sort of off-line reasoning that makes no special appeal to causation. *Ad* (c) One might say that what we assume unknown in those hypothetical circumstances are just the propositions whose truth-values or probabilities we do not hold fixed when forming agent probabilities in analogous circumstances. And my argument could then be understood as saying that there is nothing circular in this. Again the circularity seems to come from the description of the hypothetical situation in which the agent's reasoning is supposed to take place rather than the appeal to agency itself. All of this is highly unsatisfactory and shows that if MP's account is to be made plausible then a lot more work needs to be done. But my focus has been on arguing that MP's account cannot be dismissed simply on the grounds that the appeal to agency introduces a fatal circularity. And I think that that argument, if it is any good, survives the point that MP's theory needs plenty of refinement. What theory doesn't?

6.7 Conclusion

I have argued:

(1) That a variety of things might be meant by 'agency theory' and some of these things are prima facie circular.

(2) That MP's theory *is* apparently circular and that their appeal to an experience of agency is unhelpful.
(3) That the important issue is not whether we *acquire* the concept of agency independent of that of causation but whether its *exercise* is independent of that of causation.
(4) That this question can be settled by adopting an interpretational stance.
(5) That the exercise of the concept of agency is independent of causal concepts.

I conclude therefore that MP's theory *can* be absolved of circularity. If it is trivial or uninformative then it must be for some other reason.

References

Berkeley, G. (1710). *Principles of Human Knowledge*, in M. Ayers (ed.) (1975) *George Berkeley: Philosophical Works*. London: Everyman.
Collingwood, R. G. (1940). *Essay on Metaphysics*. Oxford: Oxford University Press.
Dummett, M. A. E. (1978). 'Bringing About the Past', in his *Truth and Other Enigmas*. London: Duckworth.
Evans, G. (1985). 'Things Without the Mind', in his *Collected Papers*. Oxford: Oxford University Press.
Hausmann, D. (1997). 'Causation, Agency and Independence', *Philosophy of Science*, 64, Supplementary Volume: s15–s25.
Menzies, P., and Price, H. (1993). 'Causation as a Secondary Quality', *British Journal for the Philosophy of Science*, 44: 187–203.
Michotte, A. (1963). *The Perception of Causality*. London: Methuen.
Nozick, R. (1990). 'Newcomb's Problem and Two Principles of Choice', in P. K. Moser (ed.) *Rationality in Action: Contemporary Approaches*. Cambridge: Cambridge University Press.
O' Shaughnessy, B. (2004). 'Theories of the Bodily Will', in T. Pink and M. W. F. Stone (eds), *The Will and Human Action: From Antiquity to the Present Day*. London: Routledge.
Price, H. (1991). 'Agency and Probabilistic Causality', *British Journal for the Philosophy of Science*, 42: 157–76.
Quine, W. V. O. (1976). 'Necessary Truth' in his *Ways of Paradox*. Cambridge, Mass.: Harvard University Press.

Reid, T. (2001). 'Of Power', *Philosophical Quarterly*, 51: 1–10.
Suppes, P. (1970). *A Probabilistic Theory of Causality*. Amsterdam: North-Holland Pub. Co.
Von Wright, G. H. (1975). 'On the Logic and Epistemology of the Causal Relation', in E. Sosa (ed.) *Causation and Conditionals*. New York: Oxford University Press.
Woodward, J. (2003). *Making Things Happen: A Theory of Causal Explanation*. Oxford: Oxford University Press.

7
Pragmatic Causation

ANTONY EAGLE

7.1 Russell's Explicit Arguments Against Causation

Russell (1913) famously attacked the law of causality, and indirectly the concept of causation itself, as an inessential anachronism encumbering our proper understanding of the world disclosed to us by science. Russell's primary target is the *fundamental law of causality*, which is the principle that every actual event has an actual sufficient cause, one that is guaranteed to bring that event about and in fact did so. He proposes to argue against the law by indicating that the causal relation between events that it requires does not exist.

For Russell, the relation of causation is the relation of determination: c causes e just when c determines e to occur. This succeeds as an analysis only if the relation of determination is antecedently clear. From Russell's discussion, it is apparent that he thinks determination requires that the occurrence of c be sufficient for the occurrence of e; we shall assume, not implausibly, that c determines e just when the fact of the occurrence of c, in conjunction with the background laws $\mathcal{L}$, entails the fact of the occurrence of e.[1] Different laws give different determination relations; we assume here that the laws of interest for the purposes of analysing causation are the laws of nature.

Thanks to audiences at Princeton and at the *Causal Republicanism* conference, Centre for Time, University of Sydney. Particular thanks to Toby Handfield and Jeff Speaks; thanks also to Helen Beebee, Paul Benacerraf, Karen Bennett, John Burgess, Adam Elga, Mathias Frisch, Jason Grossman, Hans Halvorson, Gil Harman, Chris Hitchcock, Graham McDonald, Lizzie Maughan, Daniel Nolan, Huw Price, Gill Russell, Mark Schroeder, Brett Sherman, Charles Twardy, and an anonymous referee for Oxford University Press.

[1] This derives immediately from the slightly more general semantic condition that models of the laws contain the determining event only if they contain the determined event (Earman 1986).

These definitions of causation and the 'fundamental law' are not a Russellian idiosyncrasy. The idea that causation is determination of one event by another appears in Hume's 'constant conjunction' regularity analysis (if c and e are *constantly* conjoined, the occurrence of c should be sufficient for the occurrence of e). Even some later accounts of probabilistic causation, which are skeptical about the supposed 'universal law', retain the idea that where there is causality, the causes determine their actual effects (Suppes 1970). These views must therefore either deny that every event has a sufficient cause in order that the existence of causality might remain compatible with irreducibly probabilistic theories, or else join many physicists in bemoaning the 'disappearance' of causality from indeterministic quantum physics. This close connection between determination and causation remains a unifying feature of many otherwise conflicting accounts of causation—causes *make* their effects happen in some way that doesn't leave the effects (wholly) up to chance (Norton in ch. 2 of this volume, section 2.2).

Russell thinks that causation as a relation of determination between events doesn't appear in fundamental physics—indeed, cannot be rendered compatible with fundamental physics—and hence should be jettisoned from a properly scientific world view. From our perspective, the indeterminism demanded by standard quantum theory might already seem to undermine this conception of causation. Russell's arguments, if sound, would raise problems for causation even supposing a deterministic fundamental physics.

Following Field (2003), we may identify two arguments in Russell to his conclusion. The first rests on the claim that the equations of fundamental science are *bi-deterministic*. That means the state s of some system S at t fixes the whole trajectory of S through the space of possible states both before and after t, in the sense that any two systems both in s at t would share all their states at all times (Earman 1986). If we make the plausible supposition that each macroscopic event is constituted by some particular microphysical state, then fixing on some particular event in the system at a time will establish which microphysical trajectory the system is on, and hence which events will occur and have occurred. Then any event determines both its temporal antecedents and temporal succedents. But if 'c causes e' is defined in Russell's sense—namely, c determines e—then by this argument e equally well causes c. However, causation is, intuitively, asymmetric: if c brings it about that e, it is not the case that e has any causal influence on c. In bi-deterministic physics, the asymmetry of determination, and hence the asymmetry of cause and effect, is lost. There seems no place in bi-deterministic physics for the causal asymmetry, which is a fundamental feature of the intuitive concept we are attempting to analyse.

This first argument isn't very compelling, for at least a couple of reasons. First, although it might be true that the causal asymmetry is not an asymmetry of determination, causation still might be defined from a relation of determination *combined with* an asymmetrical relation, where the asymmetry comes from somewhere else. One suggestion is that the asymmetry of causation depends on the asymmetrical temporal distribution of entropy, which increases with time due to unusually low entropy initial conditions (Albert 2000).

Secondly, it may still be the case that there is a macroscopic asymmetry between causes and effects. Perhaps on a global scale, the whole state of the universe at one time determines the whole state at every other time. However, if we restrict our attention to a *localised* event c, whose character does not determine every aspect of the global state, presumably there are many global states that can involve this type of event as a constituent part. Even in a deterministic system, it is possible that both (i) all trajectories which feature this event type c as part of some global state s at t have some further type of event e as a feature of a state s' at t', and (ii) not every global state that features e at some time lies on a trajectory which involves some past state that features c. So the occurrence of c determines the occurrence of e in a way that e doesn't determine the occurrence of c. Focusing attention on events of a purely local character, rather than the entire state of the system, might very well give us an asymmetry of determination between particular events. Moreover, it is arguable that this local determination is exactly what the original notion of a cause was supposed to capture, since the typical situations we use to elicit our causal intuitions concern kinds of medium-sized events, not time slices of the whole world. This second objection, which I find very compelling, depends on our being able to provide a satisfactory account of the concept of a local event.[2] For our purposes, it seems most important that the event in question must be of restricted spatiotemporal extent. Our commonsense causal views also seem to require that the events must be *discrete* (readily distinguishable from

---

[2] This, in turn, is tied up with characterizing the physically important concept of *locality* (Lange 2002).

other events going on around them), and therefore easily subsumed within familiar and natural (to us) delineations of the categories of events. Other features may also be involved, though I am unable to do full justice to these issues here.

Russell must have been sensitive to the weaknesses of his first argument, since his second argument picks up on this conception of causation between local events. Consider some small local events c and e such that the occurrence of c determines the occurrence of e but not vice versa. Russell argues that these 'local' events will not be the kind of things we typically take to be related by cause and effect. That is, events that enter into a relation of determination will not typically both be discrete and non-gerrymandered events. But then we have saved the idea of causation as determination only at the cost of savaging our ordinary intuitions about the kinds of events that can be related as cause and effect. The physical relations between states which underlie the determination relation between local events do not hold between the kinds of events that feature in our causal intuitions, or, most importantly, in our causal reasoning and the actions that flow from that reasoning.

As an example of Russell's point, consider the causal relations that support ordinary attributions of blame and responsibility. For example, suppose we blame Slim for killing Bruce by shooting him. Rather obviously, Slim's firing of a gun F and Bruce's death D are not of the right kind to get into the determination relation, because F can occur without D: for example, if Slim had missed, or if Sharon had thrown herself in harm's way, or if the bullet had exploded harmlessly in mid-air. (Similarly, in cases of preemption D could occur without F.) The problem with local determination that this case discloses is that there is always the chance of some *interference* from outside the local area at the time of the cause. To ensure that the cause guarantees the occurrence of the effect, we shall have to hold the cause to be a very large set of events, perhaps the whole past cone of potential causal influence on the effect. So if c really determines e, c will have to be incredibly more complex and larger than causes are typically taken to be. Countless other factors must be accounted for so that the natural and salient cause can determine the effect in question.

To really determine e, we shall have to make sure that all possible interfering events don't occur, which will mean specifying the events which actually fill the location of those potential interferers. Even if we allow 'negative' events, like the non-occurrence of a potential defeater, some positive characterization of what is occurring at the potential locations of those defeaters must be possible, and that characterization is important for the derivation of the determination relations from the fundamental physics. Crucially, any genuine and robust relation between cause and effect that is derived from the underlying physics will involve some events that are intuitively not causes being counted as causes simply by virtue of their being in a potentially causally efficacious location. (Precisely analogous lines of argument hold for the generation of spurious effects.)

Indeed, we shall be unable to make a distinction as regards causal efficacy with respect to these events—in particular, we shall not be able to distinguish the genuine from the merely potential causes from a set of events each of which occupies a particular kind of location, if all we have to go on are the relations that physics gives us. All of these causes and potential defeaters are necessary for the laws of physics to determine the effect consequent upon the cause; leave any out, and the determination relation will no longer hold. But this makes no distinction between the events that were, intuitively, really the causes, and those that contributed merely by having some character or other that failed to interfere. Often this kind of problem is solved by an appeal to explanatory or contextual salience: that, strictly speaking, they are all causes, but one is singled out for explanatory purposes because of special particular relationships that we bear to it. This may well be so (indeed, I'll argue for it in section 7.4), but we must recognize that the salience of some event appears nowhere in its purely physical description, and no appeal to the underlying physics will establish any priority for the genuine causes over the spurious causes.

The importance of this emerges when we consider causal deliberation. If, as Russell's arguments suggest, we can't distinguish a genuine cause from an actual non-cause that might have been a preventer, then we shall be unable to engage in goal directed activities that depend on effectively bringing about certain states of affairs. This is partly because the set of events we shall have to manage or intervene on is far too large and inhomogeneous to deal with effectively. The real problem, however, is that we shall be misled into performing actions to bring about our goals that are not effective for achieving those goals. Or rather, physics provides no reason why altering the situation with regard to genuine causes should make any more sense than altering the spurious causes. Of course, we have intuitions that govern our causal interventions, and those intuitions serve us well. But they get no support from fundamental physics, if Russell is right. Cartwright (1979) emphasized this aspect of causation when she talked of effective versus ineffective strategies. Russell can be seen as turning this proposed justification for causal reasoning against itself: if we really want causes to determine their effects, then intuitive causal reasoning simply isn't adequate for distinguishing effective versus ineffective strategies. We need the full power of our best physics; and unfortunately, our best physics don't provide a determination relation that meshes at all well with our folk theory of causation (Norton in ch. 2 of this volume, section 2.5).

We can put the net result of these arguments as follows. Physics gives us a deterministic structure of the evolution of a system over time, so in physics the notion of a cause is trivial because it counts every past event as a cause. If we wish to apply the concept of causation to some spatiotemporally local situation, then we can only have determination if we are willing to abandon the role of causes in demarcating effective strategies for manipulating that situation, which is to say if we abandon the traditional concept of causation altogether. The genuine physical determination relation that really meets the demand for effective strategies renders any use of an additional, fallible and intuitive conception of causation redundant. If the notion is redundant in physics, it is dispensable from a properly scientific account of the world. Since fundamental science is the arbiter of genuine ontology, this relation of causation should be excised from our folk ontologies: it can't even be reduced to physics, let alone found within it.

It seems to me that this eliminativism is precisely the wrong conclusion to draw. The real fault lies in the conception of causation as requiring determination. When this is combined with Russell's observations concerning the kinds of determination relations that can be defined within fundamental physics, it follows that the relation of causation is trivialized rendered unrecognizable, forcing us to include merely pseudo-causes. The appearance of triviality doesn't show that the folk conception of causation is to be rejected; rather, it shows that one part of that folk concept—determinism—is in tension with the rest of the notion. If we could show that, we could give an analysis of this folk notion of causation that limited or restricted the demand for determination of effects by causes, we could respond effectively to Russell's eliminativism. We wouldn't have a 'universal law' of causality, that every effect has a determining cause. But that cannot be had at all.

In the following section I sketch one such account of causation that weakens the determination requirement and yet is adequate to the other tasks of a concept of causation, such as grounding the distinction between effective and ineffective strategies. Before going on, I will briefly sketch the plan for the rest of this essay. In Section 7.3, we will see that a particular argument, which I call the *causal exclusion problem*, lies beneath the details of Russell's discussion, and that argument threatens to render the reconstructed concept of causation otiose. In responding to this fundamental objection in Section 7.4, I shall appeal to broadly pragmatic considerations grounding the legitimacy of certain linguistic practices—the facts supporting the legitimacy of causal language are outlined in Section 7.4.2.

7.2 Causation as Partial Dependence or Determination

In this section, I give a sketch of my favoured analysis of causation without determination.[3] Though I think very highly of the account, my main concern here is not to defend it, but rather simply to demonstrate the possibility of providing an adequate causal relation that evades Russell's arguments. That the account sketched here can do so should count in its favour against other accounts.[4] Be that as it may, the purpose of the following is illustrative rather than an attempt to argue for the framework outlined here. Those who dispute the details of this approach should feel free to substitute into my argument at this point their own preferred account of causation, provided they can satisfy themselves it too will avoid Russell's arguments. However, it will turn out that some features of this account,

---

[3] The broad outlines of the technical aspects of this approach should be familiar from the work of Pearl (2000) and Spirtes, Glymour and Scheines (2000). The philosophical development of a counterfactual theory in the causal modelling framework is basically similar to Hitchcock (2001) and Woodward (2001), though in this context I stress—for obvious reasons—the role of counterfactual dependence as being essential to avoiding the problems raised for full determination by Russell's arguments. For more philosophical detail I recommend Woodward's very rich recent book (2003). There may be an interesting connection also with Lewis' recent account of causation as influence (2000).ABewis' account requires that for c to influence e, there must be a range of relevant alterations of c that are associated with relevant alterations of e. The concomitant variation of effect variables on cause variables in the account may capture this, as well as yielding counterfactuals which give the influence a uniform treatment within the counterfactual framework.

[4] For instance, I suspect that the conserved quantities theory of Dowe and Salmon will have difficulties in meeting Russell's challenge (Dowe 2000, Hitchcock 1995).

such as its emphasis on manipulation and intervention, fit beautifully with the eventual pragmatic justification for causation in Section 7.4.2.

7.2.1 Counterfactuals and Partial Determination

This account begins by proposing that we analyse non-determining causal dependence between local events in terms of relations of counterfactual dependence between propositions concerning those events.[5] The starting point is Lewis' (1973a, 1979b) analysis of causal claims in terms of certain counterfactual conditionals. Lewis proposed, roughly, that c causes e just in case the counterfactual claim '$\neg c \,\square\!\!\rightarrow \neg e$' ('if it hadn't been the case that c, then it wouldn't have been the case that e') is true.[6] This analysis captures the intuitive idea that causation is a species of *difference-making*: causes make the difference as to whether their effects occur or not, such that their absence would entail the absence of the effect, other things being equal. These claims about difference-makers are most plausibly rendered as counterfactual claims, and any satisfactory account of causation must make sense of the intimate connection between facts about causation and the counterfactuals made true by those facts (regardless of whether that account takes this connection to exhaust the facts about causation). We may then legitimately use these counterfactuals to explain various features of causation; in our case, we use counterfactuals to explain how causes might at least partially determine their effects. Note that we also need to impose the further constraint that the counterfactuals we consider are not backtracking (Lewis 1979b, pp. 32–5)

Earlier, we assumed that the determination relation involves the occurrence of the determined event being implied by the occurrence of the determining event, given the laws. Ordinary conditionals, such as those characterizing the implication relation, satisfy the following inference rule: $\alpha \rightarrow \beta \vdash (\gamma \wedge \alpha) \rightarrow \beta$ (the rule of 'strengthening the antecedent'). If the conditional appearing in statements of determination is a standard conditional, as we assumed, then the relation of determination holds regardless of what else happens to be true. That means α must fix every fact relevant to the occurrence of β in order for the determination relation to hold: the

---

[5] It is especially nice because it preserves a kind of defeasible determination, in a way that, for example, probabilistic analyses of causation do not.

[6] I have used the same notation for the event c and the proposition which says that c occurs, the latter appearing in the counterfactual conditional. I trust no problematic confusion will occur.

only facts it need not fix are those irrelevant to the holding of the relation, like γ in the inference rule. This feature of the logic of the determination relation gives rise to Russell's problem of spurious causes: every relevant potential cause must already be fixed or entailed by α, and hence counted as part of the cause.

Counterfactual determination might appear to share this feature with standard conditional analyses of determination, especially if we attend to the gloss 'if it were the case that c, then it *would be* the case that e'. But that appearance is misleading. Counterfactuals do not support strengthening the antecedent, as the following series of counterfactual claims shows. Imagine that Slim shoots Bruce. Were Slim not to have fired his gun, Bruce wouldn't have died. But were Slim not to have fired his gun *and* had Sharon fired her gun, Bruce would have died. But were Slim not to have fired his gun, and Sharon had, *and* Sharon had missed, Bruce wouldn't have died. And so on. All of these counterfactual claims are true, both intuitively and on the standard Lewis–Stalnaker semantics (Lewis 1973b). We can see therefore that counterfactual dependence is *not* invariant if we place additional conditions in the antecedent. These additional considerations are typically potential events that we didn't consider in the initial attribution of causal effectiveness to the antecedent event. If we analyse causation as counterfactual dependence, then we may have true statements of partial determination of effect by cause while also leaving open the possibility that there may be a recherché situation where the cause is compatible with the non-occurrence of the effect. This kind of determination relation, I submit, is the best we can have that manages to avoid the Russellian spurious causes argument, retain the connection between causation and determination, and preserve the intuitions surrounding determination. Causation, on this view, involves partial or defeasible determination, as captured by counterfactual dependence.

One situation where full determination does occur even on this analysis is if the antecedent and consequent events are *maximal* global states of a system, such that either ϕ or $\neg\phi$ is fixed by the state Γ for any ϕ. Then $\Gamma_1 \square\!\!\rightarrow \Gamma_2$ cannot be defeated by affixing additional events, for there aren't any. Russell's first argument can be restated: given the fundamental physics we have, for any two global states such that $\Gamma_1 \square\!\!\rightarrow \Gamma_2$, there is a true counterfactual $\Gamma_2 \square\!\!\rightarrow \Gamma_1$. There is thus no asymmetry of determination, and hence no causal asymmetry. Russell's second argument is that global maximal states are too large and inhomogeneous to be the relata of the intuitive causal relation we were trying to analyse.

7.2.2 A Model of Causation

I begin the task of integrating these facts about counterfactuals into our model of causation by making some trivial observations. There are some events that make no difference when added to the antecedent of a true causal counterfactual. Example: we don't think that were Slim not to have fired, and were some small event ϕ on Pluto to have occurred, then Bruce would have died. This is true regardless of what ϕ is. Some events, no matter how they might have turned out, are not capable of affecting the counterfactual dependence between two other events because they are isolated from or irrelevant to those events. On the other hand, some possible events ψ on earth might well have altered the truth of the counterfactual if added to the antecedent conditions. For instance, some action ψ of Sharon's in the near vicinity of Slim's firing and Bruce's being shot might have turned out in such a way as to contribute to, or detract from, Slim's action's bringing about Bruce's death, whether or not Sharon actually performed one of the particular subclass of actions that would actually have had an impact.

To account for these platitudes, I think an analysis of causation must have three features: first, it must hold between variables rather than arbitrary events; second, it must include a variable as relevant if some of its values are counterfactually relevant; third, it must use contextual cues to rule out irrelevant variables. Let me expand on these features.

First, the causal relata. Normally, events are taken to be the causes and effects. The identity conditions for events are, plausibly, very fine-grained: two events are distinct if *any* aspect of them is different. So Bruce's death is different if he is shot by Slim *and* Sharon rather than by Slim alone. I am quite happy with finely individuated events. But since I believe Slim would have played a part in causing *something* (involving Bruce's death) regardless of Sharon's action, I had better take the relata of the causal relation to be something other than arbitrary fine-grained events. I opt for a special subset of events: those involving some *random variable* taking some value. A random variable is a function from possible events to numbers, where the numbers characterize certain features of the event. For example, I could characterize the relevant class of events that might have resulted from Slim's act, and use the random variable *Death* which takes value 1 if Bruce died, and zero otherwise (and is undefined on events that aren't relevant—where relevant will be spelled out in the next paragraph). This concept 'smooths out' variations between events which do not give rise to a different value of the random variable; however, some other random variable might be sensitive to some of those variations.[7] So in this case, the event of *Death* = 1 is caused by the event *Slim* = 1. The causal relata are not variables themselves—they are functions, not physical entities—but having variables involved means that it will not normally be the case that arbitrary disjunctive or otherwise gerrymandered events will be causally active.[8]

Secondly, what matters to the causal importance of a random variable is if at least one of its possible values can alter a counterfactual in which it features. For example, I take it that Sharon's possible actions form a relevant class of events in our example, because some of those events, if specified, alter the counterfactual dependence between *Death* and Slim's action. That is, for some of the events A that fall into that class, a counterfactual '$(\neg F \wedge A) \,\square\!\!\rightarrow D$' is true while '$\neg F \,\square\!\!\rightarrow \neg D$' also remains true). But for Sharon's activity to be *potentially* causally relevant, it doesn't matter that *actually* it was not, and she stood idly by. We might say that in such a case the variable *Sharon causally contributes* to the variable *Death*, even though the actual causal relations between actual events are insensitive to this causal relationship.

Thirdly, we must notice that the second observation puts a necessary condition on causally relevant variables. But this cannot be sufficient, otherwise all sorts of irrelevant variables will get included because in very distant worlds they alter assessed counterfactual dependence. I think that ruling out these variables is largely a matter of contextual salience, where that depends both on background information about which variables to include (only those that have an influence in situations that are serious possibilities for us (Levi 1980)), and also judgements about the distribution of values for those variables (Menzies in ch. 8 of this volume—see also section 7.4).

---

[7] This is what Field (2003) calls a 'fairly inexact variable', and is essentially the same notion of variable that appears in Hitchcock (2001), Pearl (2000), and Spirtes *et al.* (2000).

[8] So, for instance, we don't run immediately into puzzles that plague simple counterfactual theories which claim that some 'negative events' (those that occur just when some ordinary event fails to occur) have to be causally active, just because there is counterfactual dependence on a proposition stating that some positive event failed to occur.

For current purposes I think that facts about correlation and spatiotemporal connection between events that are phenomenologically salient will go a long way to explaining the choice of variables to include in a model. On this proposal, causal relations between events are very common, but frequently it is pragmatically inappropriate to state the existence of a causal relation. With respect to the values of the variables, the role of *contrast* is particularly noteworthy: judging that the hammering smashed rather than pulverized the walnut involves making quite fine distinctions between values of the variable *Hammer Force*; judging that the hammer smashed rather than failed to smash the walnut involves a coarser partition of the hammer-involving events. It is important to note that even if the variables and their ranges are chosen for pragmatic and context-sensitive reasons, the truth of the resulting counterfactuals will be a perfectly *objective* feature of those variables.

Let me deploy some new terminology to summarize these three morals about counterfactuals. Let the situation under consideration determine a contextually salient theory, specifying the set of random variables $\mathcal{V}$ that we will use to summarize the values of the events in question. The theory will encode certain counterfactual dependence claims. In particular, it will encode a pattern of mathematical dependence between parameter values for random variables. That is, it will give us facts of the form 'the value of variable V_i depends on the values of variables $V_j \ldots V_k$'. If we have a quantitative structure, the variables will be linked by functional equations that show upon which other variables the value of each variable depends. If V_i neither depends on V_j nor V_j depends on V_i, these two variables will be *independent*.

The kinds of counterfactuals we take to be true will determine exactly how the dependency is cashed out. Call a variable V a *parent* of another variable U just in case: (i) $V \neq U$; and (ii) there exists some assignment of fixed values to variables in the model such that the following counterfactual is true:[9]

Parent Were V to have some different value v, then U would have some different value u. $(V = v \,\square\!\!\rightarrow U = u.)$

Note that 'grandparent' variables are not parents: if V only acts on U through W, then the fact that W is fixed on some value will prevent the

---

[9] Thanks to Charles Twardy and Chris Hitchcock for help with this formulation.

change in V from percolating through to U. We say that a variable V is *directly causally relevant* to U if V is a parent of U (Woodward 2001). This concept allows us to construct *causal graphs* as follows. Take all the variables in $\mathcal{V}$, and put them at nodes of a graph. For each $V_i \in \mathcal{V}$, let $\mathcal{P}(V_i)$ represent its parents. For each variable V_i, draw an arrow from each $V_j \in \mathcal{P}(V_i)$ to the node V_i. We will end up with a graph something like Figure 7.1. This is a qualitative causal structure, and the parenthood relations are the most basic kind of counterfactual that should be considered in causal reasoning.[10]

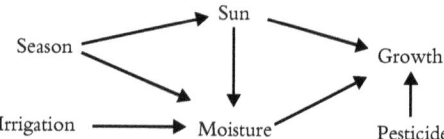

Figure 7.1. A simple causal model of plant growth and some of its factors

Consider as a simple example the model of plant growth depicted in Figure 7.1. In this model, *Season* is binary: growing season or non-growing season, which in turn influences the amount of sunlight (*Sun*) and the rainfall (*Moisture*). But moisture is influenced by sunlight (causing evaporation) and whether the crop was irrigated or not (*Irrigation*).[11] Finally, whether *pesticide* is used or not also influences the final plant growth (*Growth*). Some of these variables are binary, and some are quantities (sunlight and rainfall). Different models may choose to represent irrigation by a volume of water, not by a binary variable. Similarly, growth may be modelled by a variable taking values of 'increased', 'decreased', or 'same'.

Interestingly, some variables can cause by 'omission' (i.e. not applying pesticide can negatively influence plant growth), and it may be noted that this framework gives an easy way for causes by omissions to work—at least, after the general problem of variable selection has been solved there is no special problem of causation by omission.[12] Of course, other irrelevant variables are also omissions of things that might have impacted, and it is

---

[10] If we have in addition a probability distribution over the values of the exogenous random variables (i.e. those with no arrow leading into them), and equations which express the numerical dependence of the values of a variable on the parameters and values of its parents, we can turn this qualitative causal structure into a quantitative causal model.

[11] Irrigation and rainfall may themselves be correlated, so this model may not be perfect (since high rainfall tends to reduce the need for irrigation).

[12] Thanks to Brett Sherman and Karen Bennett for help with this.

the choice of variables to model along with judgements about salience and relevance that give content to our causal judgements.

Several things are noteworthy about this approach to a counterfactual analysis of causation. Consider Figure 7.1 again. The values of the variables *Season* and *Irrigation* are not counterfactually dependent on anything (note that the restriction against backtracking counterfactuals is crucial here: if it were not, plausibly if the value of Sun were different, then the value of *Season* would have had to have been different too). But presumably these variables do depend on something. To pretend we have isolated them, we can appeal to the contextual salience of local causes. Causal explanation has to stop somewhere, and if a certain condition on the parentless variables is satisfied, we should be prepared to stop with them. The simple condition is that parentless ('exogenous') nodes should be independent, not correlated among themselves. (I see this as a methodological condition on the construction of causal theories, not some a priori truth about systems of variables.) If X is counterfactually dependent on Y, and Y on X, neither through a backtracking counterfactual, then we should try to find another variable Z that is parent to both X and Y and screens off the counterfactual dependence in order to complete our causal model. (However, there are some cases where such a variable does not exist, for example in standard explanations of the non-local correlations in Bell-type theorems in quantum mechanics (Butterfield 1992). This is perhaps best modelled by simply keeping the two-way counterfactual dependence.)

Following naturally on from this, one can see how adding more variables can change the parental counterfactual dependencies by interpolating further intermediate causes, and by adding new parents. This can mean that contextual salience determines the causes of an event. So does the comprehensiveness of the underlying theory that supports the counterfactuals. This feature tends to support the idea that causation, as well as explanation, is often contrastive rather than absolute—it depends on the salient variables (Hitchcock 1996).

Thirdly, we must consider what kind of facts make the Parent counterfactual true. As it stands, it relies on a seemingly miraculous ability to vary the value of one variable while holding all others fixed on a certain value. This process has gone under the name of an *intervention* in the literature. In the graphical models, it can be modelled by severing the node from its parents and setting the value of the variable and other variables, effectively rendering some dependent variable independent. We are supposed to think of an intervention on X as encoded in a causal graph $\mathcal{C}$ that terminates in the fixing of some value for X independent of other values of other variables not in $\mathcal{C}$.[13] The viability of the concept of an intervention depends on the possibility of *modular* causal systems (Hausman and Woodward 1999). These are systems where each variable in the system has some independent exogenous sufficient cause.[14]

Our understanding of interventions relies on our ability to deal with quite distant counterfactual situations. Though we deploy a particular theory in reasoning about such systems, we need to be able to isolate particular parts of those systems from other parts, even to the extent of isolating the variables intervened upon from their actual causes. This will involve consideration of a set of models where the best system of laws might be quite different than the one governing the theory as a whole. How then can we get a grip on the concept of intervention? The concept of *manipulation* is important in this connection. We can begin to understand intervention by considering how human agents manipulate objects and situations in the most prosaic of circumstances. We have goals and desires, and the world is not necessarily cooperative in satisfying those desires or bringing those goals to fruition. But we have the capacity, in many cases, of doing things, of making things happen, to bring about some change in the world that makes it more amenable to our wishes. In other words, we wish sometimes to control how the world goes, and to predict what our interference might do. Once we have this simple and familiar concept of manipulation, we can begin to theorize about situations outside of our direct control, indeed those outside any human action at all. But an understanding of the way in which manipulations sever the variable intervened upon from its parents sets us up to understand how intervention more generally works, and to see manipulation as a special, particularly important, case of intervention.[15]

One problem often thought to strike accounts of causation that use the notion of intervention is circularity, because intervention itself seems a

---

[13] This isn't quite right, because expanding some graph $\mathcal{D}$ that contains X to also contain $\mathcal{C}$ would render $\mathcal{C}$ not perfectly efficacious in fixing X ($\mathcal{C}$ would no longer trump all other causal factors).

[14] Cartwright (2002) has argued that modularity generally fails. However, she seems to rely on the claim that actually non-modular systems are not possibly modular, and this claim seems false if one is willing to countenance counterfactual variation in the patterns of occurrence of instantiation of distinct variables.

[15] See, section 4.2, and also Woodward (2003: esp. s. 3.1).

causal notion. The circularity, however, is not vicious: since the variables characterizing the act of intervention itself are never included in the causal model, we may treat them as characterized solely by the counterfactuals they support. If we expand the model to include the process of intervention itself, and give a naturalized account of that particular intervention, it will no longer support the Parent counterfactual, and will appear like any other causal variable. If we take the content of the concept of causation to be given by these causal graphs, the notion of intervention we need to support the Parenthood counterfactual is not a causal one at all. However, this very strategy for defusing the charge of circularity gives rise to serious difficulties in making a global causal model that includes every variable. In such a model, nothing corresponds to an intervention conceived of non-naturalistically—once every event has a representative variable in the model, every dependence relation between events is already represented. As Pearl (2000, p. 350) remarks, 'If you wish to include the entire universe in the model, causality disappears because interventions disappear—the manipulator and the manipulated lose their distinction.'[16] Indeed, as I shall argue next (Section 7.3), Russell's arguments can be taken to be making a very similar point: once the global situation is considered, and every event encompassed by the model, the local concept of causation disappears.

Causation on the model described here is a relation between variables, and will apply to particular events only in a derivative fashion. This is because we have identified causation with a feature of abstract models of phenomena that can apply to many particular situations. Once we see the actual values of the variables as they appear by a particular situation, then we can retrodict the actual effects various potential causes had, through the light of particular assumptions about what the natural or usual range of the exogenous variables is. We make a default causal model, to use the terminology of Menzies (ch. 8 in this volume), and by using a mixture of default assumptions and evidence about actual values, we can create a restricted version of the causal model that should give us the acceptable token causes for some particular case. This will be a model that only applies to the single case in question. But which events are modelled depends on how we *coarse-grain* the space of events to give distinct values for the

---

[16] Pearl also seems to think (Pearl 2000, section 4.1) that interventions are to be connected with human free will and the causal 'unconstrainedness' of human volition.

variables. For instance, the variable *Height* could have as possible values 'less than 1m' and 'greater than 1m'; or it could be a continuous distribution over some subset of the reals, say [0,2]. Depending on how finely we divide up the events of 'having a certain height h', we get different causal models. This coarse-graining of the possible events is certainly a contextual factor in model construction, but it also plays a significant role in rendering causation a local relation. For example, in classical statistical mechanics the most fine-grained division of the space of possible events will make each distinct global state a different event, and we will be unable to avoid Russell's trivializing conclusions. Hence coarse-graining of events, perhaps to correspond to our epistemic and practical limitations, is necessary (in this theory and many others) for our deployment of the concept of causation.

Finally, what is the meaning of the arrows? An arrow between X and Y does not mean simply 'X causes Y'. Rather, it means something like 'some values of X are causally relevant to the values of Y', where this causal relevance can be at times stimulatory, at times inhibitory, and so on.[17] This seems to feed nicely into Hitchcock's (2003) recent claim that what is metaphysically primary is the multiplicity of causal connections, rather than some uniform notion of causation that is supposed to apply to all cases. Consider that some of the arrows might be purely inhibitory, some purely contributory, some a mix of both, and what constitutes the ground of the counterfactual claim might be very different in each case. Why think they can all be shoehorned into one neat causal metaphor like the 'cement of the universe'?

7.3 The Causal Exclusion Problem

At this point, I hope to have demonstrated the possibility of giving an analysis of causation that abandons the requirement that causes necessitate, or determine their effects without wholly abandoning all the intuitions about causation that led to the perceived connection between causation and determination in the first place. In particular, the analysis preserves the idea that causes are antecedent to effects and can be represented as such in

---

[17] One consequence might be that the distinction between a cause and an enabling condition will be harder to make in this framework. Thanks to Mark Schroeder on this point.

true conditional claims; but we use a counterfactual conditional rather than the strict conditional of traditional analyses of determination.

However, a serious worry remains. Indeed, I think a version of this worry underlies both of the explicit arguments that Russell gives, and hence can be identified as Russell's real challenge to causation. The problem that threatens our account is that, even if we can show how to systematically interpret causal talk so that it is *compatible* with fundamental physics, it remains true that causal talk is essentially *dispensable* once we possess fundamental physics. I call this the *causal exclusion problem*, because I think it in many ways analogous to the exclusion problem that plagues non-reductive physicalism in the philosophy of mind.[18]

To formulate the causal exclusion problem, we shall have to tease out a couple of presuppositions of Russell's arguments. I do not think they are particularly controversial, so correspondingly I believe the threat the problem poses is extremely serious, since its presuppositions are so meagre. The first is that fundamental physics provides the justification for any true physical claim. This is not the claim that fundamental physics is complete; I don't wish to presuppose the more controversial thesis of *physicalism*, that every fact is a physical fact. Here I only assume that, if we have an account of how some physical event was brought about, or produced, in terms of other physical events, the ultimate justification for that account lies in fundamental physics. Similarly, in predicting some physical event, the ultimate guide is fundamental physics. We ourselves made this presupposition when we took care to indicate how the causal models we constructed in the last section were restrictions of more complete models of the phenomena. The counterfactual claims that I argued were fundamental to causation are supposed to be supported by consideration of how the physical situation would go, were the world in the state described by the antecedent of the counterfactual. Evaluating these counterfactuals, we tacitly presupposed, involves examining how the fundamental physics tells us how the world would go, how our best physics describes the behaviour of even quite distant possible worlds.

The second presupposition is that a complete physical account of a physical event trumps a partial or approximate account. We may well

---

[18] For more on the exclusion problem in the philosophy of mind, see Bennett (2003) and the references therein.

have a serviceable default technique for reasoning about various situations, justified because it leads to no significant errors for all practical purposes, or because in a limited range of circumstances it gives precisely the correct answers even though outside that range it can be in error. But in making these claims about approximations, we appeal to the fundamental physics that describes perfectly what the default theory merely approximates. The approximation gets its value from being serviceable when compared with the fundamental theory, and only has value as long as the fundamental theory continues to be accepted. If we decide that the errors it introduces are too great, or our computational capacities increase to the point where the savings gained by using the approximation are negligible, we shall have no compunction in abandoning the approximation in favour of the fundamental theory. Unfettered by practical constraints we should have no need of approximations; and in doing ontology, particularly, such practical considerations should have no bearing on what we take to genuinely exist.

Given these presuppositions, we can state the problem. Consider some causal story in which we describe how an event e came to occur by citing one of its causes c. Is this story making an essential contribution to our understanding of the physical world, in the sense that without it, we would lack a proper account of this event e? If we answer that this story is making an essential contribution, then we deny that fundamental physics is authoritative with respect to predicting and describing the production of physical events. Moreover, we are simply denying the adequacy of physics, because we have a putative case in which one physical event brings about another and yet no physical explanation in terms of fundamental physics is forthcoming.

If we are not to deny the plausible thesis that physics is the final arbiter of physical questions, we must deny that this causal account makes an essential contribution to the correct description of the world. The alternatives are that it makes no contribution, or that it makes an inessential contribution. If the former, we are certainly better served by disposing of a putatively scientific concept that plays no useful scientific role. Let us suppose, then, that causal descriptions are inessential but legitimate contributions to scientific understanding. By our second presupposition, we would be better served by giving the more fundamental physical account which justifies and grounds the causal account, if our concern is with truth. The causal account is both inessential and less successful than the rival account given by fundamental physics. The relation of causation does no distinctive work in the giving of successful and accurate descriptions of the way the world goes. We should only accept the relation of causation if it can be rendered unproblematic: that is, if we can show that causal descriptions are able to be fitted seamlessly into a scientific picture of the world.

Russell's arguments enter at this point. I think they show that the concept of causation cannot be reduced to the fundamental physics in a seamless fashion.[19] The best way to reduce causation to physics would be if we could show that causation could be defined in terms of the fundamental physical relations, those adequate for every purely physical description and prediction of the relevant events. The upshot of Russell's arguments is that, in deterministic physics, these most fundamental physical relations are relations of determination between global states, given the physical laws. If we define causation in terms of these relations, we succeed only in trivializing the concept, or in giving a reduction that clashes spectacularly with our intuitions about causation. Neither is a satisfactory situation.

If reduction fails, as Russell suggests, then we will be forced to accept causation as an additional, autonomous concept that plays no distinctive role in a purely physical description. Combine this with the notorious difficulty in defining or identifying the precise characteristics of the causal relation, and we might well think that we are better off without it. Causal descriptions, on this view, simply obscure from view the real account of how any physical event happened. This obstruction of the truth is a reason why we should remove causal concepts from our repertoire; no powerful reason exists in favour of causal concepts. Hence, Russell adopts an eliminativist position with respect to causation: causation is superfluous, *excluded* from genuine physical significance. Even if they were to exist, causal relations would at best superfluously overdetermine which events take place; without any alternative support, the fact that they would introduce implausibly widespread overdetermination should be enough to tip the scales against causal claims.

The argument is powerful. Firstly, because it works even if causation is compatible with fundamental physics. This would render our work in Section 7.2 quite beside the point, since all the elaborate counterfactual

---

[19] If I am right about this, then Norton's claim (this volume, Section 2.4) that causal concepts are able to be generated by reduction to fundamental physics is not without difficulties.

machinery we introduce only serves as a second-class kind of description. Secondly, because the assumptions of the argument are so mild, denying any of them looks at least as bad as eliminating causal descriptions, since we can replace the causal descriptions with better though more complicated physical descriptions. Denying the first presupposition, on the other hand, and introducing extra-physical causal descriptions of physical phenomena, induces skepticism about causes. Thirdly, the features of causation we have emphasized seem only to support the exclusion argument, not rebut it. Causation is context-dependent: it is sensitive to which events or variables are included in the model, and some think it is relative to default values for the variables also. Causation is partial and local. These are precisely the features which causal accounts do not share with authoritative physical accounts, and it is precisely these deficiencies which the exclusion argument exploits. One way of putting it might be: physics has shown us how best to give an account of one event's determining another, providing a precise and detailed story where causal accounts are mysterious and imperfect. The emphasis on determination in the fundamental law of causality ironically led to the abandoning of causation as inadequate to account for determination!

Despite its many virtues, I believe the exclusion argument harbours a subtle flaw. Though causal accounts may be rendered inessential by fundamental physics, they are not thereby robbed of their distinctive contribution to *understanding*. Rather than focusing on causal descriptions of the phenomena, we should attend to causal *explanations*, which seem to play a quite different role than purely descriptive accounts. While it may well be true that the physical description of the situation is optimal with respect to the desiderata governing a good description of *how* some event came about, that same physical description needn't provide scientific understanding of *why* that event came about. Even if the global dependence relations between physical states fix entirely the physical structures, there remains an open question as to which aspects of those structures explain the resulting events. We already have prima facie reason to suspect that relations of global determination aren't adequate for explanation; simply observe that these fundamental physical relations may be bi-deterministic, and so may determine the past state given the future. Yet we should rarely, if ever, wish to accept that explanation of the past in terms of the future is acceptable.

If a causal description plays any role in explanation, as seems likely, then we shall have a strong reason to think that causal explanations are not excluded by physical descriptions.[20] Indeed, it may even be the case that no account of what explanation consists in can be given independently of the concept of causation; in which case, there could not be a problem of physical explanation excluding causal explanation.[21]

If proper attention is paid to the role of causal explanations in our conceptual economy, particularly their use in hypothetical reasoning, their importance becomes undeniable. Causal explanations can be vindicated by fundamental physics, in many cases, as the exclusion argument presupposes. If we can show moreover that these explanations provide a pragmatically essential handle on the physical facts, showing fundamental science to be continuous with our intuitions in some important sense, and supporting our self-conception as agents and our conception of the world as one among many possibilities, then we have provided an excellent reason to persevere with causation, despite the problems we have in giving an analysis of it. Russell's scruples about causation can be accommodated, even as we legitimize the pervasive use of causation in folk and scientific language. Or so I shall now argue.

7.4 Pragmatism and the Indispensability of Causation

The fundamental premise needed for Russell's eliminativism is that causal accounts are never better than accounts in terms of the functional dependence between global states given by the models of fundamental science. This premise has some initial plausibility, because of the primary role of fundamental science as providing the supervenience base for all other physical facts. But I shall argue that it is false, particularly with respect to the provision of explanations.

Let us begin by noting one premise Russell's arguments *do not* appeal to: the claim that causal explanations are incompatible with the explanations provided by fundamental physics. If this claim were true, I think it would provide an excellent reason to eliminate causal talk. But we have not relied on any such claim here (rather, presupposing it false). Nor do I think it

---

[20] See, for example, Woodward (2003, ch. 5) and Lewis (1986b).
[21] Thanks to an anonymous referee for help with this discussion.

possible to establish any such claim, since to do so would involve showing that the addition of causal claims to the body of fundamental physical truths would allow the derivation of some fundamental physical falsehood. It is clear that, on the model of causation sketched in Section 7.2, no such contradiction would be forthcoming, since each true causal claim rests on a true counterfactual claim, which in turn is evaluated by making reference to the fundamental physical models, or suitable approximations thereof. Causal models are abstractions of fundamental models; they can introduce no new claims concerning fundamental properties and relations.[22]

Russell's eliminativism turns on broadly philosophical considerations rather than logical ones. The essence of his position is that causal claims are, ontologically, freely spinning cogs in the mechanism of explanations. While causal explanations may depend on physical explanations, nothing in turn essentially depends on them. A potentially dispensable element of a scientific theory that does no genuine work should, by Occam's razor, be eliminated, especially if it is a metaphysically puzzling entity in the first place.

I will quarrel with both parts of this last claim. I will argue that causal explanations do indispensable work in the social practices of giving and requesting explanations as engaged in by creatures like ourselves. I will also argue that causation stands or falls with other modal notions, so that Russell's argument actually would succeed in undermining a huge part of our conceptual framework. Without causation and cognate notions, the world would be an alien and incomprehensible place; with it, our attempts to understand and engage with our world are useful and effective.

7.4.1 Minimalism, Legitimacy, and Ontology

Before we do this, however, we need to establish whether pragmatic claims about the role of some concept can do the metaphysical work we require of them.

If, as in the present case, we are sceptical about whether any fundamental facts neatly correspond to certain of our linguistic and cognitive practices, then we have two options. First, we can reject the language, arguing that the

---

[22] Adding causal language, then, amounts to what Field (1980) calls a *conservative extension* of the language of fundamental science, since no novel truths expressible solely in the original language become derivable. Of course, Field used a similar claim about number-language to argue *for* eliminativism about numbers.

PRAGMATIC CAUSATION   179

only legitimate basis for it must involve it representing fundamental reality in some way. Secondly, we may reject the demand that the language used in the practice be interpreted representationally, perhaps maintaining that the use of that language in a given practice might be understood in a perfectly naturalistic way without imagining that to every contentful sentence of a good practice there must correspond some part of reality that satisfies that content. Given these options, the causal exclusion argument I sketched above gives excellent reason for thinking that we cannot simultaneously maintain both of the following claims:

Representation We must understand causal talk representationally;
Ineliminability We can maintain that causal talk is, in principle, ineliminable.

But that argument does not indicate which of these claims we should abandon.

If we had the further thought that the only way to justify a practice was to show that it correctly represented Reality, then I suppose that if we wished to maintain causal language we should be forced to posit some additional ingredient of reality, not discoverable by physics (though perhaps discoverable by philosophy?), which corresponds to the content of acceptable causal claims. Some philosophers may go where this argument leads, and accept the additional ingredients; others may deny that the practice is justified. Both call for fairly radical revisions in either everyday practice or fundamental ontology, on the basis of what must seem fairly slender evidence. Better, then, to reject the further thought: a practice may be legitimately engaged in even without corresponding to some underlying reality.

To reject the further thought is to commit ourselves to what has been called *minimalism* (Johnston, 1992, section 5): the idea that sometimes a practice may be justified without that practice tracking fundamental metaphysical distinctions. Some versions of minimalism (including Johnston's) involve taking the ontology of a minimally justified practice as prima facie representational; the minimalism I will defend here will not involve this further, extremely controversial, step. Rather, the minimalism I shall defend will argue from the justification of the practice on pragmatic grounds to the claim that the language involved in the practice has a non-representational role, but that nevertheless we are entitled to use it and to act on its representational consequences. In the present section I will sketch this version of minimalism in general terms; in the next (Section 7.4.2), I will turn to causation, and argue that causal talk does have this minimal pragmatic justification. The upshot will be that causal talk is ineliminable and acceptable, yet isn't some extra non-physical ingredient of reality.

It is obvious that any adequate naturalistic explanation of the existence and persistence of a certain linguistic practice must draw upon the cognitive features of the creatures who engage in that practice. Such an explanation might involve various low level features, such as the perceptual capacities of those creatures; or it may involve features involving higher cognitive processing, for instance capacities for representation or goal-directed deliberation. Where we are the creatures concerned, it is plausible to think that there will be a complex of such features (involving our perceptual abilities, epistemic capacities, and characteristically human goals and attitudes) that will stably feature in explanations of many of our linguistic practices. Following Price (2004), we might call this stable complex of features a *perspective*, because it can be taken to characterize the viewpoint on reality that our nature forces on us (see also Price 1992, 2001). (Familiar visual perspectives originate from the constraint that we are *located* creatures.) From a given perspective, the world is most naturally modelled in ways that may not represent particularly closely the way the world fundamentally is. For example, creatures with very limited perceptual apparatus will have an internal model that will be insensitive to many features of the environment they actually inhabit.

More important for our purposes is the fact that sometimes the characteristic perspective of some creatures may make it natural to model the world as having features that it in fact lacks. Our location gives rise to an internal model which involves location-dependent properties like 'left of' and 'right of', which do not appear in a location-invariant global description (one may also consider *de se* knowledge in this connection (Lewis 1979a)). Perhaps more relevantly, our self-conception as agents seems to force on us a model of the world in which human beings are radically different in kind than the 'passive' matter that we may exercise our agency upon, despite the fact that, *qua* physical objects, we are in fact no different in kind from the rest of the population of our universe. It is quite clear in this case that our internal models of ourselves need not represent accurately any objective feature of our physical constitution. The models we choose to represent the way the world appears to creatures like us therefore may look very different to the models we choose to represent what the world is like from a global perspective.

There is no sense in which a perspective is not physical, since of course the facts about the perspective of some creatures may be studied naturalistically, by the sciences of psychology and biology. But they also provide explanatory resources that merely attending to the external world could not provide. For instance, an explanation of why some creature adopts a particular internal model need not appeal solely to the world that is being modelled, but may also appeal to the perspectival facts about the creatures who model the world in that way. It is quite clear, I hope, that if the perspective of some creature is stable enough, it may well explain the persistence of some internal way of modelling reality without requiring that the model accurately represent the external world. Since the characterizing perspective of an agent isn't a matter of choice or even conscious awareness for the most part, an appeal to that perspective may justify the internal models of that agent despite the fact that the they do not represent the external world. *A fortiori*, an appeal to perspective might justify the making of utterances on the basis of an internal model that do not represent the external world. And this, of course, is precisely what the minimalist wants.

Of course, there are non-minimalist (i.e. representationalist) ways to interpret this very same data. We may suggest that the linguistic practice correctly represents the internal model, not the external world. Or we could suggest that the practice aims to represent the external world, but fails. Both of these representationalist strategies fail to capture the distinctive role that the linguistic practice might play, because both fail to explain why the practice might persist as a successful strategy for these creatures to use in dealing with their external environment—the former cannot explain the external application, the second cannot explain the success. The non-representational minimalist interpretation, by contrast, explains both. A way of modelling reality that might be forced upon some creatures in virtue of the kind of creatures they are can be successful and persist if it it adaptive and conduces to survival. But not every adaptive trait must be explained as representing a response to some particular aspect of a creature's environment. So a practice that is explained by the nature of some creature that evolved under selective pressure can be justified without requiring that the practice is also explained by the fact that it accurately represents reality. This is particularly apparent in the case of practices that, if they represent at all, surely don't seem to represent anything that could have been exerting selective pressure. Numerical language and modal language are two prominent examples of manifestly successful practices that, if representational, represent apparently causally inert objects that could have exerted no selective pressure. As such, it can be no explanation of the success of such practices to appeal to representation. An alternative, non-representational account of the survival value of talking *as if* there were numbers or alternate possibilities is, while not without difficulties, at least some kind of explanation (Rosen 1990; Stalnaker 1984b; Yablo 2005).

On the minimalist picture, the justification for some linguistic practice, say the practice of describing ourselves as deliberative agents or the practice of number talk lies in the fact that those practices are conducive to success in creatures like us. Just how they do so is left open, though it is clear that the practice must somehow lead to predictive and explanatory strategies that facilitate our engagement with the external world. The fact that we typically engage in these practices as if they were fully representational is then explained by noting that it would undermine those strategies to regard them explicitly as false, involving us in some kind of pragmatic contradiction. However, in principle we could appeal to a naturalistic model of our behaviour, and recognize that our internal model may not be the best theory if our goal were only perfectly accurate representation with no other pragmatic considerations.

By endorsing the legitimacy of these perspective-dependent practices, we do not thereby automatically regard the objects of thought and reference in those practices as really existing in any sense. Insofar as these practices exist at all, they depend on the properties and relations of fundamental physics allowing for the existence of creatures whose constitution explains why they adopt the practices in question. It remains open to the minimalist to regard the explanatory strategies that a particular perspective-dependent practice makes available as on a par with the explanatory strategies of fundamental physics in some sense. It certainly seems to be true that these perspective-dependent practices have some degree of autonomy, since they may well persist whether or not we regard them as representing the reality they model (for example, even if there were a Platonic realm of numbers, it would play no explanatory role in the origin or structure of number discourse). We may thus get some limited devolution of ontological commitment from fundamental theories to less fundamental, perspective-dependent theories;

yet in the end I doubt, and do not need to defend for my purposes, the idea that somehow this ontological commitment is to be taken completely seriously.

7.4.2 The Causal Perspective: Counterfactuals and Deliberation

In Section 7.2, I identified a class of models that I take to characterize the core semantic aspects of causal language. Having just argued that we can sometimes give a minimalist justification for linguistic practices that need only appeal to facts about the users of those practices (Section 7.4.1), I now wish to argue that our causal language is itself a good candidate for a minimalist justification.

We begin with the obvious observation that we are *agents*. This involves having goals and projects, the capacities to accomplish some of these and the ability to deliberate about how to most effectively achieve them. It is clearly not an option for us to somehow abandon our agential status and refuse to deliberate, to act in some way as a mere object of physics. Thus our deliberative behaviour is practically inescapable. But here we may develop an observation of Ramsey's: '... from the situation when we are deliberating seems to arise the general difference of cause and effect. We are then engaged not on disinterested knowledge or classification ... but on tracing the different consequences of our possible actions ...' (Ramsey 1929, p. 158). This suggests that the justification for our causal language might come from our deliberative practices, and not from any representation of objective reality. The basic feature of deliberation that is relevant here seems to be the importance of hypothetical reasoning: making a (counterfactual) supposition and tracing what might follow from it. In some sense, entertaining such counterfactuals is constitutive of rational deliberation. When considering what to do, we should consider the possible outcomes that might ensue given our act and then weight them by their how likely we regard them as being and by how much we desire that they come about. The theory of causation I've sketched has at its core this role in hypothetical reasoning (look, for instance, at how the conditional probability distribution induced by an intervention on a single variable corresponds to the likelihood of events conditional on actions); little surprise, then, that the practical necessity of deliberating leads on to the minimalist justification of causal talk, for at base they amount to the same thing.

What is important when deliberating is that we have a relatively robust and simple means for judging which events are dependent on one another, to facilitate judgements about when an action we might undertake will be effective in bringing about our desired ends. We also wish to have a reliable and simple means of determining whether our activities will have further collateral consequences not themselves the intended goal of our activity. It should be clear that the account in Section 7.2 of causal networks of variables linked by counterfactual dependencies satisfy both of these desiderata. In virtue of these features, causal claims give us a quick and accurate way of accounting for why things happened the way they did: patterns of causal dependence summarize the relevant features of a situation that explain its overall character, as well as any event of particular interest.

Further support for the particular model of causation that I sketched above comes when we observe that our epistemic access to the external world is inherently limited to a local area. We are not able to perceive most of the possible variations in the antecedent physical states that might have some impact on the subsequent course of events, so it is important to rely on judgements of dependence that are relatively robust across a wide range of possible variations in background circumstances. Counterfactual dependence of some event on another holds only when the dependence is robustly invariant: otherwise it would not be the case that the consequent held in every situation where the antecedent held. If the events held to be counterfactually dependent on one another are relatively coarse-grained, then minor differences of detail will not be of great importance for figuring out what will come to pass.

It is crucial, then, that causal claims have some modal element, and cannot be reduced to merely actual physical connections. Robustness of a dependence relationship only makes sense if we can detect how that relation is stable under variations in the situation. This has further consequences: for example, when we make a request for an explanation, part of what we do is ask an implicitly general question: what was most responsible for the outcome, such that if we wished to reproduce it we would focus our efforts on that aspect? We rarely, if ever, request or give an explanation that accounts for an event in isolation and in such a way that the explanation does not generalize to similar cases. It turns out, therefore, that the giving of explanations depends in large part on the causal facts cited in an explanation being modally robust. Without robustness, the generalizability of causal explanations will fail. Without generalizable explanations, reasonable deliberation seems impossible. The conclusion is obvious: deliberation is not optional for creatures like us, and the only way we could undertake it is if we model reality along the lines of the causal modelling framework of Section 7.2. Our causal talk, insofar as it is accurately modelled by that framework, is thus minimalistically justified: it is a framework that we must adopt given our nature, although that adoption does not commit us to there being any objective causal or counterfactual relations that our models represent.

As it stands the deliberative aspect of our agency seems like the most plausible feature of our perspective to ground the utility of causal models, and the above account of causation neatly latches onto this ground. Physical connection models of causal language (Dowe 2000) do not have a close connection to deliberation, largely because it is implausible that modal claims may be determined by purely actual facts about causal influence. Given this, and Russellian worries about whether any candidate physical connection can play the causal role, the prospects for non-counterfactual theories of causation seem dim.

7.4.3 The Indispensability of Causation: Responding to the Exclusion Argument

At this point, the indispensability of causation starts to become apparent. Our deliberations require a modally robust relation between the action and the outcome. Causal relations fit this bill precisely. But the relation of global determination provided by fundamental physics is not robust, because in order to generalize from any one global state we need to abstract away some of the irrelevant details. As we saw earlier, any abstracting away would bring about the failure of the determination relation, because precisely in the details abstracted away might lurk a potential event that would interfere with the situation supposedly determined by the parts of the state not abstracted away. The question is moot in any case, because without a pre-existing modal/causal concept fixing which parts of the state are relevant or irrelevant to the outcome in question, there is no sense to be made of abstracting away the unimportant details. Every detail is important; any abstraction trivializes the explanation. It is the very precision and detail of the explanation in terms of fundamental physics which means it is spectacularly poor at capturing the broad outlines of a situation. On the contrary, every detail matters and none matters more than others, so any variation in even the most minor of respects leaves us unable to apply the lesson learned in the original situation (Eagle 2004, p. 395).

This is precisely where causal explanations do well, and is precisely a reason to think that fundamental explanations do not trump causal explanations, contrary to Russell's assumption and the assumption of the causal exclusion argument. Of course, every physical event has some perfectly accurate explanation in terms of fundamental physics that describes with minute precision the actual circumstances of the event. But this emphasis on the details of the actual situation precludes the generalizations that every good explanation asks for: how can we apply the lesson of this situation elsewhere? In its emphasis on the actual, fundamental physics fails to give a reasonable answer to this question. The modal aspects of causation, particularly the counterfactuals that underlie true causal claims on my account, precisely answer this question concerning generalizations. Of course, even fundamental physics provides alternative models, and hence can ground some modal claims. But the only counterfactuals the most fundamental theories give us the resources to evaluate are those of the form 'If the global state were Γ, the subsequent global state after t elapsed would be Γ_t', and these are by no means the only nor even particularly important counterfactuals that we are concerned to evaluate when deliberating. We must recognize that, as we want robust dependence relations between salient local events, causal language does a much better job of describing and managing them than the highly extrinsic and gerrymandered relations that fundamental physics provides. It would be impossible to give up causal language without also abandoning our status as agents.

This emphasis on the incapacity of fundamental theory to deal with any but the most basic modal claims actually made no direct appeal to causation. Hence the objection raised by Russell would generalize to any alternate possibility not directly represented as a fundamental global feature of some other model of the fundamental theory. Any creature, therefore, that was concerned to represent or discuss alternate possibilities would be well served to avoid the conclusion of the causal exclusion argument—for soon on its heels would follow a 'chance exclusion problem' and a 'possibility exclusion problem'. If this were to happen, we would be in serious trouble. I've argued that our status as deliberating agents depends on our being able to reason hypothetically and consider alternative situations. But there are arguments that I find extremely plausible that suggest that almost every feature of our thinking involves consideration of alternate possibilities. I am thinking here of Stalnaker's view that representation and inquiry are the fundamental practices in virtue of which we count as agents or thinkers at all (Stalnaker 1984a). Representation requires consideration of alternative possibilities in a synchronic fashion, dividing the class of possibilities into those compatible with the current situation and those incompatible, and representing the actual world as a member of the first class. Inquiry requires consideration of alternative possibilities in a diachronic fashion, as incoming evidence and reasoning narrows the class of compatible possibilities over time. Both of these practices involve, and would be unrecognisable without, alternate possibilities. Insofar as modal notions are so central, we should all wish to adopt the minimalist defence of causal models against Russell's exclusion argument.

7.5 Conclusion

The minimalist position with respect to causation is a third option between realism and eliminativism. Price (2004) characterizes his minimalism as *republican*, by analogy with the political system, because it sees the source of justification for causal claims as being neither in the world, nor nowhere at all, but rather in ourselves and in the perspective-dependent practices we endorse. In some cases, inescapable features of ourselves, combined with practices it would be inconceivable to abandon, demand an inescapable commitment of some kind to practices that do not represent fundamental reality.

So it is with causation. Though Russell's arguments unfortunately preclude a realist reduction of causation to a deterministic fundamental physics (still less to an indeterministic physics), that does not mean that causation must be excluded from our conceptual economy. Once we recognize that we are limited agents, concerned with characterizing effective and robust interventions on systems we care about, causal explanations and models have decided advantages over the explanations and models constructed using only the relations and entities explicitly found in fundamental physics. This is especially apparent once we realize that causation stands or falls with other essential modal notions. No threat to the ontological preeminence of fundamental physics need follow from a commitment to the essential legitimacy of causation for agents like ourselves.

The minimalist picture fits naturally with a (hermeneutic) *fictionalist* interpretation of the discourse in question (Kalderon 2005). This would involve regarding causal language as semantically continuous with the rest of language, but that the utterance of declarative sentences about causal relations does not have the force of an assertion that the semantic content of those sentences is literally true. (It may be that they are to be understood as assertions of something else, or not assertions at all.) The analogy holds however we understand these 'quasi-assertions': utterances within a given perspective, just as utterances in a fiction, are to be interpreted semantically as if the content of the fiction were true, but not as representing that the content obtains actually. When giving a naturalistic model of creatures as users of a particular theory, what we do is answer the question why these creatures should find it congenial or inescapable to adopt the fiction in question. I tried to answer that question for causation above (Section 7.4.2).

Though I myself prefer a fictionalist theory of causal discourse, nothing in what I have argued above relies on a fictionalist position. The narrow minimalist point required to avoid Russell's impossible conclusion is that causal explanations are practical necessities for agents like us, which provides a reason for keeping causation which is more than powerful enough to overmatch whatever reasons Russell adduces to get rid of causation. How we should then go on to understand the ontological status of causal relations—whether as having 'derivative reality' (Norton in ch. 2 of this volume, Section 2.4), or being real but irreducible entities, or taking a fictionalist stance—is a different and much larger question that I cannot answer here.

References

Albert, D. Z. (2000). *Time and Chance*. Cambridge, MA: Harvard University Press.
Bennett, K. (2003). 'Why the Exclusion Problem Seems Intractable, and How, Just Maybe, to Tract It', *Noûs*, 37: 471–97.
Butterfield, J. (1992). 'Bell's Theorem: What it Takes', *British Journal for the Philosophy of Science*, 43: 41–83.
Cartwright, N. (1979). 'Causal Laws and Effective Strategies', *Noûs*, 13: 419–37.

─── (2002). 'Against Modularity, the Causal Markov Condition, and Any Link Between the Two: Comments on Hausman and Woodward', *British Journal for the Philosophy of Science*, 53: 411–53.

Dowe, P. (2000). *Physical Causation*. Cambridge: Cambridge University Press.

Eagle, A. (2004). 'Twenty-One Arguments Against Propensity Analyses of Probability', *Erkenntnis*, 60: 371–416.

Earman, J. (1986). *A Primer on Determinism*. Dordrecht: Reidel.

Field, H. (1980). *Science Without Numbers*. Princeton: Princeton University Press.

─── (2003). 'Causation in a Physical World', in M. J. Loux and D. Zimmerman (eds). *Oxford Handbook of Metaphysics*. Oxford: Oxford University Press, 435–60.

Hausman, D. M. and Woodward, J. (1999). 'Independence, Invariance and the Causal Markov Condition', *British Journal for the Philosophy of Science*, 50: 521–83.

Hitchcock, C. (1995). 'Salmon on Explanatory Relevance', *Philosophy of Science*, 62: 304–20.

─── (1996). 'The Role of Contrast in Causal and Explanatory Claims', *Synthese*, 107: 395–419.

─── (2001). 'The Intransitivity of Causation Revealed in Equations and Graphs', *Journal of Philosophy*, 98: 273–99.

─── (2003). 'Of Humean Bondage', *British Journal for the Philosophy of Science*, 54: 1–25.

Johnston, M. (1992). 'Constitution is Not Identity', *Mind*, 101: 89–105.

Kalderon, M. E. (ed.). (2005). *Fictionalism in Metaphysics*. Oxford: Oxford University Press.

Lange, M. (2002). *An Introduction to the Philosophy of Physics: Locality, Fields, Energy and Mass*. Oxford: Blackwell.

Levi, I. (1980). *The Enterprise of Knowledge*. Cambridge, MA: MIT Press.

Lewis, D. (1973a). 'Causation', in D. Lewis (1986a), *Philosophical Papers*, vol. 2. Oxford: Oxford University Press, 159–213.

─── (1973b). *Counterfactuals*. Oxford: Blackwell.

─── (1979a). 'Attitudes *De Dicto* and *De Se*', in D. Lewis (1983). *Philosophical Papers*, vol. 1. Oxford: Oxford University Press, 133–59.

─── (1979b). 'Counterfactual Dependence and Time's Arrow', in D. Lewis (1986a). *Philosophical Papers*, vol. 2. Oxford: Oxford University Press, 32–66.

─── (1986b). 'Causal Explanation' in D. Lewis (1986a) *Philosophical Papers*, vol. 2. Oxford: Oxford University Press, 214–40.

─── (2000). 'Causation as Influence', *Journal of Philosophy*, 97: 182–97.

Menzies, P. (ch. 8, this volume). 'Causation in Context'.

Norton, J. D. (ch. 2, this volume). 'Causation as Folk Science'.

Pearl, J. (2000). *Causality: Models, Reasoning and Inference*. Cambridge: Cambridge University Press.

Price, H. (1992). 'Agency and Causal Asymmetry', *Mind*, 101: 501–20.

──── (2001). 'Causation in the Special Sciences: the Case for Pragmatism', in M. C. Galavotti, P. Suppes, and D. Costantini (eds). *Stochastic Causality*. Stanford: CSLI Publications, 103–21.

──── (2004). 'Models and Modals', in D. Gillies (ed.). *Laws and Models in Science*. London: King's College Publications, 49–69.

Ramsey, F. P. (1929). 'General Propositions and Causality', in F. P. Ramsey (1990). *Philosophical Papers*. Cambridge, UK: Cambridge University Press, 145–63.

Rosen, G. (1990). 'Modal Fictionalism', *Mind*, 99: 327–54.

Russell, B. (1913). 'On the Notion of Cause', in B. Russell (1963). *Mysticism and Logic*. London: George Allen and Unwin, 132–51.

Spirtes, P., Glymour, C., and Scheines, R. (2000). *Causation, Prediction and Search*. Cambridge, MA: MIT Press, 2 ed.

Stalnaker, R. C. (1984a). *Inquiry*. Cambridge, MA: MIT Press.

──── (1984b). 'Possible Worlds' in Stalnaker (2003). *Ways a World Might Be*. Oxford: Oxford University Press, 25–39.

Suppes, P. (1970). *A Probabilistic Theory of Causality*. Amsterdam: North-Holland.

Woodward, J. (2001). 'Probabilistic Causality, Direct Causes and Counterfactual Dependence', in M. C. Galavotti, P. Suppes, and D. Costantini (eds), *Stochastic Causality*. Stanford: CSLI Publications, 39–63.

──── (2003). *Making Things Happen*. Oxford: Oxford University Press.

Yablo, S. (2005). 'The Myth of the Seven', in Kalderon (2005), pp. 88–115.

8
Causation in Context

PETER MENZIES

8.1 Introduction

Bertrand Russell (1913) argued that the concept of cause should be extruded from the philosophical vocabulary because it is inextricably bound up with misleading connotations.

All philosophers, of every school, imagine that causation is one of the fundamental axioms or postulates of science, yet oddly enough, in advanced sciences such as gravitational astronomy, the word 'cause' never occurs... the reason why physics has ceased to look for causes is that, in fact, there are no such things. (1913, p. 180)

In this chapter I am more interested in the target of Russell's arguments than their cogency. Exactly what doctrine was he criticizing? Though he singles out several philosophers for their crude formulations of the 'law of causality', Russell does not explicitly state the target of his criticisms. Nonetheless, it is reasonable, I think, to see Russell as attacking a doctrine that might be called *causal realism*—the doctrine that causation is a feature of objective or mind-independent reality.

Actually, this is not a single doctrine, but rather a family of doctrines, varying according to what are taken to be the fundamental constituents of 'objective reality'. The particular version that Russell seems to be criticizing is one that assumes that causal relations are among the basic constituents of reality described by our most advanced physical theories. His criticism in the passage above is that this cannot be so since the advanced physical sciences do not make use of the causal concept. Put this simply, the criticism is not completely persuasive. For a defender of the causal concept could well argue that, even though causation is not explicitly mentioned in fundamental physics, it is implicitly present in the picture of reality given in fundamental physics, since causal relations supervene on the pattern of fundamental physical facts and physical laws. This more sophisticated doctrine is one Russell certainly never formulated because he did not have the concept of supervenience to hand.

In the period since Russell wrote, causal realism has become philosophical orthodoxy. The currently popular versions state that causal relations supervene on objective, mind-independent structures, though they differ with respect to what these structures are. For example, regularity theorists like J. L. Mackie (1974) think that causal relations supervene on patterns of regularities in occurrent fact. Counterfactual theorists like David Lewis (1973a, 2000) take causal relations to supervene on a network of events ordered by transitive, asymmetric relations of counterfactual dependence. Probabilistic theorists like Paul Humphreys (1989) and Ellery Eells (1991) believe that causal relations supervene on relations of probabilistic dependence between events. Process theorists like Wesley Salmon (1994) and Phil Dowe (2000) believe that causal relations supervene on patterns of causal processes and interactions that involve the conservation or exchange of physical quantities. Though these theories differ in detail, they all subscribe to the doctrine that causal relations depend completely on a substructure of mind-independent relations. To be sure, a commitment to the existence of this objective substructure often goes with a grudging admission of some minor pragmatic elements in the causal concept. For example, most causal realists are prepared to allow that pragmatic principles of 'invidious selection', as Lewis calls them, govern the way in which we select as 'the cause' a salient part of the vast network of events leading up to an event. However, setting aside such 'minor pragmatic complications', they claim that the causal concept has completely objective truth-conditions which can be stated in terms of conditions holding of the mind-independent substructure.

Causal realism is the target of this chapter, as it was of Russell's paper. One way to criticize this doctrine would be to show that there are features of causal claims that do not map onto any kind of objective substructure of events and relations. This kind of criticism would have to proceed on a case by case basis, as there are many versions of causal realism, differing in what they take to be the basic substructure of causation. Here I shall adopt a more general strategy. What I attempt to show is that the concept of causation is context-sensitive. Take any objective substructure of events and relations, whatever it may be, this pattern cannot determine the truth-conditions of a causal judgement, because its truth-value can vary from one context to another, depending on how a certain contextual parameter is set. The very same pattern of objective relations, viewed from within one context, may support a causal judgement, but, viewed in another context, may fail to support the judgement.

This style of argument is familiar in the case of other context-sensitive expressions. For example, Peter Unger (1975) has argued that the concept of flatness is sensitive to a certain contextual parameter that might be called a standard of flatness. This standard may vary from one context to another so the very same surface may truly be said to be flat in one context but not in another. The observation that the knowledge concept is context-sensitive in this way is at the heart of contextual theories of knowledge (de Rose 1992, Lewis 1996). I shall argue that the causal concept is subject to a similar kind of context-sensitivity. The objective facts of a situation do not determine whether one event causes another. No more so than the objective facts about the evenness of a surface determine its flatness, or the objective facts about the kind of evidence possessed by a person determines whether that person has knowledge. In all these cases, the truth of an attribution of causation, flatness or knowledge depends on how a contextual parameter is set.

This conclusion about causation might be seen as supporting Russell's position against causal realism. If the causal concept is so riddled with context-sensitivity, then it has no use in providing an account of objective reality. (Bas van Fraassen 1980 argues for this conclusion.) Such an *error theory* about causation accepts the causal realist's characterization of reality as mind-independent, but denies the causal realist's claim that causation is part of this objective reality. But contextualism about the causal concept does not necessarily support an error theory. For another way to dispute causal realism is to reject at the outset the realist's characterization of reality in terms of certain privileged mind-independent facts. This relatively unfamiliar position might be called *perspectival realism*. The perspectival realist acknowledges that the truth-value of causal judgements does not depend entirely on the mind-independent structures. The context-sensitive character of causal judgements indicates that their truth value is perspective-relative. Nonetheless, this does not detract from the reality of the causal relations they describe. This relatively unexplored position represents, in my view, a very promising alternative to an error theory about causation. However, in this chapter I will not try to adjudicate the merits of these two positions opposed to causal realism.

8.2 Evidence for Contextualism

Here below I cite some evidence in support of the view that the truth-conditions of causal statements are context-sensitive. The examples I cite as evidence are, for the most part, familiar from the philosophical literature on causation. They have been discussed as counterexamples to regularity, counterfactual or probabilistic theories of causation—the currently popular forms of causal realism. However, my interest in them lies in the fact that they demonstrate that the truth-value of causal statements can vary from one context to another. Even when the patterns of regularities, counterfactuals and probabilistic dependence are held constant, the truth-value of the causal statements can vary from one context to another.

(A) The Indian Famine

Pre-theoretically, we draw a distinction between causes and background conditions. The context-sensitivity of this distinction has been discussed at length by H. L. A. Hart and A. Honoré (1985).

> The cause of a great famine in India may be identified by an Indian peasant as the drought, but the World Food Authority may identify the Indian Government's failure to build up food reserves as the cause and the drought as a mere condition. (pp. 35–6)

In one context, it is appropriate to judge that the following judgement is true:

(1) The drought caused the famine and the failure to stockpile food reserves was a mere condition of the causation.

Yet in another context a contrary statement seems to be true:

(2) The failure to stockpile food reserves caused the famine and the drought was a mere condition of the causation.

This variation in judgement occurs despite the fact that the same regularities, counterfactuals, and probabilistic dependences hold in both contexts. For example, it is true in both contexts that if the drought had not occurred or if the government had stockpiled food, the famine would not have occurred.

(B) The Cricket Ball and Window

Michael McDermott (1995) describes the following example:

> A cricket ball is hit with substantial force towards a window. A fielder reaches out and catches the ball. The next thing along in the ball's direction of motion is a solid brick wall. Beyond that is the window. Did the fielder's catch prevent the ball hitting the window? (p. 525)

Several causal judgements about this situation appear reasonable. On the one hand, because the wall would have prevented the ball from hitting the window even if the fielder had not caught the ball, we are inclined to judge that:

(3) The fielder's catch did not prevent the ball from hitting the window.

On the other hand, because the ball was actually intercepted by the fielder and not the wall, we are inclined to judge that:

(4) The fielder's catch prevented the ball from hitting the window.

Again note that our judgements vary from one context to another even though the regularities, counterfactuals, and probabilistic dependences remain constant. For example, irrespective of whether we judge that the fielder prevented the ball from hitting the window or not, it is true that if the fielder had not acted, the ball would not have hit the window, and if the wall had not been there and the fielder not acted, the ball would have hit the window.

John Collins (2000) notes that our intuitions in pre-emptive prevention cases of this kind can vary depending upon the nature of the backup preventer. If it is transient thing like a second fielder, we are more inclined to judge that the first fielder's catch prevented the window from shattering. If it is a permanent thing like a wall, we are less inclined to make this judgement. Finally, if the window is on the moon (so that the earth's gravitational field is a very permanent backup preventer), we are very disinclined to make this judgement. Following Collins, Lewis (2000) concludes that our judgements of causation depend upon which possibilities we deem to be too far-fetched. It is not so far-fetched that both the fielders would miss the ball, somewhat more so that the ball would avoid the wall to smash the window, and absurdly far-fetched to suppose that the ball could evade the earth's gravitational field.

(C) Contraceptive Pills and Thrombosis

Our third example was originally proposed by Hesslow (1976) as a counterexample to probabilistic theories of causation. He assumed an indeterministic setting, but I shall adapt the example to a deterministic setting.

Betty is a young, fertile, sexually active woman who is capable of becoming pregnant. She takes contraceptive pills, which are known to cause thrombosis among women with a certain causal factor X. As it turns out, Betty has the factor X and so develops thrombosis. Let us assume that as she is taking the pill, she will avoid pregnancy, but if she were not to take the pill or use any other contraceptive method, she would become pregnant. It is a good thing that she will avoid pregnancy because when women with factor X become pregnant they inevitably get thrombosis.

It would seem that there are two causal judgements we can make about this case. On the one hand, since Betty did not become pregnant her getting thrombosis must be due to her taking the contraceptive pill. So it seems straightforwardly true that:

(5) Betty's taking contraceptive pills caused her thrombosis.

On the other hand, taking contraceptive pills has a negative effect on thrombosis in women with the factor X, since taking contraceptive pills prevents pregnancy, which would otherwise cause them to get thrombosis. So it seems plausible to judge that:

(6) Betty got thrombosis despite the fact that she took the contraceptive pill.

But finally because these two causal effects cancel each other out, it seems plausible to judge that:

(7) Betty's taking contraceptive pills made no overall difference to her getting thrombosis, since she would have got thrombosis whether or not she had taken the pills.

Christopher Hitchcock (2001b) cites this example as illustrating an ambiguity in the concept of causation, which he describes in the following terms. The consumption of oral contraceptives affects a woman's developing thrombosis along at least two different routes. By analogy with the concepts of net and component forces in Newtonian mechanics, he says that contraceptive pills have two distinct effects upon thrombosis. Along one route, the one that bypasses pregnancy, the effect of contraceptive pills on thrombosis is positive, as reflected in judgement (5). Along the other route—the one that includes pregnancy or its absence—the component effect is negative: by preventing pregnancy, contraceptive pills prevent thrombosis, which is reflected in judgement (6). However, the net effect of contraceptive pills on thrombosis is neutral, as reflected in judgement (7). So he says that when we are asked whether birth control pills cause thrombosis, we can interpret this as a question about one or the other component effect, or about the net effect.

(D) The Golfer

The following example, due originally to Deborah Rosen, has been discussed at length by Wesley Salmon (1984, ch. 7). The counterexample was intended to be a counterexample to probabilistic theories of causation and is usually presented in an indeterministic setting. However, I will present the example under the assumption that causation is deterministic.

An experienced golfer is about to drive his ball onto the green. Given his position, his level of skill and the prevailing conditions, he has an excellent chance of his holing the ball. Indeed, the only crucial variable is the angle and force of his drive: if these are within his normal range, his chance of holing out is one. But, as it happens, the player is tense and slices the ball, which veers away from the green. But then the ball hits a tree near the green and bounces back onto the green and into the cup.

Again, we are inclined to make apparently inconsistent causal judgements about this situation. On the one hand, since we can trace a causal pathway from the golfer's slicing the ball to its falling into the cup, we are inclined to judge that:

(8) The golfer's slicing the ball caused the ball to fall into the cup.

On the other hand, since the ball was veering away from the green when it hit the tree, we are inclined to think that:

(9) The ball holed out despite the fact that the golfer sliced the ball.

One way to reconcile this apparent conflict would be to follow Elliott Sober (1984) in saying that while statement (8) is true of token causation, statement (9) reflects a judgement that is true of type causation: a golfer's slicing the ball so that it veers away from the green generally hinders the ball from holing out. If we suppose that type and token causal claims have different semantics, we could then reconcile the apparent conflict between these causal statements. But Hitchcock (1993) has argued that this diagnosis cannot be correct, as statement (9) describes a token causal relation just as much as statement (8).

Clearly, we make these various token causal judgements because we find different features of the example salient in different contexts. Nonetheless, no matter which feature is salient in a given context, the same pattern of regularities, counterfactuals and probabilistic dependence hold true of the situation described in the example. For example, it is true that if the golfer had not sliced the ball, the ball would have holed out; and also true that if the ball had not hit the tree, it would have veered away from the green, and so on.

8.3 Two Orthodox Responses: Ambiguity and Pragmatics

Most philosophers find it hard to accept the conclusion that our judgements about token causation are essentially context-sensitive. In response to examples such as those of the last section, they tend to adopt one or other of two responses.

One response is to say that such examples do not impugn the idea that causal claims have context-invariant truth conditions, but rather demonstrate ambiguities in the concept of causation. This is the line of thought pursued by Hitchcock (2003), which also discusses many of the examples of the last section. According to Hitchcock, it is a mistake to look for truth-conditions for *the* concept of causation because there is simply no single unitary concept: such examples as those cited demonstrate that the verb 'cause' has many different meanings, with each meaning having its own context-invariant truth-conditions.

Contra Hitchcock, it seems to me to be remarkably implausible that the verb 'cause' should be a homonym like the verb 'bank', especially so in view of many different meanings that would have to be postulated for the verb. In the last section we considered four different, unrelated kinds of examples. If we had to posit a new ambiguity for each example, we would have four orthogonal ambiguities. However, it is implausible to suppose that our judgements about these examples depend on a sophisticated mastery of four different kinds of ambiguity. Indeed these are only a small handful of the plethora of examples that could be used to illustrate the way our judgements about causation display uncertainty and ambivalence. These considerations, taken by themselves, are hardly conclusive, I concede. But they become more compelling when one shows, as I aim to do later in this paper, that a single set of truth-conditions, albeit ones that are relativized to a contextual parameter, can explain all the different judgements prompted by these examples.

There is another standard response to examples like those of the last section that seem to illustrate the context-sensitivity of the causal concept. It is to say that there is a univocal concept of causation that has context-invariant truth-conditions and that the different uses of the causal concept in the examples are to be explained in terms of context-sensitive pragmatic principles operating on this univocal context-invariant concept. The univocal concept is not straightforwardly evident in ordinary usage because it is partially disguised by the operation of the pragmatic principles of conversation. We can, however, recover this context-invariant concept by deliberately suspending the pragmatic principles in our philosophical discourse. Sometimes this view is expressed by using separate words for the context-invariant concept and for the compound concept that results from the operation of the pragmatic principles on the context-invariant concept. Under this regimentation, the former concept is called 'causation proper' and the latter is called 'causal explanation'.

We can only assess the merits of this view by investigating whether it is successful in explaining all the diverse causal judgements we are disposed to make. To be sure, the view seems to be successful in providing a plausible explanation of the causes and conditions distinction. (See, for example, van Fraassen 1980, ch. 5.) A rough outline of the explanation goes like this. Assume that there is a network of objective causes leading up to any given event. This network consists of all those events related to the given event by some objective relation of the kind embraced by causal realists. The network will be vast, especially if the temporal ordering of events is dense, the structuring relation is transitive, and non-occurrences as well as occurrences are allowed. It is crucial that judgements about the objective causes located in this network display no context-sensitivity. In contrast, judgements about what causally explains the event in question are highly context-dependent, as they depend on the kind of explanatory question being asked. More specifically, we might give different causal explanations of the same event in different contexts because those contexts pose different contrastive why-questions. For example, in connection with the example about the Indian famine, one might ask: Why did the famine occur *at this time* rather than some other time? A good answer to this question is to mention the drought that differentiates the present time from other times. The failure of the government to stockpile reserves of food is a mere background condition that is common to all the times and so not counted as a differentiating factor. Alternatively, one might ask: Why did the famine occur *in India*, rather than in other countries that frequently experience droughts? A good answer to this question is to mention the failure of the Indian government to build up reserves of food, which distinguishes that country from others. The occurrence of a drought is a mere condition that is common to all the different countries and so does not differentiate between them.

Perhaps this account in terms of the relativity of causal explanations to contrastive why-questions can be spelt out in detail to explain the distinction between causes and conditions. (However, I express doubts about this in Menzies 2004a.) Nonetheless, the account does not seem applicable to any of the other examples previously cited. These other examples seem to involve a kind of context-sensitivity that cannot be explained in terms of the relativity of causal explanations to contrastive why-questions. A systematic explanation of all these examples is needed. Until it is provided, the view that the context-sensitivity of the causal concept can be explained in terms of pragmatic principles operating on a univocal context-invariant concept is nothing more than a promissory note.

8.4 Causes as Difference Makers

Many different approaches to causation—for example, the regularity, counterfactual and probabilistic approaches—draw their inspiration from the idea that a cause is something that makes a difference to its effects, though these approaches articulate the idea in slightly different ways. (In my view, all such theories face serious difficulties dealing with cases of pre-emption and overdetermination. However, to make our discussion manageable, I shall set to one side the question of how this idea might be articulated to deal with such difficult cases.) What is the precise content of this idea of a cause as a difference-maker? It is useful to try to express its content as clearly as possible to gain insight into how to articulate it most informatively. In my view, the philosophers who have best expressed the central idea that causes are difference-makers are Hart and Honoré:

Human action in the simple cases, where we produce some desired effect by the manipulation of an object in our environment, is an interference in the natural course of events which *makes a difference* in the way these develop. In an almost literal sense, such an interference by human action is an intervention or intrusion of one thing upon a distinct kind of thing. Common experience teaches us that, left to themselves, the things we manipulate, since they have a 'nature' or characteristic way of behaving, would persist in states or exhibit changes different from those we have learnt to bring about in them by our manipulation. The notion that a cause is essentially something which interferes with or intervenes in the course of events which normally take place, is central to the commonsense concept of cause... Analogies with the interference by human beings with the natural course of events in part control, even in cases where there is literally no human intervention, what is identified as the cause of some occurrence; the cause, though not a literal intervention, is a difference to the normal course which accounts for the difference in outcome. (1985, p. 29)

There seem to be three elements to Hart and Honoré's model of the way the commonsense causal concept works. The first element, which is implicit in Hart and Honoré's description, is that the application of the causal concept to a particular situation depends upon conceptualizing the situation as involving a certain kind of system. In applying the causal concept to a particular situation, we abstract and generalize by interpreting the situation in terms of the way a particular kind of system behaves. The second element is that we typically suppose that systems of the given kind, when left to themselves, display a characteristic way of behaving. In other words, we typically suppose that systems of the given kind follow a course of evolution that is 'normal' or 'natural' for systems of that kind when they are not subject to outside interference. The third and final element is that, when the system actually exemplified in the particular situation has deviated from its normal course of evolution, we search for something that made the difference—an event or state that is analogous to an external intervention or intrusion into the system.

In my view, Hart and Honoré's model captures one important application of the causal concept, but not the only application. (See Menzies 2004a, 2004b for discussion of another important application that is relevant to cases involving so-called 'immanent causation', i.e. cases in which the states of an unchanging object are causally connected.) However, to simplify my exposition, I shall focus on this application, and try to articulate it precisely so as to reveal the different ways in which it is sensitive to context. In the remainder of this section I shall try to articulate the application of the causal concept described by Hart and Honoré within the structural equations (SE) framework. Though there is a long tradition of the use of this framework in the social sciences, especially econometrics, the state-of-the-art presentation of the framework is Judea Pearl's *Causality* (2000). I shall take over the essentials of Pearl's framework, as it has been expounded by Hitchcock (2001a) and James Woodward (2003, ch. 2). However, I shall introduce modifications of my own to this framework which make my treatment of token causation quite different from those of Pearl, Hitchcock and Woodward.

One distinctive feature of the SE framework is that it relativizes the truth of a token causal claim about a particular situation to a causal model. This relativization corresponds to the first element of the Hart and Honoré's picture of the way the causal concept is applied. A causal model specifies the kind of system in terms of which we conceptualize the causal structure of the particular situation. Informally, a causal model represents a certain kind of system in terms of a set of variables representing the relevant dimensions of change for systems of the given kind and in terms of a set of generalizations governing the behaviour of systems of the given kind. More formally, a causal model is an ordered triple $<U, V, E>$. Here U is a set of exogenous variables whose values are determined by factors outside the model; V is a set of endogenous variables whose values are determined by factors within the model; and E is a set of structural equations.

The set E contains a structural equation for each variable, which appears on the left-hand side of its equation. The form of the equation depends on whether it is an exogenous or endogenous variable. If it is an exogenous variable, the equation simply states its actual value. But if it is an endogenous variable, the equation takes the form:

$$Y = f_Y(X_1, \ldots, X_n),$$

where $X_1, \ldots, X_1$ are all and only the variables from the sets U and V that play a role in determining the value of Y. It is important to note that the structural equations are not standard symmetric equations. In these equations side matters: the values of the variable on the left-hand side are *determined* by the values of the variables on the right-hand side. Different theorists understand the structural equations in slightly different ways. Pearl understands them to represent the basic causal mechanisms governing the behaviour of the system of the given kind, with each equation representing a distinct causal mechanism.

A causal model and its set of structural equations can be depicted in a graphical representation. The variables in the sets U and V form the nodes of a graph. These nodes are connected by directed edges according to the following rule: an edge is drawn from X to Y if and only if X appears in the right-hand side of the structural equation for Y; in other words, if and only if the values of X play a role in determining the values of Y. In this case, X is said to be *a parent of* Y. An exogenous variable is one without any parent in the graph for its model.

In relativizing causal claims to a causal model, the SE framework clearly introduces several degrees of freedom in representing the causal structure of a particular situation. As we shall see, this will be crucial to explaining the context-sensitivity of causal claims. However, it is important to emphasize that this relativization should not be seen as introducing an excessive degree of subjectivity. For example, philosophers who study causal models sometimes remark that the causal structure implied by a model depends on the choices made by the modeler—for example, choices about how to represent the situation in terms of variables, and about whether to represent them as binary or many-valued variables. In contrast, Pearl strongly dissents from such remarks. As I read him, he believes that it is a completely objective matter how a certain kind of system is to be represented in a model. The set of variables U and V of a model express all and only the objective dimensions of change for systems of the given kind, where these are joints in nature, discovered rather than constructed by us. He has correspondingly objectivist views about the set of structural equations E of a model. As remarked above, for him these represent the basic causal mechanisms that govern the behaviour of systems of the given kind. Their existence and nature are completely mind-independent matters that are settled on the basis of objective experimental and observational methods. I shall follow Pearl in his objectivist construal of the way models represent the kinds of systems that are implicitly invoked in causal talk.

It is probably best to explain the SE framework by means of a simple example. Let us adapt an example introduced by Hitchcock (1996) for a different purpose. (We shall consider Hitchcock's original purpose eventually.) Let us suppose that a person is given a certain drug, 'curit', in order to cure him of a disease from which he is suffering. He can be given different doses of the drug: no dose, a moderate 100 mg dose, or a strong 200 mg dose. The drug is known to be effective in large doses, but the cost and the risk of side-effects make it impractical to give a large dose to this patient; and so he is given a moderate dose of 100 mg. As it happens, the patient recovers; and we ask 'Did taking the moderate dose make a difference to the patient's recovery?'

To answer this question within the SE framework, we have to choose a causal model that specifies the kind of system we are investigating. Let us model the patient as a human physiological system, but let us dramatically oversimplify this kind of system by supposing it can be characterized in terms of two variables: an exogenous variable C, which can take three values 0 mg, 100 mg, and 200 mg corresponding to the different dosage levels; and an endogenous variable R, which takes the value 1 if the patient recovers and 0 if he does not recover. Finally, let us suppose that the structural equations of this model are as follows:

$C = 100$ mg;
$R = f(C)$, where $f(C) = 1$ if $C \geq 100$ mg and 0 otherwise.

The first equation is the structural equation for the exogenous variable C. The second is the structural equation for the endogenous variable R: it states that a patient recovers if and only if he is given a dose of the drug curit greater than or equal to 100 mg. It is simple matter to calculate that the actual value of the variable R is 1. The graph for this model is depicted in Figure 8.1

$$C \longrightarrow R$$

Figure 8.1

In order to answer the question 'Did taking the moderate dose of curit make the difference to recovery?' within the SE framework, one has to assess whether there is a counterfactual dependence between his taking the moderate dose and his recovery. For the framework construes the difference-making relation in terms of counterfactual dependence. Indeed, it is one of the nice features of this framework that it provides an elegant method for evaluating counterfactuals. To evaluate a counterfactual whose antecedent specifies the value of a variable, whether exogenous and endogenous, we simply replace the equation for the relevant variable with one that stipulates the new value of the variable, while keeping all the other equations unchanged. For example, to calculate what would have happened if the patient had been given no dose or a strong dose of curit, we simply replace the structural equation $C = 100$ mg with either $C = 0$ mg or $C = 200$ mg and recalculate the value of the endogenous variable R. When we do this we can see that the following counterfactuals are true:

$C = 0$ mg $\square\!\!\rightarrow R = 0$;
$C = 200$ mg $\square\!\!\rightarrow R = 1$.

Within the SE framework, the action of replacing an equation by another stipulating a new value represents the way in which the value of the relevant variable might be set by an intervention from outside the system. Of course, this is a figurative way of expressing the matter. But it is hard to avoid the use of metaphorical talk. For example, Lewis' possible worlds semantics for (non-backtracking) counterfactuals invokes a similar analogy when it stipulates that the closest deterministic worlds in which a counterfactual antecedent is true are ones in which the antecedent is realized by a miracle. The action of setting the value of a variable by a surgical intervention has special significance in the case of an endogenous variable. Replacing the equation for an endogenous variable with an equation stipulating a new value is in effect treating the endogenous variable as an exogenous one. Graphically, the edge that is directed into this variable is removed while all the other edges remain intact. As the proponents of the SE framework acknowledge, this technique for evaluating counterfactuals depends on certain non-trivial theoretical assumptions. First, it depends on the assumption that the system of equations is *modular* in the sense that one can surgically change the equation for one variable without thereby disturbing the equations for the other variables. Secondly, it depends on the assumption that the equations of a system are individually *invariant* in the sense that they continue to hold under interventions that set the values of some of their variables.

This technique for evaluating counterfactuals enables one to capture the idea of counterfactual dependence. The most natural definition within the SE framework is this:

Definition 1: A variable Y *counterfactually depends* on a variable X in a causal model if and only if it is actually the case that $X = x$ and $Y = y$ and there exist $x' \neq x$ and $y' \neq y$ such that the result of replacing the equation for X with $X = x'$ yields $Y = y'$.

As noted above, this notion of counterfactual dependence is supposed to capture the idea of difference-making. The value of one variable *makes a difference* to the value of another variable precisely in the sense that if the first variable had been wiggled to change its value, then the second variable would have changed its value too.

However, we now come to the problem that Hitchcock originally used the example to illustrate. Unfortunately, one cannot apply the definition of counterfactual dependence to the example at hand to determine a unique, unequivocal answer to the question whether giving the moderate dose makes a difference to the patient's recovery. For there are two different counterfactual cases that contrast with the actual case in which the patient is given the moderate 100 mg: the case in which he is given no dose of curit and the case in which he is given the strong 200 mg dose. The problem lies in the fact that these different contrast cases give conflicting answers to the question about counterfactual dependence. If we suppose that the hypothetical case that contrasts with the actual case is one in which the patient is given no dose of curit, then patient's recovery counterfactually depends on his taking moderate dose. However, if we suppose that the contrasting hypothetical case is one in which the patient is given the strong dose, the patient's recovery does not counterfactually depend on his taking the moderate dose. Clearly these results cannot both be correct.

The moral that Hitchcock draws from this kind of example is that we are mistaken in thinking that causation is a binary relation between a cause and effect. It is this assumption that forces us into choosing one or other result. If we think of causation as a ternary relation between a cause C and effect E relative to an alternative cause C', we can accept both results. Indeed, causal language has devices that make it possible for us to express the matter clearly. We can say the following: administering the moderate dose of curit, as opposed to no dose, made a difference to his recovery, whereas administering the moderate dose, as opposed to the strong dose, did not. Hitchcock argues that such contrastive clauses enable us to capture the extra relatum that is usually unexpressed in causal statements. Contrastive stress also serves the same purpose. For example, we might express the above idea in this way: the patient recovered because he was given *some dose* of the drug, not because he was given the *moderate* dose. This use of contrastive stress indicates that the alternative cause in the first case is 'no dose of the drug' whereas the alternative cause in the second case is 'the strong dose of the drug'.

I am inclined to think, however, that a more far-reaching moral can be drawn from this example. Let us reconsider the model of difference-making described by Hart and Honoré. The SE framework captures some of its main ideas. By relativizing causal discourse to a model, it captures the idea that causal claims involve an implicit relativity to a kind of system. It also captures the idea of a cause as a difference-maker in terms of notion of counterfactual dependence. On the other hand, the SE framework, in its present form, does not capture the guiding idea that a cause of some effect in a system is something *analogous to an intervention* from outside the system that makes a difference to the *normal course of evolution* of that system. I propose to introduce some modifications into the SE framework to capture these missing elements.

I shall modify the framework by allowing that the description of the model used to interpret a particular situation may fix the values of exogenous variables at non-actual values. In standard presentations of the framework, the values of the exogenous variables that are taken as the baselines for the calculation of counterfactual dependences are always the actual values of the variables. I propose a reversal of this usual order. Let us suppose that in framing the model relevant to a particular situation, we are permitted to set the values of exogenous variables at possible values to represent what I shall call default states of the system. Informally, the default values of the variables represent the *normal* or *natural* state of the system in question.

I shall explain this notion in more detail when I apply it to particular examples. I shall call a causal model with its exogenous variables set at their default values *a default causal model*.

We cannot apply the usual definition of counterfactual dependence to a default causal model because the definition anchors a counterfactual dependence to the actual values of the variables. We have to redefine the central notion of a difference maker as follows:

Definition 2. A value of a variable X *makes a difference* to the value of another variable Y in a default causal model if and only if plugging in the default values of the variables in the structural equations yields $X = x$ and $Y = y$ and there exist actual values $x' \neq x$ and $y' \neq y$ such the result of replacing the equation for X with $X = x'$ yields $Y = y'$.

It is easy to see the implications of this modification of the SE framework by applying it to the curit example. Suppose we say that the normal state of the patient is one in which he receives no dose of curit. So we fix on a default model in which the variable C is given the default value 0 mg, and we reason accordingly that the patient's taking the moderate dose of the drug made a difference to his recovery. On the other hand, if we suppose that the normal state of the patient is one where he is given the strong dose of the drug, so that the variable C is given the default value 200 mg, we can reason that the patient's taking the moderate dose made no difference. In this way, relativizing the assessment of a causal claim to a default causal model is roughly similar to Hitchcock's strategy of a rephrasing causal claim as a ternary relation between a cause, an effect, and a contextually determined alternative cause.

However, it is important to note the differences between the proposed strategy and Hitchcock's strategy. One difference is that Hitchcock's strategy is presented as a stand-alone strategy intended to deal with a specific difficulty. In contrast, the strategy proposed above is meant to be part of a larger strategy for formalizing the Hart and Honoré model of causation within the SE framework. As such, the strategy gains its intelligibility and its plausibility from this larger attempt to capture the insights of the interventionist or agency model that Hart and Honoré describe. For example, it is central to this model that a cause is seen as analogous to an intervention into a system that makes a difference to the system's normal course of evolution. Hitchcock's strategy makes no reference to this requirement on a cause. Another difference between the strategies concerns the scope of the context-sensitivity of causal claims. On Hitchcock's strategy, context points to an alternative possible cause to be contrasted with the actual cause, while, on the strategy proposed above, context points to a whole set of alternative possible worlds that realize the alternative cause. In other words, the proposed strategy involves the idea that context picks out a whole set of alternative background conditions as well as a contrasting cause. It is this insight—that context affects not just the choice of a contrast case, but also the set of alternative possible worlds that realize the contrast case—that provides the key to understanding the examples in Section 8.2.

8.5 The Examples Revisited

Let us return to the examples described in Section 2 to see how the modified SE framework can explain our various causal judgements.

(A) The Indian Famine

We might represent this example in terms of a model employing the following variables:

$D = 1$ if a drought occurs in India, 0 if not.
$R = 1$ if the government stockpiles reserves of food, 0 if not.
$F = 1$ if a famine occurs in India, 0 if not.

The structural equations of this model will consist of the following equations:

$D = 1$
$R = 0$
$F = D \ \& \ \neg R$

(Here logical symbols are used to represent the obvious mathematical functions on binary variables: $\neg X = 1 - X$, $X \vee Y = \max \{X, Y\}$; $X \ \& \ Y = \min \{X, Y\}$.) The first two equations state the values of the exogenous variables D and R. The third equation states the value of the endogenous variable F as a function of the variables D and R. The graph corresponding to this model is depicted in Figure 8.2.

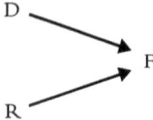

Figure 8.2

If we were to try to determine the difference-making relations according to the standard version of the SE framework, we would fix the exogenous variables D and R at the actual values 1 and 0; and then we would use those values as the baseline for assessing whether various counterfactual dependences hold. On this way of proceeding, it turns out that the famine counterfactually depends on the drought, but also depends on the failure of the Indian government to stockpile food reserves. These counterfactual dependences do not vary from one context to another and so are unsuitable for explaining how the drought makes the difference to the famine in one context, and the failure to stockpile food reserves makes the difference in a different context.

However, I have argued that it is necessary to introduce a modification to the SE framework to capture an extra dimension of context-sensitivity. This modification allows us to set the values of some variables at non-actual default values and to use them as baselines for the calculation of the difference-making relations. The default values of variables represent the states of the system that are deemed to be normal or natural in some sense.

Returning to the example at hand, let us suppose that the normal situation in India is one in which there is no drought and that the Government does not stockpile food. We can create a new default model M_1 from the old one by resetting the values of the exogenous variables at $D = 0$ and $R = 0$. Applying Definition 2 to this model, we see that the drought makes a difference to the famine, but the failure to stockpile food reserves does not. As I shall show later, when the drought makes a difference to the famine in a default model M_1, the pair of causal conditionals in (10) will hold true. I index these conditionals by M_1 to signify that they are relative to a particular default causal model M_1.

(10) $(D = 1 \,\square\!\!\rightarrow_{M1} F = 1) \,\&\, (D = 0 \,\square\!\!\rightarrow_{M1} F = 0)$

This is just the result we would expect: the drought makes a difference with respect to the famine and the government's failure to stockpile food reserves is a mere background condition that does not make a difference.

Now let us make a different assumption about the default values of the exogenous variables D and R. Let us suppose the normal situation is one in which drought regularly occurs and the government takes it upon itself to build up food reserves against the possibility of drought. So let us assume the default values of the exogenous variables are $D = 1$ and $R = 1$ in a default model M_2. Then it is possible to see that in this new default model, the failure of the government makes a difference to the famine, but the drought does not. Consequently, the pair of causal conditionals in (11) hold true:

(11) $(D = 1 \,\square\!\!\!\rightarrow_{M2} F = 0)$ & $(D = 0 \,\square\!\!\!\rightarrow_{M2} F = 0)$

Once again we have the result that agrees with intuition: in this context the government failure makes a difference with respect to the famine, while the drought is a mere background condition.

(B) The Cricket Ball, Fielder and Window

We can model this example using the following variables:

$B = 1$ if a ball is flying in direction of window, 0 if not
$F = 1$ if fielder catches the flying cricket ball, 0 if not
$W = 1$ if the wall is present, 0 if not.
$S = 1$ if cricket ball shatters the window, 0 if not.

The structural equations of the model with actual settings of exogenous variables are:

$B = 1$
$F = 1$
$W = 1$
$S = B \:\&\: \neg F \:\&\: \neg W$

The graph for this model is depicted in Figure 8.3.

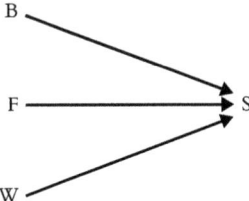

Figure 8.3

Let us consider how to convert this model into an appropriate default model. Let us assume that the system we are modelling is a flying-ball-plus-window-protected-by-a-wall and that its normal state is one in which a ball is flying through the air, the fielder does not catch the ball but the wall is present to protect the window. Accordingly, we set the default values of the exogenous variables at $B = 1, F = 0$ and $W = 1$ in a default model M_1. In this case, it is easy to check that the fielder's catching the ball makes no difference to the window's not breaking:

(12) $(F = 1 \,\square\!\!\rightarrow_{M1} S = 0 \ \& \ F = 0 \,\square\!\!\rightarrow_{M1} S = 0)$

In contrast, let us suppose that the system we are modelling is a flying-ball-plus-unprotected-window and that its normal state is one in which a ball is flying through the air, the fielder does not catch the ball, and the wall is not present. (It is easier to think of such a default model the more readily we can detach the back-up preventer from the system. This is difficult to do if the back-up preventer is a wall; and much easier to do if it is a second fielder. Compare this diagnosis with that of Collins and Lewis.) Accordingly, let us set the default values of the exogenous variables at $B = 1, F = 0$ and $W = 0$ in a default model M_2. Then it is easy to see that the fielder's catching the ball makes a difference to the window's not shattering, as reflected in the truth of the pair of causal conditionals:

(13) $(F = 1 \,\square\!\!\rightarrow_{M2} S = 0 \ \& \ F = 0 \,\square\!\!\rightarrow_{M2} S = 1)$

So our intuitions about the causal structure of this situation depend on what we take to be the system in question and what we take to be the normal state of this system. We are more inclined to see the fielder's catch as preventing the ball from shattering the window when we think of the system in question as a flying-ball-plus-unprotected-window whose normal state the does not involve the presence of the wall. (See a similar diagnosis of this example in Maudlin 2004.)

(C) Contraceptive Pills and Thrombosis

It is natural to model this example using the following variables:

$X = 1$ if Betty has factor X, 0 if not.
$C = 0$ if the Betty does not take contraceptive pills, 1 if she does, and 2 if she uses other contraceptive means.

CAUSATION IN CONTEXT   213

$P = 1$ if Betty is pregnant, 0 if not.
$T = 1$ if Betty gets thrombosis.

The causal model with the actual values of the exogenous variables has the following structural equations:

$X = 1$.
$C = 1$
$P = 1$ if $C = 0$ and 0 otherwise
$T = 1$ if $(P = 1$ and $X = 1)$ or $(X = 1$ and $C = 1)$; 0 otherwise.

The graph for this example is depicted in Figure 8.4.

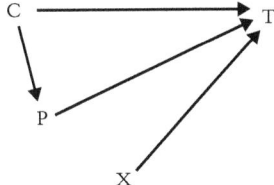

Figure 8.4

Now let us consider how this model can be modified in the light of different assumptions about the system being modelled and about its normal state. Suppose that the system we are modelling is a fertile-sexually-active-woman-with-factor-X-who-does-not-take-contraceptive-pill so that the default values of the exogenous variables of this system can be set at $X = 1$ and $C = 0$. Then it is easy to see that within the default model M_1 with these default settings, Betty's taking the pill makes no difference to her getting thrombosis. For if she were to take the pill, she would get thrombosis because she has factor X; and if she were not to take the pill, she would get thrombosis because she has factor X and would become pregnant. The following causal conditionals hold true:

(14) $(C = 1 \,\square\!\!\rightarrow_{M_1} T = 1)$ & $(C = 0 \,\square\!\!\rightarrow_{M_1} T = 1)$

Now let us make a different assumption about the system being modelled. Let us suppose that it is a fertile-sexually-active-woman-with-factor-X-who-does-not-take-contraceptive-pill-but-uses-other-contraceptive-means so that the default values of the variables are $X = 1$ and $C = 2$.

214 PETER MENZIES

It is easy to check that in the default model M_2 with these settings, Betty's taking the contraceptive pill makes a difference to her getting thrombosis:

(15) $(C = 1 \,\square\!\!\rightarrow_{M2} T = 1) \,\&\, (C = 2 \,\square\!\!\rightarrow_{M2} T = 0)$

This corresponds to the positive causal judgement that Betty's taking the pill caused her thrombosis.

(D) The Golfer

It is natural to model this example using the following variables:

$D = 0$ if the golfer does not drive at all, 1 if he drives with normal angle and force, 2 if he slices the ball.
$T = 1$ if tree is present near green, 0 otherwise.
$R = 1$ if golf ball ricochets off tree towards green, 0 otherwise
$H = 1$ if ball holes out, 0 otherwise

The structural equations for this model with actual values for the exogenous variables are:

$D = 2$
$T = 1$
$R = 1$ only if $D = 2$ and $T = 1$; and 0 otherwise.
$H = 1$ only if $D = 1$ or $(D = 2$ and $R = 1)$; and 0 otherwise.

The graph for this model is depicted in Figure 8.5.

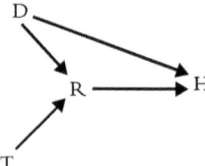

Figure 8.5

In order to capture the various intuitions about this example we have to convert this model into various default models. Let us model the situation as a system consisting of a golfer-plus-tree-near-green and let us suppose the normal state of this system is such that $T = 1$ and $D = 0$. Then we can deduce that golfer's slicing the ball made a difference to the ball's holing out in this default model M_1:

(16) $(D = 2 \,\square\!\!\rightarrow_{M1} H = 1)$ & $(D = 0 \,\square\!\!\rightarrow_{M1} H = 0)$

In order to capture the intuition that in some sense the ball holed out despite the fact that the golfer sliced it, we must model the system in a different way. Suppose we think of the system as golfer-plus-green-without-tree, reflecting the idea that the tree is an accidental and not an essential part of the system. Accordingly, let us set the default value of T to 0, keeping the default value of D at 1 in a default model M_2. Then we can see that the pair of causal conditionals hold:

(17) $(D = 2 \,\square\!\!\rightarrow_{M2} H = 0)$ & $(D = 1 \,\square\!\!\rightarrow_{M2} H = 1)$

Of course, such a despite-causal claim gets its point precisely when $D = 2$ makes a difference to $H = 0$ but in fact, due to other causal factors, it happens to be the case that $H = 1$.

8.6 Causal Conditionals

I have employed causal conditionals in the previous section to illustrate various judgements about the difference-making relation. How are these conditionals to be understood? Do they have a coherent semantics? And how do they relate to standard counterfactuals?

Let us tackle the last question first. Traditionally, philosophers have developed semantics for counterfactuals in terms of similarity relations between possible worlds. One classic treatment is David Lewis's (1973b) possible worlds semantics. (Pearl in his (2000, section 7.4) shows that the axioms of Lewis's theory follow from the axioms of his own structural semantics.) A central feature of Lewis's semantics is that it uses a system of nested spheres of possible worlds centreed on the actual world. A sphere represents a set of possible worlds that are equally similar to the actual world: the smaller the sphere the more similar to the actual world are the possible worlds within it.

Built into this semantics is the Centreing Principle to the effect that there is no world as similar to the actual world as the actual world itself. In terms of this system of spheres the truth condition for a counterfactual is stated as follows: $P \mathbin{\Box\!\!\rightarrow} Q$ is true if and only if Q is true in every P-world in the smallest P-permitting sphere. It follows from the Centreing Principle that $P \mathbin{\Box\!\!\rightarrow} Q$ is true if P and Q are true.

I propose a modified semantics for the causal conditionals that are relevant to the assessment of the difference-making relation in the modified SE framework. This semantics differs from Lewis's in two ways. The first difference is that the similarity relation is specified by reference to a contextually salient default causal model. Such a causal model determines the relevant respects of similarity to be considered in evaluating a given counterfactual. The second difference is that the system of spheres of possible worlds need not be centreed on the actual world, but on a set of what I call default worlds. I characterize default worlds as follows:

Definition 3: A *default causal model* $<U, V, E>$ of an actual system generates a sphere of default worlds that consists of all and only worlds w such that:

(i) w contains a counterpart system of the same kind whose exogenous variables in U are set at their default values;
(ii) w evolves in accordance with the structural equations in E without any further intervention from outside the system.

The intuitive idea is that the default worlds generated by a causal model exemplify a course of evolution that is normal or natural for a system of the given kind. More particularly, they represent the way that a system of the given kind would evolve from its initial default state without intervention or interference from outside the system. A crucial notion here, of course, is that of the default settings of the exogenous variables of a model, about which I shall say more in the next section.

Let us consider how this definition would apply to an example. What would a default world generated by a causal model for example 1, say, look like? As we have seen, the exogenous variables are D and R and their default values are determined in a context-sensitive way. We have seen that in one context in which it is normal for there to be no drought and no government stockpiling of food reserves, they have the default values of 0 and 0 respectively. The default worlds generated by this default model will be ones that evolve from these initial values in conformity with the relevant structural equation in such a way that there is no famine. However, in another context in which droughts and government stockpiling food reserves are the norm, they have the default values of 1 and 1 respectively. The default worlds generated this default model will be ones that evolve from these initial states in such a way that there is no famine.

The sphere of default worlds generated by a model is tied, in some sense, to the actual world. For worlds earn their membership in the sphere by virtue of their resemblance to the way the actual system under consideration would evolve in conformity with the structural equations. Nonetheless, it is important to note that the actual world need not itself belong to the sphere of default worlds. For these worlds represent how a normal system would evolve in conformity with the structural equations in the absence of outside intervention. In many cases, these worlds are ideal ones. The actual world may be very far from ideal in that the actual system may not be normal and its course of evolution may be affected by external interferences. For example, in the Indian famine example, the actual world is one where there is drought and no government stockpiling of food reserves, whereas the default worlds generated by the two models we considered lack one or other of these features. The sphere of default worlds within this framework count as the closest worlds to the actual world. The fact that the actual world need not belong to this sphere means that the Centreing Principle fails in this framework. This has some surprising implications for the logic of causal conditionals, including the failure of Modus Ponens, but these must be explored elsewhere.

So far we have attended to the question of which worlds count as the default worlds generated by a default causal model. But we need to provide truth-conditions for all causal conditionals, and so we need to specify which will be the closest-antecedent worlds for any causal conditional. In some cases, the antecedent of the conditional will overlap with the sphere of default worlds, and so the closest antecedent-worlds are specified as those antecedent-worlds that belong to this overlap. In other cases, however, the antecedent of the causal conditional will not overlap with the sphere of default worlds and the closest antecedent-worlds must be specified in some non-obvious way. Here I propose one way in which a default causal model might create an ordering of spheres of possible worlds, adapting an idea of Pearl's (2000, p. 241):

Definition 4: $\{S_0, \ldots, S_n\}$ is *a system of spheres* ordered by the model $<U, V, E>$ if and only if S_0 is the sphere of default worlds generated by the model and S_i is a sphere of worlds such that a default world in S_0 is transformed into a world in S_i by a maximum number of i interventions in the structural equations of the model.

It is easy to see that this method of ordering the spheres of possible worlds ensures that they are centreed on the sphere of default worlds and are nested within each other.

On the basis of this ordering of spheres, the truth conditions for the causally relevant counterfactuals can be formulated in general terms as follows:

Definition 5: $P \,\square\!\!\rightarrow_M Q$ is true in the actual world relative to a system of spheres ordered by a default causal model M if and only if Q is true in all the P-worlds in the smallest P-permitting sphere of the system of spheres of default worlds generated by the model.

It can be shown that these definitions provide a coherent semantics for the causal conditionals used to illustrate the difference-making relations discussed in the last section.

8.7 Default Worlds

The preceding discussion has focused on one method (but not the only method) for evaluating causal relations. According to this method for determining causes, we try to conceptualize a particular situation in terms of a certain kind of system that has a characteristic normal course of evolution. If we are successful in doing this, and we also encounter some behaviour of the system that deviates from this normal course of evolution, we judge something to be a cause of this deviation if it has made a difference to the normal course of events with respect to the behaviour.

Many philosophers will recognize this form of reasoning as Aristotle's natural state model of scientific reasoning. Elliott Sober (1980, 1984) has written that the model provided Aristotle with a technique for explaining the great diversity found in natural objects. Within the domain of physics, there are heavy and light objects, ones that move violently and ones that do not move at all. How is one to find some order that unites and underlies all this variety? Aristotle's hypothesis was that there is a distinction between the natural state of a kind of object and the states that are not natural. The latter are produced by subjecting the object to an interfering force.

In the sublunar sphere, for a heavy object to be in its natural state is for it to be located where the centre of the Earth is now. But, of course, many heavy objects fail to be there. The cause for this divergence from what is natural is that interfering forces act on the objects to prevent them from achieving their natural state. Aristotle's metaphysics reflects the kind of default reasoning that I believe is central to causal reasoning.

But this form of reasoning is not just a fossilized piece of folk science. It is common to many forms of advanced scientific reasoning. Tim Maudlin (2004) has argued that causal reasoning is especially straightforward and clear when we have laws of a particular form, illustrated by Newton's laws of motion. The first law, the law of inertia, states that a body at rest will remain at rest and a body in motion will continue in motion at a uniform speed in a straight line, unless some force is put on it. The first law specifies inertial motion, that is, how the motion of an object will evolve if nothing acts on it. The second law then specifies how the state of motion of an object will change if a force is exerted on it: it will change in the direction of the force, and proportionally to the force, and inversely proportionally to the mass of the object. The structure of these laws is especially well suited to identifying causes, Maudlin claims. Where we encounter an object deviating from its inertial state of motion (e.g. the Earth orbits the Sun rather than travelling at constant speed in a straight line), we look for a force that explains the deviation (e.g. the gravitational force produced on the Earth by the Sun). In this form of reasoning the second law is parasitic on the first: the first specifies the inertial motion, deviation from which requires explanation in terms of a force. If the inertial motion of the Earth were to orbit the Sun, its actual motion would not require a cause in the form of a force.

Maudlin argues that this form of causal reasoning is more general and not peculiar to physics. He writes:

In judging causes, we try to carve up the situation into systems that can be assigned inertial behavior (behavior that can be expected if nothing interferes) along with at least a partial specification of the sorts of things that can disturb the inertial behavior, analogous to the Newtonian forces that disturb inertial motion. Let us call the things that can disturb inertial behavior 'threats': they are objects or events that have the power—if they interact in the right way—to deflect the system from its inertial trajectory. We then think about how the situation will evolve by expecting inertial behavior unless there is interaction with a threat, in which case we see how the threat will change behavior. (2004, p. 436)

Evidently, Maudlin is making the same point I made at the beginning of this section. Where he talks of the inertial behaviour of a system, I talk of its normal course of evolution or the default worlds that exemplify its normal behaviour; and where he discusses threats that explain changes from the inertial behaviour, I discuss interventions or intrusions in the system that make a difference to its behaviour.

It would be good to have a detailed account of the notion of a default world generated by a model of a system and the correlative notion of the default setting of the exogenous variables of a causal model, especially in view of the reliance of our discussion on these notions. In very general terms, the default values of the exogenous variables represent the initial state of a system of a given kind that is normal or to be expected or is taken for granted because it requires no explanation; and the corresponding default worlds represent the normal course of evolution of the system from this initial state—normal in the sense of conforming to the structural equations in a way that is free from intervention from outside the system.

However, it is very hard to frame more detailed, positive characterizations of these notions since they are so subject to subtle idiosyncratic contextual cues, as we have seen in the examples above. Nonetheless, it is possible to make some very rough and ready generalizations about how the default worlds are determined for a system.

First, known laws or regularities clearly influence the expectations of what is the normal course of evolution for a system of the given kind. (See Toulmin 1961, chs 3 and 4). This applies in both everyday and scientific explanations. If every car passing down the street in front of my house has been reasonably quiet, then the expectation based on that regularity will determine what I think calls for explanation. A car backfiring, for example, will be something that is anomalous with respect to that expectation and will require explanation. To take a scientific example: if every planet in the solar system has been observed to conform to an elliptical orbit predicted by Newton's laws, then that expectation will form the basis of what counts as an anomalous phenomenon. When a planet is observed to deviate from its predicted elliptical orbit, an explanation will be sought in terms of the gravitational influence of a yet-to-be discovered planet.

Secondly, the default worlds are not restricted to the regular course of events unaffected by human intervention. Because nature can be harmful unless we intervene, we have developed customary techniques, procedures, and routines to counteract such harm. (See Hart and Honoré 1985.) These become a second nature. For example, the effect of drought is regularly neutralized by government precautions in conserving water; disease is neutralized by inoculation; rain by the use of umbrellas. When such procedures are established, they can determine the default worlds for the relevant system. When some harm occurs in violation of the expectations set up, a cause is often identified as an omission or failure on the part of some agent to carry out the neutralizing procedures, as the example of the famine illustrates.

Finally, another factor that can determine the default worlds for a given kind of system is what is regarded as the proper functioning of that system. For example, doctors have expectations about the normal or healthy functioning of human physiology based on their beliefs about the proper functions of the body. Their search for the causes of deviations from healthy functioning will also be influenced by these expectations as well. Similarly, biologists' expectations of the default or normal course of development of a biological trait will be influenced by their views of the adaptive function of the trait.

I am sure that it is possible to add to and improve upon these rough and ready generalizations. There may, indeed, be laws about the way in which the human mind forms conceptions of the inertial behaviour of systems and the default worlds they exemplify. However, I doubt whether any of such laws, which might improve on the rough and ready generalizations given above, could provide a general *positive* characterization of inertial behaviour that applies to every kind of system. For I suspect that the notion of the inertial behaviour or a default world for a kind of system has to be understood *negatively*: the inertial behaviour of a Newtonian system is simply the behaviour of the system when *no external force interferes*; and more generally, the default world for a system exemplifies the behaviour it would display if *there were no external causes at work*. It might be thought to be a deeply unsatisfying feature of this model of causal reasoning that no positive characterization can be provided of the central notion of inertial behaviour or default world. Nonetheless, it would be inadvisable to reject this model for this reason. For as we have seen, the kind of default reasoning I believe to be central to causal reasoning is also common to many of our best scientific theories. As already pointed out, this form of causal reasoning is embedded in Newtonian mechanics in the form of its first and second laws.

Sober (1980; 1984) also points out that the same model of reasoning is used in population genetics under the rubric of the Hardy-Weinberg law. This law specifies the equilibrium state for the frequencies of genotypes in a population when the evolutionary forces of mutation, migration, selection and drift are not at work. Even in general relativity, the geometry of space-time specifies a set of geodesics along which an object will move as long as it is not subjected to a force. Consequently, the rejection of the kind of default causal reasoning described in this paper would necessitate the rejection of the reasoning embedded in some of our best scientific theories.

References

Collins, J. (2000). 'Preemptive Prevention', *Journal of Philosophy*, 97: 223–34.
De Rose, K. (1992). 'Contextualism and Knowledge Attributions', *Philosophy and Phenomenological Research*, 52: 913–29.
Dowe. P. (2000). *Physical Causation*. Cambridge: Cambridge University Press.
Eells, E. (1991). *Probabilistic Causation*. Cambridge: Cambridge University Press.
Hart, H. and Honoré, T. (1985). *Causation in the Law*, 2nd edn. Oxford: Oxford University Press.
Hesslow, G. (1976). 'Discussion: Two Notes on the Probabilistic Approach to Causality', *Philosophy of Science*, 43: 290–2.
Hitchcock, C. (1993). 'A Generalized Probabilistic Theory of Causal Relevance', *Synthese*, 97: 335–64.
——— (1996). 'Farewell to Binary Causation', *Canadian Journal of Philosophy*, 26: 267–82.
——— (2001a). 'The Intransitivity of Causation Revealed in Equations and Graphs', *Journal of Philosophy*, 98: 335–64.
——— (2001b). 'A Tale of Two Effects', *The Philosophical Review*, 110: 361–96.
——— (2003). 'Of Humean Bondage', *The British Journal for the Philosophy of Science*, 54(1): 1–25.
Humphreys, P. (1989). *The Chances of Explanation*. Princeton: Princeton University Press.
Lewis, D. (1973a). 'Causation', *Journal of Philosophy*, 70: 556–67.
——— (1973b). *Counterfactuals*. Oxford: Blackwells Publishing.
——— D. (1996). 'Elusive Knowledge', *Australasian Journal of Philosophy*, 74: 549–67.
——— D. (2000). 'Causation as Influence', *Journal of Philosophy*, 97: 182–97.
Mackie, J. (1974). *The Cement of the Universe*, Oxford: Oxford University Press.

Maudlin, T. (2004). 'Causation, Counterfactuals, and the Third factor', in J. Collins, N. Hall. and L. Paul (eds), *Causation and Counterfactuals*. Cambridge, Mass.: MIT Press: 419–43.

McDermott, M. (1995). 'Redundant Causation', *British Journal for the Philosophy of Science*, 40: 523–44.

Menzies, P. (2004a). 'Difference Making in Context' in J. Collins, N. Hall and L. Paul (eds), *Causation and Counterfactuals*. Cambridge, Mass.: MIT Press: 139–80.

——(2004b). 'Causal Models, Token-Causation, and Processes', *Philosophy of Science*, 71: 820–32.

Pearl, J. (2000). *Causality*. Cambridge: Cambridge University Press.

Russell, B. (1913). 'On the Notion of Cause', *Proceedings of the Aristotelian Society*, 13: 1–26.

Salmon, W. (1984). *Scientific Explanation and the Causal Structure of the World*. Princeton: Princeton University Press.

——(1994). 'Causality Without Counterfactuals', *Philosophy of Science*, 61: 297–312.

Sober, E. (1980). 'Evolution, Population Thinking, and Essentialism', *Philosophy of Science*, 47: 350–83.

——(1984). *The Nature of Selection*. Cambridge Mass.: MIT Press.

Toulmin, S. (1961). *Foresight and Understanding: An Enquiry into the Aims of Science*. Bloomington: Indiana University Press.

Unger, P. (1975). *Ignorance: A Case for Skepticism*. Oxford: Oxford University Press.

Van Fraassen, B (1980). *The Scientific Image*. Oxford: Oxford University Press.

Woodward, J. (2003). *Making Things Happen*. Oxford: Oxford University Press.

9

Hume on Causation: The Projectivist Interpretation

HELEN BEEBEE

9.1 Introduction

Hume's views on causation are notoriously hard to pin down. The traditional interpretation takes Hume to be a naive regularity theorist: one event *a* causes another, *b*, just if *a* is prior to and contiguous with *b*, and events similar to *a* are constantly conjoined with events similar to *b*. Causation, on this view, just is regular association: there is no 'tie' or connection of any sort between *a* and *b*. The traditional interpretation of Hume has, of course, spawned an entire philosophical tradition, running from the logical positivists, through Quine, Davidson, and others, to David Lewis's thesis of 'Humean supervenience' ('Humean' because Lewis, in line with the traditional interpretation, regards Hume as a 'denier of necessary connections' (1986, p. ix)), and the industry of providing a reductive analysis of causation.

In tandem with the waning in popularity of that tradition in the latter part of the twentieth century, the sceptical realist interpretation of Hume has been gathering support, and has been championed by, among others, John Wright (1983), Edward Craig (1987, ch. 2), Galen Strawson (1989), and Stephen Buckle (2001). According to the sceptical realist interpretation, Hume held that regular association is all we can *know* about causation (hence 'sceptical'), but he did not question the existence of real, mind-independent causal powers in nature (hence 'realist').

Simon Blackburn gestures towards a third interpretation—projectivism—in various places (1984, pp. 210–12; 1987, pp. 55–7; 1988, pp. 178–80; 1990, pp. 107–11). According to the projectivist interpretation, Hume holds that our causal thought and talk is an expression of our habits of inference. On observing a, we infer that b will follow, and we 'project' that inference onto the world—the inference being the source of our idea of necessary connection. Thus to say that a caused b is neither to say merely that a and b are regularly associated, nor is it to assert the existence of any mind-independent relation of necessary connection.

Neither Blackburn nor (so far as I know) anyone else has articulated the projectivist interpretation in any great detail.[1] Indeed, given the lack of literature—both expository and critical—it is unclear whether one should even so much as claim that there is, at the moment, any such thing as 'the projectivist interpretation'. The purpose of this paper is to begin to remedy this deficiency: to put some flesh on the bones. Everything I say is, I think, consistent with everything Blackburn says about Hume on causation, but I may of course be wrong; and in any case there way be different ways of cashing out a broadly projectivist interpretation, just as there are different ways of cashing out a broadly sceptical realist interpretation (cf. Wright 1983 with Strawson 1989).

A full-blown defence of the projectivist interpretation would, of course, require a great deal more work: I make no attempt to defend the projectivist interpretation over rival interpretations in this paper, nor do I respond to the many possible objections that could be raised against the projectivist interpretation itself. My overall aim is simply to say enough about what a projectivist interpretation might look like to persuade the reader that such an interpretation is at least a prima facie viable alternative to the traditional and sceptical realist interpretations.

I shall proceed as follows. In Section 9.2, I introduce the projectivist interpretation by showing how it provides one way of resolving a problem Barry Stroud (1993) raises for Hume: the problem of how we can coherently think of objects as causally related when our so thinking of them appears to involve having an idea—the idea of necessary connection—which could not possibly represent how things really, mind-independently, are.

---

[1] Plenty of commentators hold that Hume takes 'projection' to have *something* to do with the genesis and/or meaning of our causal thought and talk; for example, Kemp Smith (1941), Stroud (1977) and Wright (1983)). However, none of these authors attributes to Hume a 'projectivist' view in Blackburn's sense.

In Section 9.3, I spell out the projectivist interpretation in more detail by making some (controversial) claims about Hume's conception of causal reasoning and showing how those claims fit with the claim that causal thought is a projection of causal reasoning. In Section 9.4, I argue that there is enough of a parallel between Hume's (arguably projectivist) view of ethical thought and what he says about causation to make the projectivist interpretation of Hume on causation a viable option. In Section 9.5, I address a different problem raised by Stroud (1977), concerning whether Hume has the resources to explain why the having of the idea of necessary connection plays an important role in our understanding of the world. I argue that the projectivist interpretation provides an answer to Stroud's worry by showing how the having of the idea allows us to conceive of our inferential habits as rational responses to the order of nature.

9.2 Stroud's Problem

Stroud's problem starts from the assumption that for Hume, the content of moral, aesthetic, and causal beliefs derives, somehow or other, from our 'gilding or staining' the world with sentiment (in the moral and aesthetic cases) or an impression of reflection (in the causal case). Stroud wants to think of Hume as 'holding that we do really think of objects as causally or necessarily connected, or as evil or vicious, or as beautiful' (1993, p. 21). In other words, he wants to think of Hume as holding that we are capable of believing—and hence of thinking—*that c* caused *e*, or that *X* is beautiful, or that what *Y* did was vicious. But to be capable of having such thoughts, Stroud thinks, the relevant ideas—of necessary connection, of beauty, of viciousness—must be capable of *representing* the world as being a certain way: the idea of necessary connection must be capable of representing *c* and *e* as bearing that relation to each other, the idea of beauty must be capable of representing *X* as being beautiful, and so on. But, given that the content of those ideas is given by something internal—an impression of reflection or a sentiment—it seems that they cannot be capable of representing the *world* as being a certain way at all. As Stroud puts it for the moral case:

In explaining his view of morals Hume is careful to point out that: 'We do not infer a character to be virtuous, because it pleases: But in feeling that it pleases after such a particular manner, we in effect feel that it is virtuous' (*T* 471).² But again, that a given character is virtuous is on Hume's view not something that is or could be so as things 'really stand in nature'. If we could have a feeling that a certain character is virtuous, it would have to be because we are already capable of intelligibly predicating virtuousness of some of the actions or characters we observe or think about. Simply feeling or thinking that an action pleases us in a certain way does not involve projecting or 'spreading' anything on to the action. But feeling or thinking that the action is virtuous does. The 'gilding or staining' operation which is supposed to lead to such thoughts could not therefore start from just such a feeling or impression. It must start from a feeling or impression which is 'of' something, or has an object, in the 'intentional' sense; but it cannot be 'of' any object or quality or relation which could be part of the way things 'really stand in nature'. If it were, no 'gilding or staining' would be necessary. (1993, p. 27)

Stroud's problem, then, is a problem about meaning. To hold that 'we really do think of objects' as causally connected, or whatever, is (prima facie at least) to hold that those thoughts are capable of representing how those things really are. But Hume seems to think that once we trace the impression-source of the idea of necessary connection, we will see that our causal talk and thought cannot represent how things really are, because the relevant impression itself is not an impression *of* any feature that objects as they really are could possibly possess: it is simply an impression that arises when we infer that one thing will happen on having observed another thing to have happened.

Before seeing how the projectivist interpretation resolves Stroud's problem, we first need to get clear on what a projectivist conception of causation amounts to. Following Blackburn, let's say that 'we *project* an attitude or habit or other commitment which is not descriptive onto the world, when we speak and think as though there were a property of things which our sayings describe, which we can reason about, know about, be wrong about, and so on' (1984, pp. 170–1). (Blackburn goes on to claim that projecting so defined is 'what Hume referred to when he talks of "gilding and staining all natural objects with the colours borrowed from internal sentiment"' (1984, p. 171).) To be a projectivist about causation is thus to claim that we speak and think as though causation were a mind-independent relation, even though in fact our so speaking and saying really involves projecting

---

² Hume's *Treatise* (1739–40) and *Enquiries* (1748/51) are referred to by '*T*' and '*E*' respectively.

some sort of attitude or habit or commitment onto the world. Moreover, that projection is 'not descriptive', which is to say that it does not involve representing the world either as containing mind-independent causal relations, or as being such that it *produces* that attitude or habit or commitment in us.

What attitude or habit or commitment is it that we project onto the world? Well, for Hume, in the most basic case, our thinking of one event, *a*, as cause and another, *b*, as effect arises in the first instance when, on observing *a*, we infer that *b* will follow; and we will do that just if we have had experience of the past constant conjunction of events similar to *a* (the *A*s) with events similar to *b* (the *B*s): 'when one particular species of event has always, in all instances, been conjoined with another, we make no longer any scruple of foretelling one upon the appearance of the other, and of employing that reasoning, which can alone assure us of any matter of fact or existence. We then call the one object, *Cause*, the other, *Effect*' (E 75). The impression of necessary connection, from which the idea of necessary connection derives, is the 'feeling' we get from the 'customary transition of the imagination from one object to its usual attendant' (*ibid.*)—that is, from the inference we draw from the impression of *a* to the belief that *b* will occur. So clearly, if Hume is a projectivist about causation, the relevant 'attitude or habit or commitment' will be something like the habit of inductively inferring that a *B* will occur on observing that an *A* has occurred.

What is it to 'speak and think as though' causation were a mind-independent relation? It is important to realise that, on a projectivist view, this does not involve our mistakenly *assuming that* there are mind-independent causal relations. The non-descriptive semantics of our causal talk would rule out the possibility of our even being capable of making this assumption: to think *that* there are mind-independent causal relations, in the representational sense, requires that the meaning of 'causal relations' is descriptive, which of course is what is being denied. What *would* be a mistake, on a projectivist view, would be to hold that the meaning of 'cause' is descriptive, and hence to hold that our causal talk so much as purports to be talk *about* mind-independent relations.

On a projectivist view, to 'speak and think as though' causation were a mind-independent relation is to speak and think in such a way that the habit or commitment in question—roughly, for Hume, the inductive habit—takes on what Blackburn calls 'propositional behaviour' (1987, p. 55). In the case of ethics (to put it rather crudely), 'Murder: Boo!' and 'Murder is wrong' express the same ethical attitude, but the second, and not the first, has propositional form. Similarly, the attitude expressed by 'Manslaughter: Boo! Murder: *BOO!*' takes on propositional form when one says instead, 'Manslaughter is bad, but murder is worse.' Such propositions, Blackburn says, 'stand at a needed point in our cognitive lives—they are objects to be discussed, rejected, or improved upon when the habits, dispositions, or attitudes need discussion, rejection or improvement. Their truth corresponds to correctness in these mental states, by whichever standards they have to meet' (*ibid.*).

Here is another way to put the point, this time borrowing from Huw Price (1998). Imagine teaching a novice speaker to use colour language. Two habits need to be instilled. The first is simply the habit of using the word 'red' (say) in prima facie appropriate circumstances, namely circumstances in which the novice speaker has red-experiences. The second is—as Price puts it—'the habit of taking redness to be something that falls under the objective mode of speech'. Price continues:

Against the general background of assertoric practice, the way to combine these lessons will be to teach novices to describe their redness experiences in terms of the notions of perception and belief—ordinary, world-directed perception and belief, of course, not any introspective variety. In treating the distinctive redness response as defeasible perceptual grounds for a corresponding belief, we open the way to such comments as 'You believe that it is red, but is it *really* red?' This in turn may call into play the standard methods of rational reassessment. In virtue of their acquaintance with the objective mode *in general*, speakers will be led into the practice of subjecting their colour judgements to reflective scrutiny by themselves and others. The objective mode brings with it the methods and motives for rational enquiry.   (1998, pp. 125–6)

What about the causal case? Well, as with the ethical and colour cases, the basic idea would be that our coming to speak and think in causal terms–our expressing our inductive habits in propositional form, or adopting the 'objective mode of speech'—brings with it the resources for thinking of those habits as susceptible to critical scrutiny: as habits that can be refined, rejected, warranted or unwarranted, and so on. In Hume's case, the relevant standards against which the expressed commitment is to be judged will be his 'rules by which to judge of causes and effects' (*T* 173–5).

Care is needed here, for it might seem as though to say that we cannot so much as think *that* there are mind-independent causal relations, as I did above, is tantamount to giving up on the thesis that we 'do really think of objects as causally or necessarily connected', when part of the point of a projectivist interpretation of Hume is precisely that it allows him to uphold that thesis. The two claims are not really incompatible, however. To say that we cannot so much as think *that* there are mind-independent causal relations means, in this context, to say that we cannot genuinely think of or say two events that they stand in a mind-independent relation of causation to one another. As Hume says, we are 'led astray... when we transfer the determination of the thought to external objects, and suppose any real intelligible connexion betwixt them; that being a quality, which can only belong to the mind that considers them' (*T* 168). By contrast, to say that we do really think of objects as causally or necessarily connected is to say that we are *not* led astray—we are not making any kind of mistake—when we 'speak and think as though' causation were a mind-independent relation, in the sense just described. For to speak and think so is merely for the expressed commitment to take on 'propositional behaviour'. On the projectivist view, the propositional behaviour of our causal talk and thought does not amount to our genuinely representing the world as being a world of mind-independent causal relations, but it *does* amount to our really thinking of events as causally related.

Given all this, it should be pretty clear how a projectivist interpretation of Hume on causation would resolve Stroud's problem. Stroud presupposes that our 'thinking of objects as causally or necessarily connected' is a matter of our *representing* the world as being a certain way; and the problem is that of saying how it is possible for an idea whose origin lies in an impression of reflection to represent the world in any way at all. The projectivist interpretation resolves the problem by denying that Hume takes our thinking of objects as causally or necessarily connected to be a matter of representation in the first place. Once we have rejected that assumption on Hume's behalf, the impression from which the idea of necessary connection derives no longer needs to be an impression 'of' anything, in the representational sense (in Stroud's words, the 'intentional' sense), in order for us to be able coherently to think of objects as causally or necessarily connected.

9.3 Causal Reasoning and the Image of God Doctrine

One way to characterize Hume's overall project, so far as causation is concerned, is as the wholesale rejection of the idea that our beliefs about the causal structure of the world are, or could in principle be, a species of a priori knowledge. Edward Craig argues that this conception of our epistemic access to the world is motivated by what he calls the 'Image of God' doctrine: the thesis that the human mind is the same kind of thing as (though of course less perfect than) the mind of God (1987, ch. 1). The epistemological upshot of the Image of God doctrine—what Craig calls the 'Insight Ideal'—is the thesis that the human mind can in principle have access to true beliefs in a way that is analogous to the way in which God can.

What must the world be like in order for the Insight Ideal to hold? Well, the human understanding is at its most perfect when engaged in demonstrative reasoning. So it would be natural, given the Insight Ideal, to think of the relationship between events in the world—that is, causation—as analogous to, or perhaps the same as, the relationship between stages in a logical or mathematical proof. The metaphysical upshot of the Insight Ideal is thus the view that causal relations are, as it were, the worldly correlates of inferential relations: causes necessitate their effects, or guarantee that those effects occur, in a way that is somehow analogous to, or perhaps even identical with, the way that premises in an argument necessitate or guarantee the truth of their conclusion.

Hume shows that the epistemological consequences of the Image of God doctrine are completely untenable: a priori reasoning cannot supplement sensory experience to deliver any substantive knowledge about the world at all. We cannot penetrate into the essence of objects in such a way as to reveal anything analogous to an entailment relation between causes and effects: nothing at all in our experience reveals the world to have the quasi-logical structure suggested by the Image of God doctrine. Rather, our sensory impressions can deliver no more than a succession of events which—at least insofar as they are represented by those impressions—are 'entirely loose and separate' (*E* 74). Beliefs about what is not currently available to sensation are delivered not by a priori reasoning or by penetration into the essences of objects, but by a brute associative mechanism: by the inductive habits we share not with God but with other animals.

For present purposes, what is important about the metaphysical picture prompted by the Image of God doctrine is that it takes there to be an intimate connection between the nature of causation on the one hand, and the nature of *inference* (at least in ideal circumstances) concerning matters of fact on the other. Causation is conceived as a relation such that grasp of its nature licenses inferences from one matter of fact to another. So it is the (alleged) epistemology of the inference that drives the metaphysics: we start with a thesis about the nature of the inference, and end up with a thesis about the nature of causation. Now, suppose that we take Hume to have roughly the same conception of the explanatory order as do upholders of the Image of God doctrine, so that the story about what it *is* for one thing to cause another derives from the story about how it is that we infer one matter of fact from another. What follows most naturally is a projectivist interpretation.

Here's why. The difference between Hume and the upholders of the Image of God doctrine is that Hume rejects the Insight Ideal and replaces it with the claim that inferences from one matter of fact to another are due to the operation of an associative mechanism: they are a result of habit rather than the operation of a special faculty of reason. (I shall call the associative mechanism the 'associative mechanism of causation': it is one of the mechanisms by which the mind is naturally 'conveyed from one idea to another' (T 11).) Given the supposition just made on Hume's behalf about the explanatory order, it is the nature of the inference that leads him towards projectivism. The associative mechanism of causation itself is, by definition, a *mental* mechanism: a mechanism by which one idea 'attracts' another (see T 12) as a result of custom or habit. So, unlike upholders of the Image of God doctrine, Hume cannot hold that there is any feature of the world that *corresponds* to the operation of that inferential mechanism: the world cannot literally operate in a way that mirrors the associative mechanism of causation. (Billiard balls do not act on one another out of custom or habit.) So if we are to think of the world as somehow reflecting and justifying our inferential habits, that can only be because we project those habits onto the world. According to the Image of God doctrine, our inferential habits reflect pre-existing, mind-independent, inference-justifying relations. According to Hume *qua* projectivist, we impose our inferential habits onto the world. When we come to think of two events *a* and *b* as causally related, we take ourselves to be justified in inferring future *B*s from *A*s, and we do so in virtue of our taking them to be so related. But this is only because it is that very inferential habit that we projected onto them in the first place, in coming to think of them as causally related.

Should we take Hume to think that the story about what it is to think of two events as causally related derives from how it is that we infer one matter of fact from another? From a common-sense point of view, such a view might seem to be a reversal of the natural explanatory order: ordinarily, I think, we take it that causal reasoning—reasoning from causes to effects—takes causal belief as part of the input, so that a paradigm case of causal reasoning would look like this:

(CR1) (P1) *A*s cause *B*s
 (P2) An *A* has occurred
 (C) A *B* will occur

On this view, we *start out* with a conception of what it is for *A*s to cause *B*s, and that conception will enable us to see why (CR1) counts as a legitimate inference. For example, we might hold that causation requires universal constant conjunction, in which case (CR1) turns out to be valid and hence, obviously, a legitimate inference.

This is not Hume's conception of a paradigm case of causal reasoning, however. For in order to deploy (CR1), we need antecedently to believe (P1). But belief in (P1) could only come about by deploying an inference that takes us from beliefs about the past to beliefs about what will happen in the future–that is, an inference that generates *expectations*: beliefs about what is not present to the senses or memory. And it is that very inference that (CR1) is supposed to capture: (CR1) is supposed to explain how it is that we move from what is present to the senses or memory—the impression or belief that an *A* has occurred—to an expectation (the expectation that a *B* will occur).

So for Hume, causal reasoning in general cannot start from an antecedently-held belief in some causal claim. Instead, Hume holds that it is causal reasoning *itself*—the operation of the associative mechanism of causation—that delivers our capacity to think of event *a* as a cause of event *b* (and hence of *A*s as causes of *B*s generally). For Hume, I think, the paradigm case of causal reasoning is just this:

(CR2) (P1)   *a* has occurred
       (C)   *b* will occur

Such reasoning is *causal* reasoning because we will infer (C) from (P1) (that is, infer *b* from *a*) just when we think of *a* and *b* as cause and effect. And in the first instance—on the first occasion on which the associative mechanism generates the inference from *a* to *b*—the mechanism is *also* what makes us think of *a* and *b* as cause and effect. As Hume says, 'when one particular species of event has always, in all instances, been conjoined with another, we make no longer any scruple of foretelling one upon the appearance of the other, and of employing that reasoning, which can alone assure us of any matter of fact or existence. We then call the one object, *Cause*; the other, *Effect*' (*E* 75). On the projectivist interpretation, our so thinking of them is not a matter of our performing any further cognitive feat—our ascertaining that the truth conditions for '*A*s cause *B*s' are met, say; rather, it is simply a matter of our projecting the inference itself onto the events.

There is a parallel here between causation and entailment that might make this point clearer. I just claimed that, for Hume, paradigm causal inference does not proceed by starting out with a causal claim as a premise. The parallel claim is true for entailment: if I infer, say, 'P' from 'P & Q', the inference does not proceed like this:

(E1) (P1)   'P & Q' entails 'P'
      (P2)   P & Q
      (C)   P

Rather, it proceeds like this:

(E2) (P1)   P & Q
      (C)   P

The fact that 'P & Q' entails 'P' is what makes the inference from the former to the latter a legitimate inference; but it does not make the inference legitimate by functioning as an additional premise. My grasping the fact that 'P & Q' entails 'P' just *is*, in a sense, my grasping the fact that (E2) constitutes a valid inference.

This, I claim, is a pretty close parallel to Hume's conception of *causal* reasoning. My thinking of *a* and *b* as cause and effect is not an additional premise, *belief* in which legitimizes the inference. Instead, my thinking of *a* and *b* as cause and effect just *is*, in a sense, my thinking of the inference from *a* to *b* as a legitimate inference.

There are two important disanalogies between a priori reasoning and entailment on the one hand, and causal reasoning and 'causes' on the other. First, entailment holds between mental items—ideas—whereas causation is a relation between events in the world.³ This is precisely where projection comes into the story: in thinking of *a* and *b* as causally related, we project the inferential relation between the *idea* of *a* and the *idea* of *b* onto *a* and *b* themselves.

Second, a priori reasoning is guaranteed to be truth-preserving, whereas causal reasoning is not. Of course, the question of where this leaves the epistemological status of causal reasoning, according to Hume, is a thorny one. It has recently been argued by several authors that Hume is not, as has been traditionally supposed, an inductive sceptic, in the sense of believing that no belief at all about the unobserved is justified (see for example Owen 1999, ch. 6). There are good reasons to think that this is right. Hume draws clear distinctions between good and bad causal reasoning, for example when he lays down 'rules by which to judge of causes and effects' (*T* 173–6). Moreover, he describes the project of the *Treatise* and the *Enquiry* as a 'science of man', which is 'the only solid foundation for the other sciences' (*T* xvi). It would be very peculiar indeed for Hume to say this, knowing full well that within a hundred pages or so he would be concluding that no belief based on empirical investigation is any more reasonable than any other (see Owen 1999, p. 146 and Baier 1991, p. 55).

This non-sceptical interpretation of Hume's attitude towards inductive inference fits well with the projectivist interpretation. Recall that, according to projectivism about causation, the expression of our habits or commitments takes on 'propositional behaviour'—we adopt the 'objective mode of speech'—because of the need to conceive of those habits or commitments as susceptible to critical scrutiny. If Hume is no inductive sceptic, then he holds that our inferential habits *are* susceptible to critical scrutiny. According to the projectivist interpretation, our accepting or rejecting causal claims just *is* the projection of our attitudes towards

---

³ Of course, it is controversial whether entailment really does hold between mental items; one might think that entailment is a relation between propositions, and conceive of propositions as mind-independent, worldly entities. This controversy is beside the present point, however, since Hume himself does not suggest that there are such mind-independent entities as propositions.

particular inferences: in accepting that *a* caused or will cause *b*, we endorse the inference from *a* to *b*, and in rejecting the claim that *a* caused *b*, we reject the inference.

9.4 Causal and Ethical Projectivism

A large part of the motivation for the projectivist interpretation comes from seeing a parallel between Hume's views on causation and his views on ethics and aesthetics, where a projectivist interpretation more clearly has some textual support, most notably in Hume's claim that '*taste*... gives the sentiment of beauty and deformity, vice and virtue. [It] has a productive faculty, and gilding or staining all natural objects with the colours, borrowed from internal sentiment, raises in a manner a new creation' (*E* 294). In this section, I shall argue that there is enough of a parallel between the causal and ethical cases to make a projectivist interpretation of Hume on causation at least a prima facie viable option.

Hume's views on ethics are, of course, no less a matter for interpretative dispute than are his views on causation. However the difference between the viable interpretative positions in the case of ethics is more subtle than in the case of causation: it is generally agreed that Hume is what Mackie calls a 'sentimentalist' about ethics (Mackie 1980, ch. 5), and this rules out the possibility that Hume intends either a reductionist analysis of vice and virtue or a realist view according to which vice and virtue are mind-independent features of characters or actions. Hence in the case of ethics a projectivist interpretation, being a variety of sentimentalism, is more obviously a serious interpretative option, along with (among other options) a secondary-quality view (what Mackie calls 'dispositional descriptivism'), emotivism, and what Mackie calls 'the objectification theory', according to which moral features are 'fictitious, created in thought by the projection of moral sentiments onto the actions (etc.) which are the objects of [moral] sentiments' (Mackie 1980, p. 74).

Stroud (1977, ch. 8) argues, in effect, that Hume holds the objectification theory, and that this interpretation 'coheres better than any alternative with [Hume's] general philosophical aims' (1977, p. 185):

More of Hume's aims would be served by a theory of moral judgments that follows the same general lines as I suggested for the case of necessity. I contemplate or observe an action or character and then feel a certain sentiment of approbation towards it. In saying or believing that X is virtuous I am indeed ascribing to X itself a certain objective characteristic, even though, according to Hume, there really is no such characteristic to be found 'in' X. In that way virtue and vice are like secondary qualities. In saying that X is virtuous I am not just making a remark about my own feeling, but I make the remark only because I have the feeling I do. In 'pronouncing' it to be virtuous I could also be said to be expressing or avowing my approval of X. Hume thinks that approval is a quite definite feeling, so for him it would be expressing my feeling towards X. (1977, p. 184)

As we saw in Section 9.2, Stroud later expresses the worry that the view thus ascribed to Hume cannot, in the end, be made to work. The problem, remember, is that we cannot intelligibly 'ascribe to X itself a certain objective characteristic'—that is, *represent* X as possessing that characteristic—if our doing so involves 'gilding or staining' X with a feeling or impression. The projectivist interpretation was supposed to solve that problem in the case of causation by denying, on Hume's behalf, that 'ascription' amounts to representation; and of course the same move can be made in the ethical case. In fact, Stroud comes slightly closer to projectivism in the ethical case than in the causal case, because he holds that in the ethical case, 'in "pronouncing" it to be virtuous I could *also* be said to be expressing or avowing my approval of X' (my italics). But he rejects an 'emotivist' interpretation on the grounds that Hume 'thinks of a moral conclusion or verdict as a "pronouncement" or judgment—something put forward as true' (1977, p. 182). Again, a projectivist interpretation solves this problem, since according to projectivism the projection of sentiment does indeed involve our putting forward moral conclusions, pronouncements or judgements: that is what 'adopting the objective mode of speech' (to use Price's phrase) amounts to.

Of course, a full defence of a projectivist interpretation of Hume on ethics requires more than merely showing that it solves Stroud's problem. But suppose that Stroud's argument for the claim that the objectification theory coheres better with Hume's aims than do any of the other alternatives that he canvasses is correct. (Stroud does not consider the projectivist interpretation.) If so, we do have reason to think that the projectivist interpretation is the best interpretation. This is partly because it solves Stroud's problem, but also because it allows Hume not only to make sense of our moral pronouncements, but to endorse them. According to the objectification theory, our moral pronouncements are all, strictly speaking, false: we ascribe to actions or characters features which they do not possess. But Hume does not suggest that he thinks this. He does not suggest that there is anything defective about our moral pronouncements, or that, from the perspective of concern for truth and falsity, no moral pronouncement fares better than any other. Instead, he seems straightforwardly to endorse some moral pronouncements and reject others. Hence, at least prima facie, the projectivist interpretation makes very good sense of what Hume says about our moral thought and talk.

Unfortunately, there appear to be some significant disanalogies between what Hume says about virtue (and beauty) on the one hand, and what he says about causation on the other. So, even assuming the viability of a projectivist interpretation in the moral case (and the aesthetic case), there are prima facie grounds for suspicion that Hume does not endorse a projectivist view of causation. I shall argue, however, that the differences are not as great as might be thought.

Perhaps the most significant problem is that Hume explicitly *contrasts* the 'boundaries of *reason* and of *taste*' (*E* 294). Reason 'conveys the knowledge of truth and falsehood' while taste 'gives the sentiment of beauty and deformity, vice and virtue'; reason 'discovers objects as they really stand in nature', while taste 'has a productive faculty' which 'raises in a manner a new creation'; the standard of reason is 'founded on the nature of things', while the standard of taste arises 'from the internal frame and constitution of animals' (*ibid.*). ('Reason' here is to be understood to include reasoning from experience.) If Hume had more or less the same—projectivist—view about causation as he has about beauty, deformity, vice and virtue, then he would surely have to hold that causal beliefs or judgements stand on the side of taste rather than reason. But surely Hume holds that causal beliefs stand on the side of reason rather than taste: surely he holds that causal beliefs are beliefs about 'matters of fact', while moral judgements are not. Indeed, if 'reason' is supposed to include reasoning from experience, then it includes *causal* reasoning, since that is just what Hume takes reasoning from experience to be. So it seems that causal beliefs stand on the side of reason by definition.

A second problem, related to the first, is that Hume asks, right at the beginning of his discussion of ethics in the *Treatise*, '*[w]hether 'tis by means of our ideas or impressions we distinguish betwixt vice and virtue, and pronounce an action blameable or praise-worthy'* (*T* 456); and his answer is 'impressions'. This sets up a second apparent difference between his treatment of vice and virtue on the one hand and necessary connection on the other. As Stroud says:

> In the case of necessity we are said to have an *idea* of necessity that we employ in formulating our *belief* that two events are necessarily connected, but Hume nowhere mentions a corresponding *idea* of virtue or goodness and he never talks explicitly about moral *beliefs*. (1977, p. 185)

Instead, Stroud points out, Hume tends to talk about moral 'pronouncements' and 'judgements'.

So it seems that Hume is drawing a clear distinction between, on the one hand, belief, reasoning, and matters of fact; and, on the other, 'pronouncements' and matters of taste. And it seems that he intends causation to fall into the first category, while vice and virtue fall into the second category. I shall argue, however, that it is not at all obvious that Hume *does* take causation to fall into the first category.

Consider, first, Stroud's claim that 'we are said to have an *idea* of necessity in formulating our *belief* that two events are necessarily connected'. Well, Hume undeniably does say that we have an idea of necessity. But, so far as I can tell, he nowhere talks about the 'belief' that two events are necessarily connected. Instead, he says that 'we call the one [event] *cause* and the other *effect*' (*T* 87; see also *E* 75), and that we 'pronounce...two objects to be cause and effect' (*T* 87; see also *E* 75).

In fact, given Hume's restrictive sense of 'belief', this is just what we should expect: Hume restricts 'belief' to what is inferred on the basis of a present impression (of sensation or memory) together with past experience of constant conjunction (see *T* 94-8). Whatever it is we do when we 'pronounce' *a* to be a cause of *b*, the only candidates for being the objects of *belief* here are *a* and *b*. One might therefore object that Hume's unwillingness to talk about our believing *that a* caused *b* is merely a by-product of his somewhat idiosyncratic use of the term 'belief'. But this would be unwarranted, for Hume's notion of 'matter of fact' is similarly restricted. 'Reasoning concerning matters of fact', for Hume, is reasoning *from* one matter of fact *to* another; that is, from cause to effect. He does not need to think, and nowhere says, that *a*'s *being* a cause of *b* is an *additional* matter of fact.

What I am suggesting here is that Hume's explicit contrast between reason and taste can be read as a contrast between the *objects* of reason (including causal reasoning) on the one hand—what reason leads us to *believe*—and the 'objects' of taste on the other. Reason leads us to form beliefs about matters of fact: beliefs about 'objects as they really are in nature'. In the case of ethics, by contrast, there are no such matters of fact to represent. There are no 'objects' of moral thought, since such thought does not attempt to represent matters of fact; it does not attempt to capture objects as they really are in nature. Moral thought, unlike reasoning, does not deliver belief in matters of fact; rather, it involves moral 'pronouncements', in which sentiment plays an ineliminable role. *This* contrast between reason and taste is one that makes no implicit claim about causal thought—as opposed to the objects of causal reasoning—at all. Hume need not think of causation as an *object* of reason, in the sense that our causal pronouncements or judgements are themselves beliefs that purport to represent matters of fact. So the contrast is entirely compatible with the claim that causal pronouncements, like moral pronouncements, are not beliefs in matters of fact, do not discover objects as they really stand in nature, and 'raise in a manner a new creation'.

This still leaves us with a version of the difference noted by Stroud, however. While I have denied that Hume holds that we have causal *beliefs*, strictly speaking, Hume nonetheless does appear to think that the *idea* of necessary connection plays a role in our causal thought, whereas in the case of moral thought, it is the *impressions* of vice and virtue that are supposed to play the role: in his discussions of ethics, he does not talk about the *ideas* of vice and virtue.

However, we can resolve the apparent discrepancy between causal and moral thought by pushing Hume towards the view that there *are* ideas of vice and virtue, which play a role in moral thought, even though he does not say that there are. There are three reasons for thinking that Hume can be pushed towards that view. First, at the beginning of the *Treatise*, Hume talks quite freely about 'the ideas of passion and desire' (*T* 7) and 'the idea of pleasure or pain' (*T* 8). And he goes on to say that these ideas in turn produce 'the new impressions of desire and aversion, hope and fear' which 'again are copied by the memory and imagination, and become ideas' (*ibid.*). If Hume thinks we have ideas of desire, aversion and the like, he has no principled reason to deny that we have ideas of vice and virtue.

Second, as Stroud points out, the fact that Hume denies that the idea of virtue or goodness plays a role in our moral pronouncements presents him with a problem: 'what could a moral "pronouncement" be? It would seem to consist only of an impression or feeling, but how do we employ that very feeling in formulating a "pronouncement"?' (1977, p. 185). Stroud goes on to say that what Hume *ought* to do here is to say what he says in the case of necessary connection: 'I make the distinction [between vice and virtue] on the basis of my impression or feeling, but I use an *idea* of viciousness or virtuousness in making my pronouncement' (1977, p. 186).

Third, Hume *needs* to hold that we can *think* about vice and virtue without actually having the relevant impressions—in which case we must do so by deploying the *ideas* of vice and virtue rather than the impressions—even if he holds that we cannot *pronounce* a person or action to be virtuous without having the corresponding impression. For otherwise we would not be able even to entertain the possibility that our moral judgements are mistaken, or wonder what the appropriate moral attitude in a particular case is. For example, I might judge that a certain politician is deplorably insincere and manipulative, and do so because of the moral sentiment I feel when I consider his actions; but I am still capable of considering the possibility that I have misjudged him. Or I might, on meeting someone for the first time, form no moral view of her at all; but this does not stop me wondering whether or not she is considerate, selfish, generous, dishonest, or whatever. Again, feeling no moral sentiment whatever towards her, I am not in a position to *judge* her to be any of these things; but I can perfectly well imagine that she might be.

The claim I have been trying to establish is that *if* projectivism is a viable interpretative position in the ethical case, then it is also a viable interpretative position in the causal case. Admittedly, I have not given much by way of argument for the claim that, in the case of ethics, a projectivist interpretation is a viable option. But I think it is at least pretty clear that a projectivist interpretation does the best justice to Hume's claims that taste 'raises in a manner a new creation' (*E* 294). Insofar as there are parallels between Hume's treatment of the ethical case and his treatment of causation (as Stroud (1993) argues), we thus have at least some reason to think that the projectivist interpretation is a serious contender when it comes to Hume's conception of causation.

9.5 What Does the Idea of Necessary Connection Add?

Finally, I want to address a different worry, which Stroud raises (1977, pp. 224–34), concerning whether, given Hume's views, our having the idea of necessary connection really adds anything to our understanding of the world. He brings the worry out by considering how we differ from hypothetical beings whose minds work just like ours do, *except* that they lack the impression—and hence the idea—of necessary connection. Let's call such beings 'connectionless beings'. Stroud notes that, since their minds operate according to just the same associative principles as ours do, connectionless beings come to have just the same expectations, on the basis of past regularities and current experience, as we do: they too would infer, and be just as certain as we are, that the black ball will move on seeing the white ball make contact with it.

Connectionless beings, however, 'would presumably differ from us in never saying or believing that certain things *must* happen, or that two sorts of things come together *of necessity*' (1977, p. 227). Stroud's worry is that Hume is not in a position to think that this difference amounts to anything very significant:

> it would seem that the notion of necessity does not serve to describe or refer to some objective feature of the world that we, but not they, have discovered. All their beliefs about the actual course of their experience would be the same as ours. And although our minds do differ from theirs in 'possessing' the idea of necessary connection, surely we are not actually describing or referring to that difference, or to anything else in our minds, when we use the word 'must' or attach the idea of necessity to something we believe. What then is the difference? According to the theory of ideas, we, but not they, are simply the beneficiaries of an additional mental item that forces itself into our minds on certain occasions, and we then go through the otherwise empty ritual of adding that unanalysable idea of necessary connection to some of our beliefs. (*ibid.*)

The worry, then, is that, according to Hume's theory of ideas, the having and deploying of the idea of necessary connection can be no more than a mere 'empty ritual'. The mere possession of a mental object—the 'having' of the idea of necessary connection—cannot explain the important role that thinking of the world in causal terms has in our judgement and reasoning. What 'needs to be understood before Hume's programme can succeed', Stroud says, is 'how it is possible for us to think about more than the actual course of events in the world, or what is involved in our accepting statements whose modality is stronger than "existence" or what is actually the case' (1977, p. 230).

Stroud sees this problem with the idea of necessary connection as part of a much wider problem whose root lies in the theory of ideas—in the view that the ability to think about the world, to deploy concepts, is merely a matter of the presence in the mind of a 'mental item'. I shall not discuss whether or not Stroud is right about this in general; rather, I want to argue that, in the case of the idea of necessary connection, there is no real problem given the projectivist interpretation.

Note first that Stroud takes it for granted that on Hume's view, connectionless beings' 'beliefs about the actual course of their experience would be the same as ours'. There is a sense in which this is true. Once we have been persuaded by Hume's arguments that there is no *sensory* impression of necessary connection, our *beliefs* about the actual course of *sensory* experience will be the same as those of the connectionless beings. But there is still, on Hume's view, a *phenomenological* difference between us and the connectionless beings, since our sensory experience is accompanied by the impression of necessary connection, and theirs is not. I shall argue later that this phenomenological difference is important, even though it generates no difference in beliefs about the actual course of experience.

Stroud also seems to think that it is relevant that, on Hume's view, 'the notion of necessity does not serve to describe or refer to some objective feature of the world that we, but not they [that is, connectionless beings], have discovered'. So, now, suppose that the notion of necessity *does* serve to describe or refer to genuine, mind-independent necessary connections which we 'discover'. Suppose, in other words, that the view that Hume is attacking—the view that we detect genuine, mind-independent, a priori inference-licensing necessary connections between events—is correct. For current purposes, I'll call that view 'causal realism'. Stroud appears to think that if causal realism is true, the problem does not arise.

Why might this be? Well, Stroud seems to demand two related things of an adequate account of the 'having' of the idea of necessary connection. First, it must explain how we are able to think of the world as being such that something *must* happen, or that one event happens *because* another event happens, and so on. And second, it must explain how we are able to think about more than what *actually* happens: how we are able to 'go beyond beliefs about the course of all actual events, past, present and future' (1977, p. 229), which is what we do when we engage in counterfactual reasoning: when we come to believe that if an *A* had occurred, a *B* would have occurred. We can see how causal realism succeeds on both counts. It succeeds on the first count because, according to causal realism, our causal thought (and experience) unproblematically *represents* the world as being such that, given one event, another *must* happen, or such that one event happens *because* another happens. And it succeeds on the second count because if we are capable of believing that *A*s and *B*s as necessarily connected, then presumably we are also perfectly capable of believing that, had an *A* occurred, it *would* have been necessarily connected to a *B*—and hence believing that if an *A* had occurred, a *B* would have occurred.

In fact, we can add a third requirement on an adequate account of the having of the idea of necessary connection—one that lies at the heart of Hume's interest in causation. (Whether this is a requirement that *we* ought to endorse is a controversial question, but I don't think there is much doubt that Hume would have endorsed it.) The account must explain how an impression of, or a belief in, one matter of fact can be a *good reason* to believe in some other matter of fact. That is, the idea of necessary connection must be such that it allows us to conceive of our inferences from causes to effects as *rational* inferences. Again, causal realism satisfies this requirement: if our sensory experience reveals one event—the cause—to be such that another event—the effect—is *guaranteed* to follow, then of course our having of the impression (and idea) of necessary connection explains why an impression of the cause constitutes a good reason to believe that the effect will follow. Indeed, if we think of the issue in terms of the Image of God doctrine, it is the whole *point* of causal realism that it satisfies this requirement: the whole point of holding that the world is a world of detectable, a priori inference-licensing necessary connections is precisely that their detection licenses a priori inference.

So causal realism satisfies the three requirements for an adequate theory of what is involved in having the idea of necessary connection. What about Hume's own view? Is Stroud right to say that *that* view fails to satisfy the requirements? Well, there are very large differences between Hume's view, *qua* projectivist, and the causal realist view he is attacking. *Qua* projectivist, Hume rejects the epistemological thesis that inference from causes to effects is a priori inference, and he rejects the corresponding metaphysical thesis that necessary connection is the relation, or a feature of the cause, that makes such a priori inference possible. He also rejects the semantic thesis that our thinking of events as causally or necessarily connected is a matter of *representing* them as standing in such a relation.

Despite these differences, however, there is a close connection between causal realism and the projectivist view I am attributing to Hume; for according to both views, our deployment of the idea of necessary connection is inextricably linked with our conceiving of causes as *grounds* of our expectations. Our having the idea of necessary connection just *is* a matter of our conceiving of the world as a world of causal relations: as a world whose causal structure is revealed by and serves to justify our inductive inferences. Of course, the major difference between causal realism and projectivism is that on the projectivist view, our inductive inferences only 'reveal' the causal structure of the world because that causal structure is itself a projection of our inferential habits. But this (from a projectivist perspective) does not make our conception of the world as causally structured any less central to our conception of our inferential habits as rationally constrained by the world. Given all this, projectivism seems to me to meet the three requirements just as well as causal realism does; the fact that, on the projectivist view, our thinking of events as causally or necessarily connected is a matter of projection rather than representation does not make the having of the idea of necessary connection any more of an 'empty ritual' than it is on the causal realist view.

What about the connectionless beings? Does the projectivist interpretation provide a conception of what the having of the idea of necessary connection amounts to, which makes us importantly different to connectionless beings? Is there an important difference between being able to think or say that the black ball *must* move, and being able to think or say only that the black ball *will* move? I think so. For, in saying or thinking that the black *must* move, we conceive of ourselves of having good *reasons* for thinking that the black will move. In his discussion of Hume on inductive inference, Stroud says: 'To say that the murderer *must* have only four toes on the left foot is to indicate that what you already know is good or conclusive reason to believe that about the murderer, and not just that he does have only four toes on the left foot' (1977, p. 63). On the projectivist interpretation, the causal case—one's thinking that the black *must* move—is just the same. Indeed, Stroud's case *is* a causal case—at least for Hume, given that he holds that *all* reasoning concerning matters of fact is causal reasoning. Our having good or conclusive reason to believe that the murderer has only four toes on the left foot, and our consequently coming to hold that the murderer *must* have only four toes on the left foot, is a matter of reasoning from effects to causes. The inference from crime-scene evidence—footprints in the sand, say—to facts about the murderer's anatomy just *is* a matter of thinking of the footprints as effects, and drawing a conclusion about what caused them. In other words, Hume would, I think, deny that there is any special epistemic, as opposed to causal, sense of 'must' at work in the claim that the murderer must have only four toes on the left foot: connectionless beings would be no more able to think or say that the murderer must have four toes than they are able to think or say that the black ball must move.

It does not follow from any of this that we are in a *better* position than are the connectionless beings, if our ultimate aim is to track the regularities in nature—that is, to make, and have confidence in, predictions that turn out to be true. As Stroud notes, it is 'implausible to suggest that [connectionless beings] would differ in being less *certain* than we are about, say, billiard balls, falling bodies or death. If their minds worked according to the [associative principle of causation], there is no reason to suppose that less force and vivacity, and therefore less certainty, would be transmitted from impression to idea in their case than in ours' (1977, p. 227). And of course connectionless beings can 'indicate' that they have good reason to believe that the murderer has four toes on the left foot by saying so, rather than saying (as they cannot, because they lack the idea of necessary connection) that the murderer *must* have four toes on the left foot. On the other hand, it is no part of Hume's thesis that we are in a better position than connectionless beings; it is no part of his thesis that we, armed as we are with the idea of necessary connection, will be better able to get around in the world, or will be better scientists, or whatever, than connectionless beings. Because we project our habits of expectation onto the world and they do not, we think of inductive reasoning as causal reasoning—as reasoning from causes to effects and vice versa—while they do not. But there is a sense in which it is the having of the habit, and one's thinking of the habit as legitimate or justified, that is important, and not the ability to project the habit onto the world in such a way that one gets to think of the world as a world of causes and effects.

Having said that much, there is still a sense in which Hume can hold that we are better off than our connectionless counterparts. Connectionless beings have the associative mechanism of causation (though of course they would not call it that). That mechanism will generate just the same expectations as it does in us, and will track nature's regularities just as successfully. But, because the impression of necessary connection arises from the operation of the mechanism, we, but not they, can track the operation of the mechanism much more easily than they can; and we will therefore find it much easier to conceive of the inferences generated by the operation of the mechanism as rational.

To see why, consider what is needed in order for connectionless beings to conceive of the inferences generated by the associative mechanism of causation as rational. The mechanism will generate expectations, given observed constant conjunction and a present impression as input, just as ours does. Connectionless beings will be able to tell that there is such an associative mechanism, because they will be in a position to notice that sometimes an expectation will naturally arise thanks to their having previously observed the relevant constant conjunction, and so they will come to realise that an associative mechanism, with experienced constant conjunction of As and Bs and a present impression of an A as input, generates belief that a B will occur. And they will be able to consider the expectations generated by the associative mechanism, as opposed to those generated by some other means (superstition or education, say), as justified, just as we are. So far, so good. But none of this will be *obvious* to connectionless beings. The expectations generated by the associative mechanism are accompanied by no special phenomenology. A given expectation—that the black will move, say—will simply appear in the mind *as* an expectation; it will not, as it were, wear its genesis in the associative mechanism on its sleeve. In order to think of a given expectation that a connectionless being finds herself with as rational, she will have to consciously think about how that expectation arose—about whether it arose thanks to the associative mechanism, or whether it is due to some other, less reliable mechanism: education, say.

We, on the other hand, thanks to the impression of necessary connection, do not have to go through any such laborious procedure. When *we* come to expect that the black will move, that expectation *does* wear its genesis on its sleeve, for it is accompanied by a phenomenology that is lacking in cases where expectation is generated by, say, education. It is that phenomenology, and the corresponding projectivist semantics, that allows us automatically and legitimately to think of ourselves as rationally responding to a causally structured world, rather than to a world of loose and separate events.

Consider how things are with dogs. Dogs' expectations, according to Hume, are generated by the same associative mechanism as ours (see *T* 176–9). But dogs, unlike us, are not capable of caring about *how* their expectations are generated: they are not capable of conceiving of one expectation as more or less rational than another. Dogs thus have no use for an impression of necessary connection; for them, such an impression would be (or perhaps is) merely a 'feeling' they get when they expect *walk soon* or *dinner now*. Connectionless beings, unlike dogs, *are* capable of caring about how their expectations are generated, and they are capable of conceiving one expectation as more or less rational than another. But their ability to do so is hampered by their lack of an impression, and hence an idea, of necessary connection. If they had such an impression, they would be able to think of themselves as reasoning from causes to effects—rather than succumbing to superstition, say—without having consciously to consult past experience in order to work out what generated their expectation on a given occasion. We, unlike connectionless beings, can do just that. As Hume says, albeit in a slightly different context: 'Those, who delight in the discovery and contemplation of *final causes*, have here ample subject to employ their wonder and admiration' (*E* 55).

References

Baier, A. (1991). *A Progress of Sentiments: Reflections on Hume's Treatise*. Cambridge, Mass.: Harvard University Press.

Blackburn, S. (1984). *Spreading the Word*. Oxford: Oxford University Press.

—— (1987). 'Morals and Modals', in Blackburn (1993), 52–74.

—— (1988). 'How to be an Ethical Anti-Realist', in Blackburn (1993), 166–81.
—— (1990). 'Hume and Thick Connexions', in Read and Richman (eds) (2000), 100–12.
—— (1993). *Essays in Quasi-Realism*. New York: Oxford University Press.
Buckle, S. (2001). *Hume's Enlightenment Tract*. Oxford: Oxford University Press.
Craig, E. (1987). *The Mind of God and the Works of Man*. Oxford: Clarendon Press.
Hume, D. (1739–40). *A Treatise of Human Nature*. L. A. Selby-Bigge and P. H. Nidditch (eds), 2nd edn (1978). Oxford: Clarendon Press.
—— (1748/51). *Enquiries Concerning Human Understanding and Concerning the Principles of Morals*. L. A. Selby-Bigge and P. H. Nidditch (eds), 3rd edn (1975). Oxford: Clarendon Press.
Kemp Smith, N. (1941). *The Philosophy of David Hume*. London: MacMillan.
Lewis, D. K. (1986). *Philosophical Papers, vol. II*. Oxford: Blackwell.
Mackie, J. L. (1980). *Hume's Moral Theory*. London: Routledge.
Owen, D. (1999). *Hume's Reason*. Oxford: Oxford University Press.
Price, H. (1998). 'Two Paths to Pragmatism II', in R. Casati and C. Tappolet (eds), *European Review of Philosophy* 3: 109–47.
Read, R. and K. Richman (eds). (2000). *The New Hume Debate*. London: Routledge.
Strawson, G. (1989). *The Secret Connexion*. Oxford: Oxford University Press.
Stroud, B. (1977). *Hume*. London: Routledge.
—— (1993). '"Gilding or Staining" the World with "Sentiments" and "Phantasms"', in Read and Richman (eds) (2000), 16–30.
Wright, J. P. (1983). *The Sceptical Realism of David Hume*. Manchester: Manchester University Press.

10
Causal Perspectivalism

HUW PRICE

10.1 Foreign Metaphysics and the Benefits of Travel

As objects go, foreigners are a pretty respectable bunch. They are not figments of our collective imagination, or social constructions, or useful fictions. They are not mind-dependent, and they do not disappear when we don't keep an eye on them. Our 'folk theory' about foreigners is not subject to some global error, and the term 'foreigner' certainly manages to refer. Some of our beliefs about foreigners are mistaken, no doubt, but only by failing to accord, case-by-case, with the objective reality to which they are certainly answerable. There are many facts still to be discovered about foreigners, such as their precise distribution in space and time. Moreover, these are matters for scientific study. And so on. In a nutshell, foreigners are as real as we are.

Yet think of the discovery each of us made when, minds broadened by travel, we realized that foreigners themselves use the very same concept, but apply it to us! What we learnt (at that unsettling moment) was that the distinction between them and us, foreigners and compatriots, is not as objective as we had assumed. It is a distinction drawn 'from a perspective'—that of a *local* speech community, embedded in a tribal population. There are objective divisions in the world, of course, but not the asymmetric distinction that each side sees from where it stands. God sees us as Afghans, Zimbabweans, and many things in between, perhaps, but not as

---

I am indebted to many people for comments and discussion of this material, among them David Braddon-Mitchell, Amit Hagar, Chris Hitchcock, Carl Hoefer, Jenann Ismael, Doug Kutach, Peter Menzies, Brad Weslake, and Jim Woodward, and participants in conferences in Sydney in July 2003, and Venice in May 2004. I am also grateful to the Australian Research Council, for research support.

locals and foreigners.¹ So, the reality of foreigners notwithstanding, there's a sense in which foreignness is a less objective matter than we used to think.

Perspectivity of this kind raises important philosophical issues. Some are general issues: How is the relevant notion of perspective best characterized? Is it one phenomenon or several? And how far does it extend—how many of our conceptual categories are perspectival, in whatever sense or senses the notion turns out to encompass? Others are local issues: For some particular concept or group of concepts of philosophical interest, is that concept or group of concepts perspectival, in any interesting sense? And if so, how, and how can we tell? In this chapter, I am interested in one of these local issues: roughly, the question whether our *causal* concepts are perspectival.

I began with the local–foreigner distinction with three points in mind. First, it is a striking and familiar example of the general phenomenon I am interested in, and illustrates well that noticing perspective is often a matter of learning to see things 'from another point of view'—from the viewpoint of someone who uses the same concept as we do but applies it differently, in virtue of a difference between their circumstances and ours. Sometimes, of course, it can be very hard to make this imaginative shift, especially when the alternative standpoint simply doesn't occur within our own linguistic community—when 'we' all occupy the same viewpoint, in the relevant respect. (Call this a *homogeneous* perspective.) I'll argue that causation is like this. For basic physical reasons, all humans share a homogeneous perspective, in the relevant respect.

The second useful lesson is the one I started with. Perspectivity doesn't automatically lead to simple-minded forms of anti-realism or subjectivism. It may lead to more sophisticated forms of the same thing, but in virtue of being more sophisticated, these won't challenge the 'obvious' truths of common sense in the same way.

Finally, the example has a useful structural similarity to the case of causation. Why? Because the local–foreigner distinction is strongly asymmetric, a fact variously revealed in the different ways we behave towards foreigners and locals. The example thus provides a case of an *apparent* asymmetry in reality (apparent, anyway, from a sufficiently naive viewpoint) that has turned out to be merely perspectival. I want to argue that in an analogous

---

¹ I am thinking here of the philosophical 'view from nowhere' kind of god, of course. Some gods have regrettably perspectival viewpoints.

way, a perspectival view of causation makes better sense than alternatives of both the *asymmetry* of causation, and its *temporal orientation*—that is, of the intuitive difference between cause and effect, and the fact that at least in general, causes precede their effects in time.

Although the local–foreign distinction is asymmetric, it isn't temporal, and another useful way of approaching the issue of causal perspectivity is via more general questions about the conceptual manifestations of our temporal situation—the temporal aspects of the ways in which we are constructed, situated, embedded and oriented in time. In some respects, this is familiar philosophical territory. Tense, and the distinctions between past, present and future, are widely regarded as products of our own temporal perspective. On this view, clearly, there's an analogy between 'now' and 'then', on the one hand, and 'us' and 'them', on the other. True, this comparison is somewhat controversial. Some writers, such as presentists, maintain that there is an objective *now* in a sense in which there is not an objective *us*. And in any case, the two distinctions are clearly disanalogous in another respect: our location with respect to whatever fence we mark by 'us' and 'them', compatriot and foreigner, while perhaps not immutable, is very much more constant than the temporal location we mark by 'now'.

Not all parts of the temporal territory are so inconstant, however. I'm especially interested in the conceptual manifestations of our asymmetric *orientation* in time. In my view, some of our modal notions, including causation itself, have a perspectival character closely linked to the 'oriented', or temporally asymmetric aspects of our constitution, as entities or structures located in spacetime. Clearly, these aspects of our temporal situation are not things that are likely to change, either for each of us individually from time to time, or from one of us to another, across the community. In this respect, then, we humans all share the same temporal perspective (even though, as we'll see, physics might allow other possibilities). So any conceptual perspectivity grounded on our temporal orientation is likely to be homogeneous, if anything is; and therefore, presumably, very hard to see.

At this point, travel can no longer help us. We need alternative resorts, alternative metaphors. One of the most powerful is Kant's comparison of the revelation (as he saw it) that there is an anthropocentric ingredient in some of our conceptual categories to the Copernican discovery of the perspectival character of the sun's apparent motion around the earth. Transposed to the present project, Kant's analogy makes several important points: that perspective can be hard to see; that its discovery may challenge deeply entrenched intuitions about the nature of reality; and—perhaps most importantly of all, in the present context, in virtue of the role of causation and related concepts in science—that scientific virtue may lie on the perspectivalist's side.

As the case of *foreigner* made clear, unmasking the perspectival character of a concept does not lead to simple-minded anti-realism—we may continue to use the concept, and even to affirm, in a variety of ways, the objectivity of the subject-matter concerned, despite our new understanding of what is involved (of where we 'stand') in doing so. Nevertheless, there is a tendency to think that perspectivity is incompatible with good science, in the sense that science always aims for the perspective-free standpoint, the view from nowhere. In my view, it is important to see that science itself might challenge this philosophical conception of science. For suppose we came to accept a perspectival genealogy for causation and related notions (perhaps for counterfactual reasoning, for example). In essence, this would be a scientific account of a particular aspect of human linguistic and cognitive practice, explaining its origins in terms of certain characteristics of ourselves, as structures embedded in time in a particular way. A corollary would be that uses of these very concepts in science—including, indeed, in this very explanation—would themselves be held to reflect the same embedded perspective. Thus some aspects of current scientific practice would be revealed *by science* to be practices that only 'make sense' from this embedded perspective—so that if, *per impossibile,* we could step outside this perspective, these aspects of science would cease to be relevant to us.

Would this be a *reductio* of the perspectival account of causation? Or a fundamental challenge to science? Neither, in my view. On the contrary, it would be continuous with a great scientific tradition, a tradition in which science deflates the metaphysical pretensions of its practitioners, by revealing yet further ways in which they are unlike gods. For Kant, Copernicus was the giant in this tradition. For us, Darwin towers beside him. The vertiginous lesson we need continually to relearn, on these lofty shoulders, is how insignificant we are, from the world's point of view; how idiosyncratic the standpoint from which we attempt to make sense of it. But however unsettling we ourselves find the blows that science thus delivers to our metaphysical self-image, science itself has not only survived, but thrived, on this diet of self-imposed self-deprecation. I see no reason to think that the present case will be any different, if some of science's own core categories and activities turn out to be perspectival in a newly-recognised way; a way that depends on the peculiar standpoint that science's own practitioners occupy in time.

But would the relevant aspects of current science then stand revealed as bad science? Or would it be the philosopher's 'view from nowhere' conception of science that would have been shown to be mistaken—an inaccurate conception of what science is, or could be? Again, neither, in my view. The perspectivity of (some aspects of) current scientific practice turns out to be entirely appropriate, given its role in the lives of creatures in our situation. In that sense, it is not 'bad science', and doesn't need to be reformed or eliminated. In appreciating this perspectivity, however, we get a new insight into the nature of the non-perspectival world, which 'looks like this', from our particular point of view. So there is good news, too, for 'detached' science.

I will return briefly to these issues at the end of the paper. In the main, however, I am going to focus on the first-order question about the character of the causal relation itself. Is the distinction between cause and effect like the distinction between us and them—a perspectival projection onto a non-perspectival reality? Or is it better understood as non-perspectival from the start? I want to make a case for the perspectival view.

My argument relies heavily on analogy. I take the optimistic view that once the key elements relevant to the issue of the status of the causal asymmetry are laid out in a clear way, the attractions of the perspectival view are easy to see. The role of the analogies is to help to achieve the required clarity: 'Look', I want to be able to say, 'The options in the causal case are just like *that*.' With this aim in mind, I'll begin with some simple examples of perspectival and non-perspectival concepts, in use in particular arenas, and then offer a slightly more elaborate example, which has all the structure I take to be needed to provide a useful comparison with the causal case. The early sections of the paper will be devoted to developing this argument for the perspectival character of causal asymmetry.[2]

The latter part of the paper tries to take the analysis a stage further, proposing a kind of diagnosis of the perspectival character of causation. By

---

[2] As I will explain, the focus on this asymmetry plays a dual role: it provides an important explanatory puzzle, to which the perspectival view offers the best solution (or so I'll argue); and it supports a powerful 'existence proof', based on symmetry considerations, for the possibility of an alternative causal perspective that reverses the asymmetry.

identifying some key elements in those aspects of our epistemic and practical 'architecture' that seem essentially associated with causal thinking, I'll offer an abstract characterisation of what might be called the *causal viewpoint:* a distinctive mix of knowledge, ignorance and practical ability that a creature must apparently exemplify, if it is to be capable of employing causal concepts. My project is thus a kind of naturalized Kantianism about causation. It aims to understand causal notions by investigating the genealogy and preconditions of causal thinking; by asking what general architecture our ancestors must have come to instantiate, in order to view the world in causal terms.

10.2 Perspective on the Field of Play

Imagine a soccer field, viewed from a goalkeeper's perspective. There is a near end of the field and a far end of the field, and the difference makes a great deal of practical difference to the goalkeeper's role in the game: typically, goalkeepers have much more work to do when the ball is at the near end than when it is at the far end. Clearly, however, this difference is perspectival. The opposing goalkeeper has the same conceptual categories, but applies them with precisely the opposite orientation to the physical arena that separates him from his opponent—and neither, obviously, is objectively right or wrong. (The referee, in the middle of the pitch, can't settle the issue—from his perspective, the near–far distinction, as applied by the goalkeepers, is simply inapplicable.)[3]

Contrast this to the case in which, literally, the two goalkeepers are not playing on a level field. Suppose that the pitch is higher at one end than the other. This asymmetry, too, makes a practical difference, in various ways. Each goalkeeper needs to adjust his kicks, for example, to allow for the effect of the slope. But here, clearly, the situation is not symmetric: one goalkeeper needs to kick away from his own goal more forcefully than he would if the field were level; the other needs to kick away less forcefully than he would if the field were level. This lack of symmetry

---

[3] The referee can adopt a kind of second-hand perspectival vocabulary, of course, referring to the two ends of the pitch as 'the near end for goalkeeper A' and 'the near end for goalkeeper B', for example. Although parasitic on the perspectival concepts employed by the two goalkeepers, however, these descriptions are not themselves perspectival.

256   HUW PRICE

reflects the fact that the high-end–low-end distinction is objective, or non-perspectival—in contrast, therefore, to the near-end–far-end distinction. (The referee is certainly able to rule on which is the higher end of the pitch.)

In these spatial examples there are four kinds of factor in play:

(i) The role that certain concepts—*near* and *far*, *high* and *low*, for example—play in some practical game or activity, and the reasoning associated with that activity.
(ii) Some spatial asymmetries in the application of these concepts.
(iii) Some spatial asymmetries in the location or other attributes of the users of these concepts.
(iv) Some physical attributes of the environment, such as the slope or altitude of the pitch, which are of (possible) relevance to the activity in question, and which may also be distributed asymmetrically in space.

Within these simple parameters, we have already seen how natural it is to regard the spatial asymmetries of some concepts (e.g. *near* and *far*) as reflecting those of the users of the concepts; those of other concepts (e.g. *high* and *low*) as reflecting those of external features of the environment.

The relevance of these examples is that they have much of the structure associated with the case of causal asymmetry. As I will explain, we can distinguish an analogous set of four factors in the causal case—the main difference being that in that case, of course, time replaces space. My next example is intended to make the analogy even more explicit. It introduces an asymmetric binary relation, defined on pairs of spatial points, which is intended to provide a useful analogue of the cause–effect relation itself. The example is designed so that the spatial relation in question is best understood in a perspectival way; thereby (or at least this is the intention) helping to make it clear how the same might be true of the causal relation.

10.3 The Forest of Forking Paths

Imagine a network of paths, on a forested north-facing hillside. Imagine creatures who map this terrain, and choose trajectories to travel through it. These trajectories depend on the availability of paths, but not only on that, for there is a further major constraint. These creatures begin their journeys (or, as we might say, their lives) at the top of the slope. From that point, things go downhill; for, like rolling stones, these little travellers rely on gravity to make their journey.

In thinking about actual and possible trajectories, these creatures employ a binary relation, defined on pairs of spatial points. They say that point B is *accessible* from point A if there is a possible trajectory that would lead them from A to B. The notion of trajectory is thus directed and transitive, and turns out (let us suppose) to be asymmetric: if A is accessible from B, then B is not accessible from A. Suppose also that at least in most cases in which B is accessible from A, B is to the north of A. We now have several obvious candidates for what it is for B to be accessible from A.

3.1 REDUCTION TO SPATIAL DIRECTION. B is accessible from A iff B is connected to A by a suitable path and lies to the north of A.
3.2 REDUCTION TO SLOPE. B is accessible from A iff B is connected to A by a suitable path and B is at a lower altitude than A.
3.3 REDUCTION TO TYPICAL SLOPE. B is accessible from A iff B is connected to A by a suitable path and lies in the direction from A in which the terrain is *typically* lower.[4]

In addition, we have a perspectival option:

3.4 REDUCTION TO THE TRAVELLER'S PERSPECTIVE. B is accessible from A iff a typical traveller arriving at A could proceed to B (at least in principle, perhaps).

Which of these alternatives makes best sense of the use that the creatures in question make of the notion of accessibility?

Let's begin with 3.1. The first issue is whether it gets the extension of the concept right. Is it in fact true, in practice, that a point B is accessible from A only when B is to the north of A? Or do some trajectories run east–west, or even southwards, at least for short distances? If so, then these are counterexamples to 3.1.

---

[4] There are two importantly different ways to read this proposal. According to one, accessibility is a fundamentally asymmetric relation, the asymmetry resting on the fact that the relation is constituted, in part, by some relation to the prevailing slope. According to the other, accessibility is essentially a symmetric relation, to which an asymmetric labelling is affixed by convention—a convention that links in practice to the direction of the prevailing slope, though any available 'signpost' would do as well. It is the former reading we have in mind here. The latter could be made explicit, but would then play a role in what follows very similar to that of 3.1, being vulnerable to very similar objections. I ignore it, henceforth, for ease of exposition.

Suppose for the moment that 3.1 passes this first test. Let's now think about the question, 'What is it that *explains* the fact that a point B is accessible from A only when B is to the north of A?' In the light of 3.1, the appropriate answer is that this is simply what it *means* for B to be accessible from A. But the puzzle now bursts out somewhere else, of course. If this is what it means for B to be accessible from A, why is it that our creatures can only travel in an 'accessible' direction? In other words, why is accessibility relevant to their practical and cognitive behaviour, in the way that we have assumed that it is?

We are inventing the story as we go along, of course, and so we could simply stipulate that this is how it is. These creatures are simply keyed to rolling north, and that's that. This 'primitivist' strategy seems likely to be unappealing in any realistic case, however, because we don't expect space to play this direct explanatory role, and because what is perhaps the same coin we can't see any mechanism which could produce such behaviour, absent any other kind of physical property with suitably asymmetric spatial distribution.

In any realistic case, then, we expect that the fact that our creatures travel north will be explained by some spatial asymmetry in their own physical constitution, or that of their environment, or both. The prevailing slope might well provide such a feature, but in that case it seems more accurate to say that accessibility tracks slope, and that it is the further fact that the slope is towards the north that explains the fact that accessibility is towards the north. 3.1 thus defers to 3.2 or 3.3.

The merits of 3.2 and 3.3 can be assessed by looking for counterexamples. Concerning 3.2, for example, we want to know whether there are cases in which B is accessible from A, without being at a lower altitude than A. Perhaps there are small rises on the generally downhill slope, and some trajectories rise over these hills, before continuing downwards. If so, then this counts against 3.2

Concerning 3.3, similarly, we want to know whether there are cases in which B is accessible from A even though B lies with respect to A in the direction in which the terrain is typically higher, rather than lower—that is, as it happens, south rather than north of A; or, less dramatically, but also in tension with 3.3, whether there are cases in which B is accessible from A even though B is merely east or west of A. Either kind of case counts against 3.3.

Let's add some detail to the story, to settle the issue. Suppose, as above, that our creatures are rolling stones—freewheeling but intelligent rocks, who rely on gravity and momentum for their journey through this forest of forking paths. It is now easy to imagine how, even though the prevailing direction of travel must be downhill (and therefore, in the assumed environment, to the north), small variations of two kinds are possible. Our rocks can roll uphill for short distances, thanks to their momentum. And they might sometimes roll southwards for short distances, either downhill, when small variations in the prevailing slope permit it; or uphill, again, when they are able to convert kinetic to potential energy. (Perhaps they achieve this trick by first tackling a small uphill rise, and then doubling back—imagine the options available to a snowboarder!)

It now seems clear that 3.4 is closest to the truth. Accessibility is partly 'relative to the perspective of the rolling stone'. It depends not only on the external environment, but also on properties of the stone: its speed, mass and direction of travel (i.e. its momentum). There are various ways in which this might be formalized. Momentum might figure in a truth-condition, an assertibility condition, or perhaps other alternatives. This level of detail is unnecessary, however. The crucial point is simply that any satisfactory elaboration will be perspectival (or 'lithocentric'), in taking accessibility to depend on contingent attributes of the stones in question, as well as on the properties of the environment.

Accessibility is intended to be a prospective notion. Our intelligent rocks are planners, who think in advance about where their journeys might take them. What they want to know, when they consider whether B is accessible from A, is whether, if their journey takes them to A—if they arrive at A in typical fashion—they can expect or hope to reach B. In virtue of the general orientation of their hillside, and their own reliance on its slope for their locomotion, a typical way of arriving at A is from the south. Absent any special circumstances in which they are able to progress in a curving path back towards the south, B will be accessible from A only if it lies to the north, or at least not to the south.

Accessibility thus has a prevailing northerly direction, in virtue of the typical motion assumed in thinking about new cases. The explanation of the typical motion appeals to the typical slope, and this in turn explains why accessibility tends to be aligned with the downhill slope. But if the creatures have no trouble, in theory or in practice, in handling occasional small sections of level or even uphill path, then this correlation between accessibility and slope will often break down at a local level (with or without concomitant exceptions to accessibility's prevailing general northerly orientation). It may even break down in dramatic ways, at least in theory, if our rolling stones, idealizing beyond their real physical limitations, can imagine arriving with lots of momentum at the base of substantial uphill sections of path. In those cases, despite the dramatic departure from the normal downhill gradient, the uphill sections will seem accessible, at least in principle.

Imagine that some of the more far-sighted rocks, deducing on theoretical grounds that no downhill slope can continue for ever, realize that their own hillside will eventually level out. Perhaps the terrain stays flat after that, or perhaps their world is more symmetric, and their own downhill slope is matched by an uphill slope on the far side of a valley. In either case, unlikely though it might be that any actual rock should roll that far, a sufficiently idealized notion of accessibility might well extend to such regions. If so, the accessibility 'arrow' would still be taken to point predominantly to the north, even in these regions which lack the usual gradient in that direction.

Note the contrast between two ways in which the prevailing slope can be relevant. For the objectivist options, 3.2 and 3.3, slope comprises part of what is effectively a truth-condition for ascriptions of accessibility. For the perspectival option, it plays a very different role: it is something like a necessary physical precondition for the existence of creatures equipped to *occupy* the perspective in question. Roughly, their perspective depends on the fact that they are in motion, and (in the circumstances as described, at least) their motion depends on the slope. There is a big difference between these two ways of characterizing the relevance of the prevailing slope; even though it only shows up, roughly speaking, where the slope gives out. If slope is part of a truth or content condition, then where the slope gives out, so too does the extension of the concept in question. If it is a precondition for occupation of a perspective, then even the most level terrain may be viewed asymmetrically, provided only that there is enough slope, somewhere, to produce the kind of asymmetric creature who can occupy that perspective.

Thus, as our thoughtful little rocks consider their intuitions about how the notion of accessibility is properly applied, in the light of an understanding of their own nature and circumstances, they are already in a position to see that the notion is perspectival; that its application depends on an important spatial asymmetry in their own condition. Moreover, it seems to me, they are able to reach this conclusion without having the further thought that there might be creatures in which a similar spatial asymmetry had a different orientation—creatures who would hence apply the same notion with a different extension.

Nevertheless, this further thought, when it is achieved, acts as a virtual trump card. Realizing that they might live on one side of a valley, our creatures suddenly recognize that there might be rocks on the opposite side of the valley ('rock*s', perhaps) who would see things in precisely the way that they themselves do, but with the opposite spatial orientation. At this point, our creatures cross the kind of conceptual frontier that we ourselves crossed, when we realized that other people see us as foreigners—and see the most powerful reason for regarding accessibility as perspectival.

True, a proponent of 3.2 or 3.3 might argue that this case shows simply that the direction of accessibility varies from place to place—just as that of *downhill* does, for example. But we've already seen why this isn't a serious rival to the perspectival proposal. It doesn't explain as well as the perspectival view the postulated extension of the notion of accessibility to cases lacking the actual or typical slope in question (or, indeed, why slope matters in the way that it does to the practical activities of the creatures in question). And in any case, if the perspectivalist's opponent is the kind of realist who wants to regard the arrow of accessibility as a deep structural feature of reality, these reductionist alternatives provide little real comfort. So long as the symmetries of the relevant physics permit reversal of whatever asymmetry is taken to provide the reductive ground of accessibility, its arrow becomes at best contingently a global constant; and risks extinction altogether, in regions without the asymmetry in question, such as the floor of the valley that separates rocks from rock*s. This 'reversibility objection' remains very powerful, then, even for an opponent blind to the advantages of perspectivalism.

10.4 Applying these Lessons to the Causal Asymmetry

In the previous example, we had four kinds of factor in play: a certain concept, *accessibility,* playing a central role in a practical activity; a marked spatial asymmetry in the application of this concept; a spatial asymmetry in the physical characteristics of the users of the concept in question; and some spatially asymmetric attributes of the environment, of relevance to the practical activity in question. As we saw, the key issue was whether the spatial asymmetry of the relation of accessibility should be regarded as analytic (3.1), as reducible to the environmental asymmetry either directly (3.2) or on average (3.3), or as grounded on the 'intrinsic' asymmetry of the users of the concept (3.4). The example was chosen so that the perspectival option made best sense of the use of the concept in question. And the focus on spatial asymmetry played a dual role: it provided an important *explanandum*, to which the perspectival option offers the best *explanans;* but also the basis of a kind of existence proof, based on symmetry considerations, for an alternative perspective that reverses the asymmetry in question.

In the causal case, similarly, we have four factors to consider:

(i) The concepts *cause* and *effect*, with important conceptual links to certain practical activities, namely, intentional action and deliberation. In particular, an action and its intended outcome are held to be related as cause and effect: means and end are cause and effect.

(ii) Some temporal asymmetries in the application of these concepts, such as the fact that causes typically *precede* their effects.

(iii) Some time-asymmetric characteristics of the users of the concepts in question, such as the fact that they typically deliberate about *future* actions, on the basis, at least in part, of information received in the *past*.

(iv) Some temporal asymmetries in the environment, such as the prevailing thermodynamic asymmetry.[5]

Again, the issue is how best to explain the temporal asymmetry in the application of the concepts in question, and to relate it satisfactorily to the 'internal' and environmental asymmetries of (iii) and (iv). And again, there seem to be four main options:

4.1 REDUCTION TO TEMPORAL DIRECTION. B is an effect of A iff B is causally connected to A and occurs later than A.

---

[5] For present purposes I'll assume that the thermodynamic asymmetry, broadly construed, is the only time-asymmetric feature of the physical world that might plausibly be held to be relevant at this point. Thus I rule out, for example, the possibility that the asymmetry of causation might have something to do with the small T-symmetry violations known in microphysics. This assumption doesn't seem to me to be controversial, but in any case, much of the argument below would easily transpose to a proposal of the latter kind.

4.2 REDUCTION TO THERMODYNAMIC GRADIENT. B is an effect of A iff B is causally connected to A and B is at a higher entropy than A.
4.3 REDUCTION TO TYPICAL THERMODYNAMIC GRADIENT. B is an effect of A iff B is causally connected to A and B lies in the temporal direction from A in which entropy *typically* increases.[6]
4.4 REDUCTION TO THE AGENT'S PERSPECTIVE. B is an effect of A iff doing A is a means of bringing about B, from an agent's perspective—roughly, if controlling A is a means of controlling B.

Here is a sketch of a case for preferring 4.4, along the lines of the case we made earlier for preferring 3.4. Again, let's begin with 4.1. The first issue is whether it gets the extension of the concept right. Is it in fact true, in practice, that an event B is an effect of an event A only when B occurs later than A? Or is causation sometimes simultaneous, or even backwards? If so, then these are counterexamples to 4.1.[7]

Suppose for the moment that 4.1 passes this first test. Let's now think about the question, 'What is it that *explains* the fact that an effect always occurs later than its cause?' In the light of 4.1, the appropriate answer is that this is simply a consequence of what it *means* for one event to be an effect of another. But if this is true, why, then, is the distinction between cause and effect relevant to human action, in the way noted in (i)? Why are the *ends* of our actions typically *effects* of the *means* by which we achieve them?

If there is to be a satisfactory answer to this question, in the light of 4.1, it must appeal to the principle that the means–end relation itself has the appropriate temporal orientation: ends must occur later than their corresponding means. But this in itself, now, is something that calls for explanation—and it is no use simply appealing again to linguistic convention. It cannot be convention all the way down. There is a genuine temporal asymmetry in our deliberative practice—namely, roughly, the fact that our deliberations are future-directed but not past-directed—and this calls for explanation. In the end, then, this temporal asymmetry needs to be explained in terms of some temporal asymmetry in our own physical constitution, or in our environment, or some combination of the two.

---

[6] As in the case of 3.3, there are two possible readings of this proposal. I ignore the reading according to which the thermodynamic asymmetry is merely a convenient temporal 'signpost', by means of which to label the two ends of a time-symmetric relation.

[7] This is a familiar objection to Humean conventionalism about the direction of causation, according to which it is merely a terminological matter that causes precede their effects.

As in the spatial case, it is possible, at least in theory, that the relevant environmental asymmetry might be a primitive asymmetry of time itself, on which our deliberative behaviour somehow depends. In other words, it might be held that there is an intrinsic direction to time—an intrinsic distinction between past and future—independent of, or at least more basic than, any other physical time-asymmetry; and that it is somehow a necessary fact that deliberation is sensitive to this fundamental directionality, again in a manner unmediated by other physical time-asymmetries.

This primitivist view will perhaps attract more support than the corresponding view in the spatial case. Certainly, some philosophers profess to believe that time has some such intrinsic directionality—and at least in some cases, seem to think that their grounds for believing it are direct, or phenomenological, of a kind which is insensitive to the distribution and possible reversibility of the known physical time-asymmetries (such as that of thermodynamics). For present purposes, I simply want to note the theoretical cost of this view. It requires a deep link between the mental, on the one hand, and some deep and fundamental time-asymmetric aspect of physical reality, on the other—without the time-asymmetry concerned being manifest at intermediate levels.

Similar remarks would apply to attempts to appeal to a primitive asymmetry of causation itself, as a brute metaphysical fact. Again, it would be mysterious how such a fact could have the relevance it needs to have for deliberation, if the connection isn't somehow mediated by asymmetries in us or our environment. Otherwise, once again, we need some peculiar primitive link between minds and fundamental metaphysics.

If we set these primitivist proposals aside, then the time-asymmetry of deliberation needs to be explained in terms of some other temporal asymmetry, in our own physical constitution, in our environment, or in some combination of the two. The thermodynamic time-asymmetry might provide such a feature, for example, but in this case it seems appropriate to say that the direction of deliberation follows that of increasing entropy, and that it is the fact that entropy increases towards (what we call)[8] the future that explains the fact that we deliberate 'in that direction'. Thus 4.1 defers to 4.2 or 4.3.

---

[8] The qualification is needed because 'past' and 'future' may well themselves be perspectival notions. More on this later.

On the face of it, just as in the spatial case, the plausibility of 4.2 and 4.3 can be decided by looking for counterexamples to one or other suggestion. Concerning 4.2, for example, we want to know whether there are cases in which B is an effect of A, without being at a higher entropy than A. If so, then this counts against 4.2.

Concerning 4.3, similarly, we want to know whether there are cases in which B is an effect of A even though B is oriented with respect to A in the direction in which entropy is typically lower, rather than higher—that is to say, earlier rather than later than A; or, less dramatically, but also in tension with 4.3, whether there are cases in which B is an effect of A even though B is merely simultaneous with A. Either kind of case counts against 4.3.

My own view is that we have no trouble making sense of simultaneous causation; that we have little trouble in making sense of 'entropy-reducing'[9] and 'backward' causation;[10] and, as I will argue in a moment, that we certainly have no trouble in making sense of ordinary, directed causation in cases in which—in virtue of the manifest time-symmetry of the physics in question—it is simply implausible that typical slope is important (except in so far as it supports our perspective). If I am right, then, as for the rolling stones, only the perspectival view gets the intuitive extension of the concepts in question more or less right. According to the perspectival view, causal reasoning too depends on the standpoint of creatures engaged in a certain kind of journey. In this case, it is a journey into an uncertain future, a journey in which—*from the epistemic standpoint of the creatures themselves*[11]—their choices determine what path they take through a tree of forking possibilities.

I want to develop one more example, in order to illustrate the very striking advantages of the perspectival view, and also to motivate an hypothesis about the genealogy of a crucial element of the asymmetry of our causal reasoning.

---

[9] Here's a sketch of an argument to this effect. Consider a typical entropy-increasing process, such as gas molecules escaping from a bottle. Gradually reduce the number of particles. Presumably this does not change the essential causal structure of the case, and only changes by degree the extent to which it is an entropy-increasing process (however this is characterized). But reduce the number far enough, and you approach a case we could produce in reverse in a laboratory. In the reverse case—so our intuitions tell us—the causal arrow would still run from past to future; and hence, now, in the direction in which the entropy of the local system in question *decreases*.

[10] I'll mention some examples in a moment.

[11] This qualification is indispensable, in my view, and marks the second major respect in which causal reasoning is perspectival. More on this later.

10.5 Stargate Doughnut

Imagine a photon, p, which spends billions of years in intergalactic space as it travels from one distant galaxy, G_{past}, to another, G_{future}. At some point in between, at a time t, it passes through the central aperture in a tiny doughnut-shaped object, which happens to be spinning on a transverse axis, somewhere in deep space. As this doughnut spins, it periodically occludes the path the photon takes from G_{past} to G_{future}, and hence acts as what we could call a 'stargate'. At t, however, the gate is open, and the path is unobstructed.

Consider the following counterfactuals:

> PROPOSITION 1: *If Stargate Doughnut had been closed at time* t, *the photon* p *would not have been absorbed at* G_{future}.
>
> PROPOSITION 2: *If Stargate Doughnut had been closed at time* t, *the photon* p *would not have been emitted at* G_{past}.

We take Proposition 1 to be true, and Proposition 2 to be false. Accordingly, we take the orientation of the stargate to be a cause of the *later* position of the photon, but not the *earlier* position of the photon. What is the source of this time-asymmetry?[12]

On the face of it, the example seems completely independent of any thermodynamic details of the systems in question. It is far from clear that any sense can be made of the notion of the entropy of a discrete microscopic system of this kind. Even if it could, however, the example doesn't depend on the existence of a thermodynamic gradient. After all, we could quite well imagine the same kind of situation, in a universe in thermodynamic equilibrium. (We would need to replace galaxies with some photon source at the same temperature as everything else, but this makes no significant difference to the example.)

In some sense, of course, our intuitions about the case are sensitive to temporal direction. What else could we rely on in distinguishing Proposition 1 from Proposition 2, after all, given their apparent symmetry

---

[12] There is another intuitive asymmetry lurking here, related to that of Kripkean necessity of origins: we are somewhat more inclined to allow that it is the *same* photon we consider in Proposition 1 than in Proposition 2, namely, the one that actually passes through the stargate at t. I don't have space to explore the origins of this second asymmetry here, but I regard it as a very plausible target for the perspectival approach, with likely connections to the causal and counterfactual cases.

in other respects? But as we saw, however, it is no use trying to rely simply on temporal direction, to explain the difference between the two cases. If we say it is merely a conventional matter—a matter of the meaning of 'cause' and 'effect'—that causes occur before their effects, this will certainly imply that Proposition 2 (or a causal variant of it) must be false; but it doesn't explain why that is relevant to our decision behaviour, in the way that it is. (We think that we could use the stargate to influence future photon positions but not past photon positions, for example.)

The perspectival proposal is that what we bring to the case, in imagination, is the typical perspective *we* have as deliberating agents—the perspective we bring to the situation, quite unconsciously, when we think about *manipulating* the stargate. This deliberative perspective displays a very marked temporal bias, of course. Roughly, it treats the past as fixed, and only the future (or some subset of the future) as under our control. According to the perspectival view, it is this asymmetric perspective *on our part* that grounds the intuitive asymmetry between Proposition 1 and Proposition 2.

It might be objected that in introducing 'us' in this way, we implicitly 'import' the thermodynamic gradient, on which our existence depends—thus contradicting my claim that the case is insensitive to the thermodynamic details. However, there is no such conflict, as the spatial analogy developed earlier makes clear. In that case, the prevailing slope was relevant in the sense that without it, there could be no rolling stones of the kind described, and hence no creatures from whose perspective the notion of accessibility would make sense. As we saw, however, this is very different from reduction of the relation of accessibility *to* any fact about the prevailing slope.[13]

In the rolling rocks case, I suggested that in making judgements about accessibility, the creatures consider the issue of the accessibility of a point B from a point A, under the assumption that they arrive at A in a typical manner—that is, in their case, from the south, with a certain amount of momentum. We noted that momentum is easily varied in imagination, as it were, supporting a well-defined sense of 'accessible in principle'. Accessibility assessed in this manner thus comes to transcend the capacities of any actual creature.

---

[13] Compare: we need brains to be in a position to talk about anything at all, but it doesn't follow that in talking about anything at all, we are talking *about* our brains.

What is the corresponding exercise in the causal case? In my view, we get an excellent explanation of our intuitions about the stargate example if we postulate, again, that what underlies those intuitions is an exercise of imagination: we imagine *intervening,* to change the orientation of the doughnut, and assess the likely outcomes of such interventions. Roughly, anything that changes or 'wiggles' in these imagined cases, when the orientation of the doughnut is 'wiggled', is regarded as an *effect* of that orientation.

Of course, it is hardly news that some such notion of intervention plays a crucial role in our understanding of causation. Manipulability approaches to causation have a long history, and the importance of a notion of intervention to an understanding of causal reasoning has been displayed in a powerful formal way in recent years by writers such as Glymour, Pearl, and Woodward.[14] By and large, however, these authors think of their project in objectivist terms. In other words, they think of the causal structure of the world—including interventions themselves, to the extent that they comprise a crucial element of that structure—as something that exists independently of human agents. But there is a deep tension in such a viewpoint, in my view, stemming from the fact that intervention is a deeply perspectival notion. Intervention acts as a kind of Trojan Horse against objectivist approaches: when (rightly) accorded a central role, it thwarts the metaphysical ambitions of the theories in question.

Why is intervention a Trojan Horse? Because of the in-built bias I mentioned earlier—the bias manifest, *in our case,* in the fact that when we imagine intervening, we carve up the relevant aspects of reality, on broadly temporal lines, into a fixed or 'given' past and an open or mutable future. This carve-up, I claim, is perspectival: it reflects contingent features of our own circumstances, in such a way that other thinkers, differently 'situated' in the relevant respects, would carve matters up in a different way. In their hands, then, the same conceptual framework acquires a different extension, in the manner characteristic of perspectival concepts.

We cannot simply take for granted that the carve-up in question is perspectival, of course; the point needs to be argued. So far, the stargate example is simply an intuition pump. It makes it clear that we retain strongly time-asymmetric intuitions about the possibilities for intervention,

---

[14] See especially Spirtes, Glymour, and Scheines (1993), Meek and Glymour (1994), Hausman (1998), Pearl (2000), and Woodward (1997, 2000, 2003). There is also an excellent introduction in Woodward (2001).

and hence about causal direction itself, even where the 'objective' physics of the situation is manifestly symmetric. I am going to reinforce this intuition pumping in two ways. First, in Section 10.6, I will present what seems to me the most powerful argument in favour of the perspectival view. As in the rolling rocks case, it exploits symmetry considerations, to argue for the possibility of creatures with an alternative perspective on the same objective reality—roughly, creatures whose carve-up holds fixed what we hold open, and vice versa. (As I'll explain, the relevant sense of 'possibility' is rather thin: the argument is really a further, very forceful, intuition pump.)

The second kind of reinforcement, in Section 10.7, is both more concrete and more general. I'll aim for an abstract characterisation of deliberation—a characterisation that abstracts away from various contingencies about the kinds of agents we ourselves happen to be—in order to identify the contingency in question in the most general possible terms. This will demonstrate, I think, that the notion of intervention is ineliminably perspectival, and it will also provide a deeper understanding of the origins of our intuitions about the stargate case. I want to show that our intuitions about the case are just what they ought to be, if they are indeed manifestations of the viewpoint of an agent in the general sense, who (to put it somewhat figuratively) happens to be embedded in time in a particular way.

10.6 The Reversibility Argument: Causation in a Gold Universe

In arguing that the perspectival view makes better sense than the alternatives of our intuitions about the direction of causation, I have so far avoided placing any weight on the suggestion that there might be creatures who use causal concepts, but whose perspective does not coincide with ours. As I have presented it, then, the argument would be unaffected by the discovery that there were good physical reasons to exclude this possibility. As we will see, however, the dialectic is far from even-handed at this point. The reversibility argument, exploiting physical time-symmetries to make a case for the possibility of alternative causal perspectives, doesn't need to wait for physics to convince us that these perspectives are actually occupied. As in the rolling stones case, in my view, the reversibility argument therefore provides something close to a trump card for perspectivalism. Yet it is a card that remains strangely neglected, and the role of the following brief excursion into physics is to bring it into play.[15]

As I have already emphasized, the main candidate for a physical asymmetry that seems likely to be associated with the causal asymmetry, whether by the reductive or perspectival routes, is the asymmetry associated with the second law of thermodynamics—the general tendency of entropy always to increase. The behaviour the second law describes is strongly time-asymmetric, of course. As we might put it, entropy increases in one temporal direction ('towards the future'), but decreases in the other ('towards the past'). Since the late nineteenth century, physicists have puzzled about the origins of this time-asymmetry, especially in the light of the apparent time-symmetry of the underlying laws of physics.

One simple but crucial insight is that entropy would not be increasing in our region, if it were not for the fact that entropy was low, at some point in the past. After all, if entropy had had its maximum possible value in the past, the effect of the second law would simply have been to keep it at that maximum—there would be no general entropy increase. In the last 40 years, modern cosmology has thrown some remarkable light on the nature of this low entropy 'boundary condition' in the past. It now appears that matter was extremely smoothly distributed, about 100,000 years after the Big Bang. In a system in which the dominant force is the attractive force of gravity, this is a highly ordered or low entropy state: the 'natural', equilibrium-seeking behaviour of gravitating matter is to agglomerate into large, inhomogeneous clumps; and the tendency of matter to do this, coupled with the homogeneous initial state, drives the production of galaxies and stars.

The upshot seems to be that the conditions required for evolution of creatures such as ourselves depend on this cosmological low entropy boundary condition in the past. Moreover, it seems plausible that it is no accident that we are aligned in the way that we are with respect to the entropy gradient provided by the existence of this condition—that any creature is bound to 'remember' in the direction in which entropy decreases, and hence to regard that direction as the 'past'.

In fact, the suggestion that 'past' and 'future' are perspectival in this way was made already by Boltzmann in the 1890s—though against the

---

[15] For a more extended introduction to the topic, and recommendations for further reading, see Price (2004a).

background of a rather different proposal about why entropy is low in what we call the past. Following a proposal by his assistant, Schuetz, Boltzmann suggested that we might live in the kind of chance fluctuation from equilibrium that, although extremely rare, would nevertheless be inevitable eventually, in a sufficiently ancient universe. Such fluctuations would be two-sided dips in the entropy curve, of course, and Boltzmann suggested that it is no accident that we seem to live on an 'uphill' rather than a 'downhill' side of the dip—that is to say a side on which entropy is increasing, rather than decreasing. The sense of uphill or downhill depends on our sense of the direction of time, and that, Boltzmann suggested, is perspectival.

Thus Boltzmann's proposal already introduces the possibility of creatures whose temporal perspective (and hence also, presumably, causal perspective) is reversed relative to ours. However, this is not the most congenial form of this idea, because Boltzmann's hypothesis has some unwelcome consequences. It turns out that if Boltzmann's statistics are our guide, it is much easier for random fluctuations of this kind to produce fake records and memories, than the real events of which they purport to be records.[16] As a result, Boltzmann's hypothesis about the origins of the low entropy past implies that, almost certainly, all our 'records' and 'memories' are misleading. So the proposal is hard to take seriously.

However, the possibility of creatures whose temporal perspective is reversed relative to ours emerges again in a new form, in the modern cosmological alternative to Boltzmann's account of the origins of the low entropy past. One of the early contributors was the cosmologist, Thomas Gold. Among other things, Gold (1962) proposed that there might be a deep connection between the second law of thermodynamics and the (then) newly discovered expansion of the universe. Roughly, Gold's idea was that the expansion of the universe creates new possibilities for matter

[16] To give an example I have used elsewhere, suppose that God wants to leaf through possible worlds, until he finds the complete works of Shakespeare, in all their contemporary editions. According to Boltzmann's own statistical measure, it is vastly more likely that God will hit upon a world in which the texts occur as a spontaneous fluctuation of modern molecules, than that He'll find them produced by Shakespeare himself. This is simply because entropy is much higher now than it was in the sixteenth century. According to Boltzmann, probability increases exponentially with entropy. So the higher-entropy twenty-first century—'Shakespearian' texts and all—is much more likely than lower-entropy sixteenth century: almost all possible worlds that include the former don't include the latter. In Boltzmann's terms, then, it is extremely unlikely that Shakespeare and his contemporaries ever existed. And the same goes for the rest of history!

(rather like a piston withdrawing in a cylinder of gas, which makes new positions accessible to the molecules of gas within the cylinder). Gold proposed that the increase of entropy simply reflected the tendency of matter to take up these new possibilities, just as the gas expands behind the retreating piston. This proposal implies that if the universe were eventually to recollapse to a 'Big Crunch', entropy would need to decrease. After all, a recollapse is simply an expansion viewed in reverse.[17] Hence if expansion implies entropy *increase,* then recollapse implies entropy *decrease.*

Gold's suggestion seems to be flawed, however, in a way that comes to light if we make a *modus tollens* of his *modus ponens.* It doesn't seem to be the case that recollapse automatically leads to decreasing entropy. Why should the gravitational clumps in recollapsing matter gradually disperse, after all, rather than simply becoming even larger? But if recollapse doesn't imply that entropy decreases, then expansion doesn't imply that entropy increases (again, it is a simple symmetry argument).

So something else is required to explain why entropy increases: a low entropy 'boundary condition' in the past, of the kind we have already described (and cosmology has apparently discovered). But the role of this condition means that one aspect of Gold's proposal is in a sense reopened. Would entropy decrease, if the universe recollapses? Well, it depends. It would do so if there is a 'boundary condition' near the Big Crunch of the same kind as the one we have discovered near the Big Bang—a condition that ensures that entropy is low in that region.

In this sense, then, this symmetric cosmological model—the 'Gold universe', as it has come to be called—remains a live possibility, at least until we have a better sense of how to account for the low entropy condition in the past. And at this point, symmetry comes to Gold's aid. We cannot dismiss the possibility of a Gold universe on the grounds that it would require an incredibly unlikely 'fine-tuning' to make entropy decrease. For, at least in the absence of any time-asymmetry in the underlying physics,

---

[17] Gold himself argued that if the laws of physics are time-symmetric, the expanding and contracting universes cannot be distinguished—we have just two different ways of describing the same thing. (As in the Boltzmann–Schuetz case, it is presumably no accident that we see it as expanding.) As Larry Schulman pointed out to me, Gold himself did not actually consider the case in which the same universe expands and then contracts. The usual attribution of this proposal to Gold—e.g., by Davies (1974, p. 193), as well as by me (Price 1996)—depends on an extrapolation from Gold's views. But Schulman also reported that Gold was not unhappy with the extrapolation, or with the use of the term 'Gold universe' to describe such a cosmology, and I will continue to use this label.

the fine-tuning required is the same in either temporal direction. It doesn't exclude a low entropy past, so it can't exclude a low entropy future.

At present, then, it remains a live empirical possibility that the universe contains regions in which the thermodynamic gradient is reversed. In such regions, it seems likely that intelligent creatures would have a time-sense reversed relative to ours. There isn't an objective matter which side—us or them—gets it right about the direction of time, or about the direction of causation. They deliberate with respect to what we regard as earlier events. For them, the causal arrow runs directly counter to the way it runs for us.

This possibility seems to me to make a very strong case for the perspectivity of the causal asymmetry. Moreover, the strength of the case seems to depend very little on how seriously we need to take the hypothesis that our own universe might be a Gold universe. The hypothesis simply gives us an easy way to imagine the possibility that there might be creatures, elsewhere in the actual universe, whose time-sense is the mirror-image of ours. Suppose we grant that if there were such creatures, of whatever origins, then two things would follow: (i) they would think that the causal arrow is oriented in the direction that we would call future-to-past; and (ii) their perspective would be as valid as ours. Then we have all it takes to establish that causal direction is perspectival *for us*—whether they exist or not.

Point (ii) might conceivably be challenged. It might be claimed that if there were such time-reversed creatures, their reversed causal perspective would simply 'get it wrong'. But what could it be that these creatures are thus supposed to be wrong *about,* exactly, if not something grounded in the physics (which is symmetric, by assumption)? This response seems committed to a version of causal primitivism, in other words, with the attendant disadvantages. Whatever this primitive causal relation is supposed to be, how could we possibly know which side—us, or our time-reversed cousins—is right about its direction?[18] And in any case, why should it matter? Why should such a relation have any particular connection to deliberation (unlike what we might call 'quasi-causation', which would be the perspectival substitute, grounded on deliberation from the start)?

---

[18] Another possibility, suggested to me by some recent work by Tim Maudlin, would be to argue that these time-reversed 'creatures' could not really be minds or agents at all. In raising this possibility, however, Maudlin appeals to the intuition that such time-reversed states would not enjoy the causal relations to one another that the corresponding states in our brains enjoy. In the present context, then, this proposal would have to turn on the kind of primitivism just mentioned. It couldn't provide a non-question-begging argument *against* the symmetric view.

Of course, one might respond to the appeal to the Gold universe by trying to defend a non-perspectival alternative, such as 4.3 or 4.4. Such a view would imply, as a thoroughly objective matter, that causation simply changes direction in a Gold universe, in regions in which the entropic arrow reverses. As we have already seen, however, these objectivist approaches make poor sense of our intuitions about a range of cases—the stargate case, for example, and apparent cases of entropy-reversing and backward causation—and certainly do less well than the perspectival view in the task of explaining the links between causation and deliberation. And in any case, to echo a point I made in the rolling rocks case, such objectivist approaches seem to concede a large part of the game, in this context, by conceding that the direction of causation is at best contingently a global constant.

The force of the appeal to the Gold universe lies in the fact that it exploits a symmetry associated with time, in order to take us in thought (so to speak) where travel cannot take us in fact. Beginning with the conditions on which our own perspective depends, a simple temporal reflection yields conditions that support a conflicting perspective on the same reality. But although this is a powerful way of exhibiting the perspectival character of causal thinking, it doesn't get to the heart of the matter. *Why* does temporal reflection make such a big difference to our causal intuitions? I have already suggested that the answer lies in the links between causal reasoning and deliberation. To support this suggestion, I now want to make some general remarks about what is involved in deliberating, or 'being an agent'. Among other things, I want to try to identify some more basic sources of the perspectivity of the causal viewpoint—sources that seem only contingently linked to our temporal perspective, in the way that the reversibility argument exploits so successfully.

10.7 The Architecture of Deliberation

In considering the nature of the agent's perspective, we should distinguish two projects. The first aims for an abstract characterization of the structure, or functional architecture, of deliberation in general—what is essential to anything that deserves to count as an agent. The second considers the temporal aspects of that architecture, in so far as it is instantiated in us. Thus we need to distinguish 'essential' features of deliberation from contingent facts about the temporal characteristics of deliberation in our case. In principle there might be non-human agents, who, while instantiating a broadly similar functional architecture to us, *qua* deliberators, do so in a way that involves a different relation from ours to time and perhaps to space.[19]

Let's begin, then, with some general remarks about the structure of deliberation. In any deliberative process, presumably, there must be a range of things that the deliberator in question takes to be matters for deliberation: in other words, the alternatives among which she takes herself to be deliberating. For formal convenience, let's regard these alternatives as a class of propositions, denoted by OPTIONS. These are the propositions the agent takes herself to have the option of 'deciding to make true', in other words. It will be helpful to subdivide this class into DIRECT OPTIONS, comprising those matters over which an agent takes herself to have immediate control, and INDIRECT OPTIONS, comprising those ends she takes herself to be able to accomplish indirectly, by an appropriate choice from her DIRECT OPTIONS. And let FIXTURES denote everything else—all matters of fact that are not held to be a matter of choice in the deliberation in question.

FIXTURES will contain a subset, KNOWNS, comprising those facts the deliberator takes herself to *know* at the time of deliberation, and also a larger subset, KNOWABLES, comprising matters she regards as either known or knowable, at least in principle, before she makes her choice.[20] Why must KNOWNS and KNOWABLES be subsets of FIXTURES? Because it seems incoherent to treat something both as an input available to the deliberative process, at least in principle, and as something that can be decided by that process. Control trumps a claim to knowledge: I can't take myself to know that P, in circumstances in which I take myself to be able to decide whether P, in advance of that very decision.[21]

---

[19] Compare D. C. Williams's (1951) suggestion that there might be consciousnesses spread out across space, their successive mental states related like the palings in a picket fence.

[20] Note that 'knowable' is ambiguous in this context. I can believe that something not known could be known, but in a way which would then deprive me of the relevant choice. For example, I believe I can decide where my garden furniture will be placed tomorrow; but I also acknowledge that I could come to know the truth about this matter, prior to any decision on my part, say if I were to discover that someone had concreted the furniture in place, in its present location. The sense of knowability I have in mind here is not this one, however, but that of knowability without loss of control in the matter in question.

[21] Choice is a contextual defeater for a claim of knowledge, as we might put it. Note that the bilking argument against backward causation relies on this conflict, in the case in which the outcomes of an action may lie in the past.

So far, then, we have a very simple template, characterizing a deliberator's view of the world. In terms of this template, acting, or intervening, is a matter of fixing something not already fixed—of moving something from OPTIONS to FIXTURES, as it were. This deliberative template is linked to, but importantly distinct from, an even simpler epistemic template, that divides the world into KNOWNS, KNOWABLES, and everything else. The link turns on the fact that KNOWNS and KNOWABLES seem to provide a (perhaps *the*) major constraint on what goes into FIXTURES. As we have just seen, what is accessible to us cannot be something we can take ourselves to control. Yet, as we also noted, there is much flexibility: something knowable and hence in FIXTURES under some circumstances, may nevertheless be regarded as controllable and hence in OPTIONS in other circumstances.

In practice, for us, this seems especially true of states of affairs in the future. We plan under certain assumptions about what the future will be like, which we take as KNOWNS—for example, normally, that the sun will rise tomorrow. But this seems to be very context-sensitive: if we want to consider an action that involves eliminating the sun, we won't take the fact that it will rise tomorrow as a given—its rising will be in OPTIONS, not FIXTURES.

Let's think some more about the temporal characteristics of OPTIONS and FIXTURES. We have already seen that the division between the two does not in general line up in a tidy way with the future–past distinction, because much of the future will typically be regarded as FIXTURES, for the purposes of any particular deliberation. But is what is in OPTIONS always in the future? Here we need to be careful, I think. Deliberation seems to presuppose a personal or subjective time for the deliberating agent herself, and in that time, the choices that comprise the relevant DIRECT OPTIONS certainly lie in the personal future—indeed, perhaps this is what 'future' should *mean*, in this context. But this doesn't seem to imply that the material manifestations of these choices need also occur later in time than the process of deliberation itself. We can make some sense of the thought that deliberation might take place outside the temporal arena of the material world altogether—think of deliberation by a god, outside spacetime herself, but able to fix matters of fact by fiat, anywhere within spacetime.

All the same, we human deliberators seem to be firmly embedded in time, in a way that ensures that for us, the 'internal' or personal time of deliberation does line up with the external time of the material world; and hence guarantees that the immediate material manifestations of our choices occur *after* the deliberations of which they are the outcomes. For us, then, the immediate choices that we regard as DIRECT OPTIONS always do lie in the 'external' future, as well as in our personal future.

Our INDIRECT OPTIONS would have to lie entirely in the future, as well, if we took it that what lies in the past goes into FIXTURES by default. As a fact about how we ordinarily think about these matters, this does seem to be the case, or close to the case. Indeed, it seems to me plausible to regard this assumption—call it the *fixed past principle,* or FPP—as something like a piece of naive physics. Certainly, this suggestion about the presuppositions of our causal reasoning provides an immediate explanation of our intuitions about the stargate case—in effect, it is as if we assume we could know the position of the photon, before it reaches the stargate.

Calling FPP a piece of naive physics does not prejudge the issue as to whether it is perspectival. We are familiar with the fact that some parts of naive physics turn out to be more perspectival than others (colour more so than up and down, perhaps, and that more so than weight). All the same, it is easy for considerations of the status of FPP to be pulled in two different directions, with different apparent implications for its objectivity. There is no real conflict, in my view, but it is important not to let either consideration obscure the other.

On the one hand (favouring objectivity), we seem to be able to make sense of ways our world might turn out be in which FPP would clearly fail. With a little conceptual effort, we seem to be able to make sense of backward causation: cases in which the *indirect* outcomes of some of our actions lie in the past. Thus we seem to be able to make sense of versions of Newcomb problems in which one-boxing can be made to seem rational, because we can take ourselves, in effect, to *choose* what is already in the box. And more realistically, it has been suggested (e.g. by me, in Price 1996) that quantum mechanics might deserve to be interpreted in a way that gives us some kind of control over the past. Getting one's head around these suggestions certainly requires a bit of effort, but it seems do-able, at least for most of us; which suggests, as has often been noted, that backward causation is not ruled out on analytic grounds. To that extent, then, FPP seems flexible, at least at the edges, in a way that depends on how the world turns out to be.

On the other hand (favouring perspectivity), this flexibility does not extend to what we might regard as the core of FPP: the assumption that the past is *typically* fixed, whereas the future is not. However much we acknowledge that in a Gold universe our time-reversed cousins would see things differently, we can't imagine our own perspective shifting to align with theirs. Yet as the comparison makes clear, inflexibility does not imply objectivity. The core of FPP remains perspectival. It is inflexible because we can't change our viewpoint, not because there is no viewpoint involved.

Thus I suggest that to the extent that our default deliberative template embodies FPP—the principle that nothing in the past can lie in our OPTIONS, DIRECT or INDIRECT—this is a product of two kinds of contingency. The more basic contingency concerns our own 'situation' in the world. It is characterizable, albeit indirectly, as the respect in which we differ from our Gold universe cousins; and in being grounded on this contingency, FPP is perspectival. The less basic contingency concerns the world itself, and is characterizable as the absence, at least in familiar arenas, of the kinds of phenomena that would be interpreted—by asymmetric agents such as ourselves—as allowing some control of the past. As I have already noted, the hypothesis that FPP provides our default deliberative template offers a ready explanation of our intuitions about the stargate case—and this could be further confirmed, I think, by exhibiting the malleability of those intuitions in the light of proposals about backward causation.

The view that FPP is perspectival does not imply that the fact that we hold the past and not the future fixed is 'up to us', or conventional in that sense. On the contrary, it seems to be determined by factors we certainly cannot change. To a large extent, I have suggested, it seems to be a consequence of our epistemic template. We regard the past as fixed because we regard it as knowable, at least in principle. This is clearly an idealization, but one with some basis in our physical constitution. As information-gathering systems, we have epistemic access to things in (what we call) the past; but not, or at least not directly, to things in (what we call) the future.

Plausibly, this fact about our constitution is intimately related to the thermodynamic asymmetry, at least in the sense that such information-gathering structures could not exist at all, in the absence of an entropy gradient. Although the details remain obscure, I think we can be confident that the folk physics reflected in the temporal asymmetry of our epistemic and deliberative templates does originate in *de facto* asymmetries in our own temporal orientation, as physical structures embedded in time.

These latter asymmetries are not perspectival, of course—they are as objective as the fact that our rocks rolled downhill in a northerly direction.

But these objective asymmetries *in us* underlie a perspectival asymmetry in our causal concepts, in much the way that the objective asymmetry of the rocks underlies the perspectival asymmetry of accessibility. And as in that case, the perspectivity of the concepts is revealed in the fact that they continue to be applied asymmetrically, even in instances that lack any relevant intrinsic asymmetry in themselves. That was the point of the 'stargate' example, in Section 10.5.

10.8 Contingency, Ignorance and Manipulability

At the end of Section 10.5, I noted the strong emphasis on intervention in much recent work on causation. In interventionist terms, the issue of the temporal directedness of causation comes down to the question as to why dependent 'wiggles' typically occur *after* rather than *before* the wiggle on which they depend. We now have two arguments to show that this is perspectival (and thus, as I claimed, that interventions are a Trojan Horse for causal objectivists).

The first argument exploited links to temporal factors, in order, by means of a simple symmetry argument, to make sense of the idea of agents with a different perspective—agents with the opposite temporal orientation, at the other end of a Gold universe. I have offered the following diagnosis. In posing the question, 'What else wiggles, if we wiggle *this*?', we normally simply presuppose that the past does not change, as a given of the enquiry in question. The constraint comes from the needs of the deliberative standpoint, as instantiated in creatures whose typical epistemic access is to things in the past. Idealized in a natural way, I have suggested, this epistemic imbalance requires us to treat the past as fixed, for deliberative purposes. *For us,* then, the notion of a wiggle thus becomes strongly time-asymmetric, in the way revealed by our intuitions about the stargate case.[22]

The second argument puts these considerations into a more general framework. From a purely logical point of view, there's no need to constrain wiggling in this temporal way. I have already suggested that we

---

[22] Objection: Can't we simply *observe* that wiggling a variable doesn't change anything earlier than the time of the wiggle? If so, how can it be a perspectival matter? Reply: It isn't so, for the issue in question always concerns a counterfactual—what would have been the case if we had wiggled, or wouldn't have been the case if we hadn't wiggled. In assessing these counterfactuals, we don't rely simply on observation, for we never observe the counterfactual case.

can make sense of the idea of an atemporal god, able to wiggle the material world in a much less temporally-constrained manner; and in principle, such an example could be modified to 'fix' any set of facts whatsoever. But *something* has to be held fixed, for otherwise the question, 'What changes, if we change this?' has a trivial answer: 'Everything!'

So the strong temporal asymmetry of the notion of intervention—and hence, apparently, of our causal thinking in general—stems not merely from the fact that we are agents, but also from a further contingency concerning our temporal circumstances: above all, the strong temporal bias of our epistemic access to our environment. Our causal intuitions are a product both of general aspects of the architecture of deliberation, and of specific facts about the way in which that architecture in implemented in our own case.

I now want to show that there are two further respects in which the causal viewpoint is necessarily perspectival, also brought to light by an understanding of general features of the architecture of deliberation. Again, both factors have a lot to do with the epistemic constraints on deliberation, though they differ from the constraint we have already encountered in that it is ignorance, not knowledge itself, that does the crucial work.

The first new factor stems from a constraint on deliberation that may seem trivial. As we've already observed, the agent's perspective presupposes lack of prior knowledge of the outcome of the choice in question. The OPTIONS and the KNOWABLES are necessarily disjoint, as we put it earlier. In terms of interventions, this amounts to the observation that to ask in a sensible way, 'What else changes, if this changes?', there must be things we leave open, as well as things we hold fixed. So there is a contingency to the standpoint of the would-be interventionist that depends on what she is ignorant about, and hence can leave open, as well as one that depends on what she assumes known, and hence holds fixed. As I'll argue later, this apparently trivial fact turns out to have important implications for the plausibility of strong forms of realism about causation.

The second new constraint stems from the fact that interventions are supposed to be 'free variables', independent of anything except their effects. As is often noted in the recent interventionist literature, this leads to a puzzle, in cases in which the (physical or biological) agent of an intervention is part of the same closed system as the object 'wiggled'. Interpreted in an objectivist manner, then, interventionism makes the metaphysics of causation hostage to the possibility that there may be no causation, literally speaking, because there are no genuinely open systems.

The perspectival solution to this puzzle rests on an observation due to F. P. Ramsey. Ramsey is famous as a pioneer of pragmatic subjectivism about probability. In one of his last papers, he extends this subjectivist viewpoint to laws and causation. He links the asymmetry of cause and effect explicitly to the perspective we have as agents, saying that 'the general difference of cause and effect' seems to arise 'from the situation when we are deliberating' (1929, p. 146). He then goes on to identify what he seems to take to be the crux of the agent's perspective, namely, the fact that from an agent's point of view, contemplated actions are always considered to be *sui generis*, uncaused by external factors. As he puts it, 'my present action is an ultimate and the only ultimate contingency' (1929, p. 146).

I have argued elsewhere (Price 1992b) that this amounts to the view that an agent thinks of her own actions as probabilistically independent of everything except their effects—as not themselves determined by anything 'further back'. This is where causal chains *begin,* as it were, from the agent's own perspective. And this should be read in reverse, I think. We should explain the genealogy of the notions of cause and effect by noting that we apply the terms, initially, on the following basis: we say that B is an effect of A, when we think that *doing* A would be a way of ensuring B (or increasing the probability of B, in a more general version). This is entirely in keeping with the interventionist insight, of course (though without ambitions to objectivist metaphysics).

I have also argued that this approach provides the most promising basis for a probabilistic theory of causality. Among its virtues is the fact that it avoids the problem of spurious causes: correlations due to common causes don't translate into probabilistic dependencies from the agent's point of view, because the presence of the common cause is incompatible with the assumption of *sui generis* origins. The argument turns on a defence of evidential decision theory against Newcomb-style objections. Indeed, I think the viability of the approach in general depends on this defence, for it is this that ensures that the probabilities in question need only be evidential, and hence not dependent on a prior modal notion (as invoked in causal decision theories). What needs to be shown is that correlations between prior causal states and actions do not translate into evidential dependencies from the agent's perspective; and the crucial point is that in the means–end context any such subjective dependency would itself be a causal factor, so that the principle of total evidence would immediately undermine the judgement on which it was based. (For the details, see Price 1986, 1991, 1992b.) This argument shows that Ramsey's suggestion is coherent in purely evidential terms, I think. As such, it is then available to ground our causal concepts in the way that Ramsey suggests.

For the moment, the relevant point is that Ramsey's 'contingency' is entirely a product of the perspective of a deliberating agent. Its source lies not in the world, but in a certain kind of *ignorance* on the agent's part—roughly, ignorance of the causes of her own actions.[23] Again, then, it is an epistemic constraint on deliberation and intervention.

In the most general terms, then, the argument for the perspectival character of causal concepts goes something like this. Causation has a conceptual tie to intervention, and hence to deliberation. But the possibility of deliberation is epistemically constrained, in several ways—it depends on both knowledge and ignorance, and these epistemic factors are contingencies, whose limits may well vary from agent to agent. Hence causal judgements are correspondingly perspectival: they are necessarily 'situated', relative to some implicit boundaries to the knowledge and ignorance of the agent concerned.

We have now distinguished three epistemic constraints, which all contribute to the deliberative perspective. The first was a constraint imposed 'from below'—a lower bound on an agent's knowledge, providing the main constraint on FIXTURES (and apparently based, in our own case, on our epistemic access to matters in the past). The two new factors are constraints imposed 'from above'—upper bounds on an agent's knowledge, deriving from its necessary ignorance, as it deliberates, both of what it is actually going to do, and of some of the immediate precursors of its decision. Again, these epistemic constraints are more fundamental than their temporally based manifestations in our own case. As before, we can imagine a divine creature, able to intervene at will at arbitrary points of a spacetime arena. Such a creature might have an atemporal notion of causation (the 'effects' of its interventions showing up in various directions, throughout the manifold). But it must share our limitations in one respect, if it is to

---

[23] *Pace* E. F. Schumacher, then, it is ignorance, not knowledge, that makes us free, at least in this case. Knowledge of the causes of our own actions would make us mere spectators.

think of itself as deliberating at all. It must be sufficiently ignorant for the notion of choice to make sense, by its own lights.[24]

It is worth noting that the ignorance required by the two new epistemic constraints seems deeply protected by the structure of the deliberative process. Deliberation always allows a kind of feedback loop, whereby the conclusions at any intermediate stage can feed into the deliberative process itself. This feedback can easily be made self-undermining, in the face of any claim to knowledge either of the causes of the contemplated action, or of the nature of the action itself. As cognitive structures embodying this kind of feedback, there is thus no danger that we are suddenly going to prove too clever for our own good, and have to abandon the deliberative perspective altogether. But the ignorance is there all the same, no matter how ineradicable. And our sense of ability, or control, depends on these epistemic limitations.

10.9 Locating Causal Perspectivalism (I): Comparison with Chance

This conclusion has a surprising corollary: far from being omnipotent, an omniscient creature *could not deliberate at all*. Recall the playing field example from Section 10.2—there, we noted that the referee doesn't see the pitch as the goalkeepers do, asymmetrically divided into the near end and the far end. Similarly, I maintain, an omniscient creature does not see the world in causal terms—if science aims for the god's-eye point of view, then Russell was right, and science has no place for causation. (More on this in a moment.)

This might be thought a *reductio* of the perspectival view. But is it really absurd? On the contrary, I think, for it has a familiar and comparatively orthodox cousin in the contemporary philosophical landscape. Many writers maintain that if the physical world is deterministic, there are no non-trivial chances, or objective probabilities. Thus in a deterministic world all probability is epistemic on this view, and we need it only because we are ignorant. Laplacean gods, omniscient about the complete present state

---

[24] As we noted earlier, the notion of choice seems to presupposes a personal time in which choice takes place. So such an agent cannot be entirely atemporal, even if it occupies a different time dimension than the one in which its god-like interventions manifest themselves in our world.

of the world, and hence able to infer the rest, would have no use for probabilistic notions. My conclusion is that something similar is true of causation. Once we appreciate that agency depends on ignorance, we see that causation becomes epistemic in a similar way. Again, it is a way of thinking about the world that we need because we are not gods.

Indeed, the two cases go hand-in-hand. In the deterministic case, Laplacean gods omniscient about the complete present state of the world cannot be deliberators, because they have inferential access to their own interventions. So they have no more use for intervention-grounded causation than for probability. While in the indeterministic case, a stronger omniscience that gave a creature direct access to any part of spacetime would render redundant objective chance, as much as intervention-grounded causation. Chance, too, thus has a perspective-making tie to ignorance—a fact revealed, I think, as it is for causation, by the strong time-asymmetry of ordinary notions of chance and propensity.[25]

10.10 Locating Causal Perspectivalism (II): Effective Strategies

Another useful way to position causal perspectivalism in the philosophical landscape is to relate it to a well-known discussion by Nancy Cartwright, which explores similar territory in pursuit of different quarry. In 'Causal Laws and Effective Strategies', Cartwright begins with the familiar distinction between laws of association and causal laws. Noting Russell's argument that, as she puts it, 'laws of association are all the laws there are, and that causal principles cannot be derived from the causally symmetric laws of association', she goes on to argue 'in support of Russell's second claim, but against the first' (1979, p. 419). So Cartwright agrees with Russell that '[c]ausal laws cannot be reduced to laws of association', but maintains that 'they cannot be done away with' (1979, p. 419).

Cartwright's argument for the latter conclusion is that causal laws are needed to ground an 'objective' distinction between effective and ineffective strategies. She illustrates this distinction with some 'uncontroversial examples'. Thus:

---

[25] Rightly taken to be grounds for suspicion about the ontological credentials of propensities, in my view, by Carl Hoefer (2004).

[W]hat is, and what is not, a good strategy...is an objective fact....Building the canal in Nicaragua, the French discovered that spraying oil on the swamps is a good strategy for stopping the spread of malaria, whereas burying contaminated blankets is useless. What they discovered was true, independent of their theories, of their desire to control malaria, or of the cost of doing so. (1979, p. 420)

However, Cartwright argues, the 'objectivity of strategies requires the objectivity of causal laws'. In other words, 'causal laws cannot be done away with, for they are needed to ground the distinction between effective strategies and ineffective ones...[T]he difference between the two depends on the causal laws of our universe, and on nothing weaker' (1979: 420). Thus Cartwright contends that our primary reason for believing in the objectivity of causation is that it is needed to draw a distinction we need when we deliberate.

For my part, I have agreed that there is a deep conceptual link between causation and deliberation. However, I have argued that this suggests that causation is perspectival, because the deliberator's viewpoint is necessarily 'partial', or incomplete. So the needs of deliberators are not a good guide to the nature of the world, as seen by God.[26] But what, then, of the distinction with which Cartwright began, between effective and ineffective strategies? Am I committed to denying that this distinction is objective? To maintaining that it is an anthropocentric matter that oiling swamps controls malaria, but burying blankets doesn't?

Recall the lessons of the case with which we began. From the French perspective, it was certainly an objective matter that there were foreigners in Nicaragua. It was independent of French preferences in the matter, for example, or of any theories or projects conceived in Paris. Nevertheless, the distinction the French drew between themselves and '*les étrangers*' was perspectival—a distinction drawn from the French viewpoint. The fact that the presence of foreigners was an objective matter, from that viewpoint, does not imply that there wasn't a viewpoint involved.

Similarly in the case of causation, I maintain. From the homogeneous deliberative perspective that we humans all share, it is an objective fact that oiling Nicaraguan swamps is an effective strategy for reducing malaria.

---

[26] At least not directly, though of course we can still ask how the world has to be, for creatures like us to have a need for these perspectival concepts. And we can still ask the question whose analogue in the Copernican case is this one: What kind of solar system looks like this, from the perspective of observers on the surface of a rotating planet in orbit around a sun? More on this later.

Indeed, it is an objective fact that when the swamps were oiled, that oiling *caused* a reduction in malaria. But again, the objectivity of these matters, from this viewpoint, does not imply that there is no viewpoint involved. It doesn't imply that there is a viewpoint involved, either, of course. To investigate that matter, we need to do the hard work we have been doing in this chapter. In particular, we need to think about the distinctive preconditions of causal judgements, of the use of causal concepts, in search of the tell-tale contingency and variability that is a mark of perspectivity.

I have argued that that investigation reveals that causal judgements are indeed perspectival. And it seems to me that, properly understood, Cartwright's own argument provides additional support for this view. Why? Because if we accept with Cartwright that causal laws are not reducible to laws of association, we face an issue analogous to one we raised at several points for various reductivist and primitivist proposals: namely, that of explaining *why* causation matters in deliberation in the way that it does. The argument goes something like this. If we make causation an objective matter, then either its relevance to us goes via physics, or it is somehow 'direct'. The latter option is mysterious, for reasons we have already canvassed in the case of various other primitivist proposals. The former option is in tension with the plausible Humean view that physics can be cashed in terms of laws of association.

True, the realist will respond that the point of Cartwright's argument is precisely that it shows that this Humean view of physics is mistaken—that the physical world contains primitive causal facts of some kind, not reducible to matters of association. But this reply simply avoids one horn of the dilemma at the cost of the other. Causal facts become mysterious primitives once more, knowable only by their relevance to decision. This dilemma is the real 'hard problem' for causal realism, in my view.[27] It is avoided, however, by a view that *begins* with deliberation, and sees causal judgements as projections and idealizations of judgements made from the agent's perspective.

Realists are likely to argue that the first horn of the dilemma is more attractive than I have claimed. One popular strategy, comparable to David Lewis's strategy in the case of chance, is to compare the access that this

---

[27] To borrow some terminology from Hitchcock (ch. 3 in this volume), who borrows it in turn from David Chalmers.

approach gives us to causation to the kind of access we have to theoretical entities in science. In each case, what we know about the object in question is that it is whatever 'plays a certain theoretical role'—in the case of causation, a role that includes, *inter alia* and centrally, a particular significance for decision.

Because it accords such a central role to the role of causal notions in practice, this approach is in one sense very close to the kind of pragmatic perspectivalism that I favour—how close depends on the prospects for the kind of reference-fixing machinery which is supposed to make it more realistic. And the prospects are rather poor, in my view. Briefly, if the crucial notion of theoretical role is cashed in causal terms, as it is in Lewis's account of reference-fixing elsewhere in science, then the approach is viciously circular in the case of causation itself—in this sense, causation isn't just another theoretical term (see Price (2001, s. 7) for an elaboration of this point). If it is cashed in (non-causally grounded) semantic terms, then, as Stephen Stich (1998) points out in an analogous context, the metaphysical enterprise becomes hostage to a theory of reference that is nowhere in range, above or beyond the horizon. And to what end? The interesting part of the story, the part this approach presumably shares with pragmatists such as me, is a story about how creatures like us come to use causal concepts. Moreover, as we noted earlier with reference to reductionist responses to the reversibility arguments, the arguments that support perspectivalism show that if and when reference does get fixed, in some interesting non-minimal sense, it will be in a way which reveals a relativity to the circumstances of the speaker—which implies that we pick out something different by 'cause' than do our time-reversed cousins in a Gold universe. In all, then, a great deal of work, for no significant benefit—or so it seems to me.

The key point is that by *beginning* with deliberative practice, a perspectival view resolves the mystery that Cartwright's paper brings close to the surface, namely the question why causation should be relevant to deliberation in the first place.[28] Moreover, the 'cost' of perspectivalism, in so far as there is

---

[28] Lewisean realists do just as well in this respect, of course, for they begin in the same place. But they must avoid the 'dormative virtue' temptation, of thinking that their additional metaphysics buys them some deeper account of why causation matters to decision. As I have said, I think it is doubtful whether it buys them anything useful. Certainly knowledge of the causal facts can do no work in explaining or justifying our judgements about effective and ineffective strategies. Once the link between causal

one, is nicely circumscribed by Cartwright's discussion. For if Cartwright is right in thinking that the primary argument for causal realism turns on the apparent objectivity of the distinction between effective and ineffective strategies, then the main task facing perspectivalism is to find some way around that obstacle—to find some way to reconcile perspectivalism with our intuitions about the objectivity of effective strategies. And that, I've suggested, isn't all that hard to do. These are precisely the intuitions we should expect, given the homogeneous nature of the causal perspective.

10.11 Russell's Republic Revisited

I've agreed with Cartwright that causal laws cannot be done away with, because they are needed in deliberation. But for me, this reflects the peculiarities of the deliberative standpoint, rather than a perspective-independent reality. So for me, the conclusion that causal laws are indispensable comes with a qualification: *as long as we continue to deliberate*. In practice, of course, this qualification makes little difference. There is no risk that we humans will cease to be deliberators. But there is still a formal question about the implications of my view to the issue between Cartwright and Russell. Cartwright contends that Russell is wrong to think that causal laws can be eliminated from science. Where does my view stand on this issue?

The answer depends on how deeply embedded in science the causal viewpoint turns out to be, in two senses: whether it is dispensable, and if not, whether it encompasses the whole of science, or just an aspect of science. Concerning the former, it is easy to see how the causal viewpoint might be ineliminable. If explanation is an indispensable element of science, and explanation is essentially causal, then to adopt a scientific stance is necessarily to view one's subject-matter in causal terms. And if causal explanation is not only *a* goal but *the* goal of science, then we have the stronger conclusion. Causal perspectivalism would imply that the scientific viewpoint is *wholly* as well as *ineliminably* 'embedded'.

As I said earlier, this would not amount to a *reductio* of the perspectivalist view. It would do so if a perspectivalist claimed non-perspectival authority

---

facts and facts about effective strategies becomes analytic, we don't have knowledge of the former until we've decided the issue about the latter. The epistemology of causation becomes parasitic on the epistemology of rational decision, rather than vice versa.

for his own pronouncements, for the perspectivalist project is itself scientific—it treats the use of causal concepts as an aspect of human behaviour and psychology, and sets out to provide an explanation, in the same scientific spirit in which we might investigate the emotions, say, or vision. Such an explanation will presumably *use* causal concepts (and related elements of reasoning, such as counterfactuals), but there seems to be no vicious circularity here (no more so, in my view, than in the fact that we use language in theorising about the origins of language). There would only be an inconsistency if the perspectivalist claimed a standpoint more detached than that of science in general, but such a claim seems entirely unmotivated, by the perspectivalist's lights.

Accordingly, the view that all of science is interestingly perspectival would not be inconsistent, so far as I can see. All the same, I don't think that *causal* perspectivalism has this consequence. Not only is there a legitimate place for a non-causal viewpoint in science, in my view, but such a viewpoint seems implicit in the perspectivalist project itself, considered as a scientific enquiry. After all, think about the kind of explanation the perspectivalist is looking for: namely an understanding of how we humans came to employ causal concepts, in order to prosper in environments in which causation is not part of the pre-existing furniture.

As in Section 10.9, the case of probability provides a useful analogy. Think of the project of explaining the practical utility of acquiring epistemic notions of probability, for creatures in a deterministic (and therefore, on most accounts, chance-free) environment. In both cases, as I have argued elsewhere (Price 2004b), a natural way to approach the issue is in terms of what I called 'meta-models'—simplified models of idealized agents in idealized environments, intended to make sense of the question as to what difference it makes to creatures when they model their own environments in causal or probabilistic terms. This approach brings with it the idea of a non-causal or non-chancy reality, within which causal or probabilistic reasoning evolves.

In the present context, then, the point I want to stress is that to treat causal concepts as perspectival, in this naturalistic spirit, is inevitably to theorize about the non-causal world to which we apply these concepts—the bare Humean world, in the midst of which we embedded creatures come to think in causal ways. I have already noted the Kantian character of this project. In these terms, the bare Humean world plays the role of the non-causal 'thing in itself'. But unlike Kant's own *Ding an sich,* it is a realm to which empirical science—here, the naturalized Kantian enterprise of perspectivalism itself—gives us some kind of access.[29] Indeed, as I noted above, Kant's own Copernican analogy provides a model for the investigation in question: we ask ourselves, 'What kind of reality would look like *this,* from the peculiar standpoint we humans happen to occupy?'

In endorsing this enquiry, then, I side with Russell, against Cartwright, in favour of Humean science. But there's another important sense in which I disagree with Russell (though without thereby agreeing with Cartwright). By contemporary lights, it is natural to read Russell as a causal eliminativist. By these lights, the discovery that causation is a harmful relic of a bygone age, as Russell puts it, would be comparable to the discovery that there is no phlogiston, or luminiferous ether. But these are misleading comparisons for the implications of causal perspectivalism, in my view.

Why are they misleading? Because there are two different ways to discover that something isn't as real as we thought. There's the kind of discovery we made with respect to phlogiston, ether, unicorns, leprechauns, and the like; and there's the kind of discovery we made with respect to foreigners. We didn't discover that foreigners do not exist, but merely that the concept *foreigner* is perspectival. And that is where causation belongs, in my view, along with folk favourites such as up and down, night and day, and the rising sun itself.[30]

Thus although I am less realist about causality than Cartwright, I am also a less revolutionary anti-realist than Russell himself. I don't want to eliminate causation altogether from science, but merely to put it in its proper place, as a category that we bring to the world—a projection of the deliberative standpoint. Causal reasoning needn't be bad science, in my view. On the contrary, it is often an indispensable construct for coping with the situation we find ourselves in, as enquirers and especially as agents. It is bad science to fail to appreciate these facts, but not bad science to continue to use causal notions, where appropriate, having done so. Some perspectives simply cannot be transcended.

---

[29] Of course, this would be compatible with the conclusion that there is some deeper sense in which the *Ding an sich* is off-limits to science, for more general reasons.

[30] The difference between causation and these examples is mainly that it is much easier to change perspective by moving around the planet, or moving off it, than it is to alter our temporal and epistemic perspective. That's why causation *seems* more objective, and is more indispensable, in practice. But the difference is one of degree.

By offering a modest, pragmatic, agent-centered view of causation, perspectivalism thus provides an irenic third way between Russell and Cartwright. It forments revolution, but a quiet revolution, in the spirit of Kant's Copernican revolution, that avoids the mysteries of 'monarchist' metaphysics without the anarchic nihilism of causal eliminativism. It dethrones causation, certainly, but saves it, for all ordinary purposes, by revealing its human face—truly, then, a republican compromise.

References

Cartwright, N. (1979). 'Causal Laws and Effective Strategies'. Noûs, 13: 419–37.
Davies, P. (1974). *The Physics of Time Asymmetry*. London: Surrey University Press.
Gold, T. (1962). 'The Arrow of Time'. *American Journal of Physics*, 30: 403–10.
Hausman, D. (1998). *Causal Asymmetries*. Cambridge: Cambridge University Press.
Hitchcock, C. (this volume). 'What Russell Got Right'.
Hoefer, C. (2004). 'Time and Chance Propensities'. Typescript.
Meek, C. and Glymour, C. (1994). 'Conditioning and Intervening'. *British Journal for the Philosophy of Science*, 45: 1001–21.
Pearl, J. (2000). *Causality*. New York: Cambridge University Press.
Price, H. (1986). 'Against Causal Decision Theory', *Synthese*, 67: 195–212.
_____ (1991). 'Agency and Probabilistic Causality'. *British Journal for the Philosophy of Science*, 42: 15–76.
_____ (1992a). 'Agency and Causal Asymmetry', *Mind*, 101: 501–20.
_____ (1992b). 'The Direction of Causation: Ramsey's Ultimate Contingency', in D. Hull, M. Forbes and K. Okruhlik (eds), *PSA 1992: Volume 2*. East Lansing, MI: Philosophy of Science Association, 253–67.
_____ (1996). *Time's Arrow and Archimedes' Point*. New York: Oxford University Press.
_____ (2001). 'Causation in the Special Sciences: the Case for Pragmatism', in D. Costantini, M. C. Galavotti, and P. Suppes (eds), *Stochastic Causality*. Stanford: CSLI Publications, 103–20.
_____ (2004a). 'On the Origins of the Arrow of Time: Why There is Still a Puzzle About the Low Entropy Past', in C. Hitchcock (ed.), *Contemporary Debates in the Philosophy of Science*. Oxford: Basil Blackwell, 219–39.
_____ (2004b) 'Models and Modals', in D. Gillies (ed.), *Laws and Models in Science*. London: King's College Publications, 49–69.
Ramsey, F. P. (1929). 'General Propositions and Causality', in D. H. Mellor (ed.) (1978). *Foundations: Essays in Philosophy, Logic, Mathematics and Economics*. London: Routledge and Kegan Paul, 133–51.

Russell, B. (1913). 'On the Notion of Cause', *Proceedings of the Aristotelian Society*, 13: 1–26.

Spirtes, P., Glymour, C., and Scheines, R. (1993). *Causation, Prediction and Search*. New York: Springer-Verlag.

Stich, S. (1998). *Deconstructing the Mind*. New York: Oxford University Press.

Williams, D. C. (1951). 'The Myth of Passage', *Journal of Philosophy*, 48: 457–72.

Woodward, J. (1997). 'Explanation, Invariance, and Intervention', in L. Darden (ed.), *PSA 1996: Volume 2*. East Lansing, MI: Philosophy of Science Association, S26–41.

—— (2000). 'Explanation and Invariance in the Special Sciences', *British Journal for the Philosophy of Science*, 51: 197–254.

—— (2001). 'Causation and Manipulability', in E. Zalta (ed.), *The Stanford Encyclopedia of Philosophy* (Fall 2001 Edition). http://plato.stanford.edu/archives/fall2001/entries/causation-mani/.

—— (2003). *Making Things Happen: A Theory of Causal Explanation*. New York: Oxford University Press.

11

Counterfactuals and the Second Law

BARRY LOEWER

I will be discussing a kind of conditional (or counterfactual) typically expressed in English by subjunctive conditionals. Here are some examples: 'if I were to strike this match there would be an explosion', 'if there had been an explosion it is likely that I would have heard it', 'if the explosion had not occurred then the window would not have broken'. This kind of counterfactual is intimately connected with laws, explanation, causation, choice, knowledge, memory, measurement, chance, the asymmetry of past and future, and so forth—a veritable *Who's Who* of philosophically and scientifically significant concepts. Philosophers may disagree about the order of explanation among these items and counterfactuals but everyone ought to agree that we would make significant progress understanding them all if we had an account of what makes this kind of counterfactual

> This chapter arose out of hours and hours of discussions between David Albert and myself about statistical mechanics, counterfactuals, chance, and related issues. It was originally intended to be a co-authored paper. But as it emerged I ended up writing it and after consulting with David we decided that it is best if it is authored by myself alone. The ideas in the paper (as will be made clear) are built upon the account of statistical mechanics that David develops so compellingly in *Time and Chance* (Albert 2000). However, the reader should be aware that although there is broad agreement between David and myself concerning the foundations of statistical mechanics there are differences in emphasis. In the book David makes some intriguing suggestions about how counterfactuals are connected to statistical mechanics but doesn't develop the account and at this point he is unsure that he would develop them in the way I do here. It goes without saying, although I will say it anyway, that my thinking on this topic is enormously indebted to David's work. Other philosophers who have connected counterfactuals and statistical mechanics in ways similar to my approach are Adam Elga (2000) and Doug Kutach (2003). My discussion is indebted to their work and to discussions with them. I am also grateful to Frank Arntzenius, Craig Callender, Tim Maudlin, Mathias Frisch, Brad Weslake, and Laurie Paul for comments on various earlier incarnations of this paper. I am especially grateful to Frisch's penetrating criticisms of earlier versions of this paper.

statement true/false; what facts in our world (and worlds like ours) serve as the truth-makers of these counterfactuals. One answer to the analogous question for certain regions of discourse is that their basic statements express fundamental facts that do not supervene on anything more basic. But whatever attractions this kind of view may have for some subject matters (e.g. spatial relations, values, consciousness, and modality) it is off the table here. More basic facts about laws, causation, probabilities, the distribution of fundamental properties, dispositions, our interests and beliefs, possible worlds or something else determine the truth-values of counterfactuals.[1]

The counterfactuals of interest are typically temporally asymmetric. If a counterfactual's antecedent mentions a nomologically possible and relatively local event (or situation) then if that event had (not) occurred subsequent events would or would are likely to have been very different but the course of prior events (typically) would have remained pretty much the same. For example, if Hitler had been assassinated in 1943 the subsequent course of the war and world history would have been very different but history prior to the assassination would have been pretty much the same. Explanation, causation, chance, choice, and memory are also temporally asymmetric. It is tempting to think that all these temporal asymmetries have the same source. But what can that source be? The issue of where temporal asymmetries come from becomes especially puzzling once we realize that the fundamental dynamical laws of nature are—or for all we know might be—temporally symmetric. There is no 'arrow of time' to be found in them.[2] Some philosophers suggest that time itself has an intrinsic direction or orientation that underlies the apparent passage of time and the temporal asymmetries.[3] But even if there is something to this suggestion it is not at all clear how time's supposed intrinsic direction connects with counterfactuals (and the rest) in a way that can explain the

---

[1] Not everyone agrees (Lange, 2005). A referee remarked that it would be interesting to argue for the claim from more basic principles. I agree that it would be but will leave that for another occasion.

[2] More accurately there is no 'arrow of time' in the known dynamical laws that plausibly can account for these asymmetries. Certain processes involving elementary particles (e.g. neutral kaon decay) seem to involve temporally asymmetric processes as a matter of law. But no one seriously thinks that these rare processes can be the source of the temporal asymmetries of counterfactuals and so on.

[3] Maudlin, 'On the Passing of Time' (2005) argues that time possesses an intrinsic direction and suggests that this direction is connected with the direction of nomological explanation and causation.

temporal asymmetries.[4] Of course, one could simply take the direction of time as given and stipulate a way of evaluating counterfactuals (or stipulate that causes must temporally precede effects or that memories can only be of the past etc.) that necessarily supports temporal asymmetries. Rather than a *metaphysical* or a *linguistic* account the project pursued in this paper is to develop a *scientific account of* temporal asymmetries. My approach is to ground the temporal asymmetry of counterfactuals (the direction of causation and the other temporal asymmetries) in contingent facts and laws. Although the fundamental dynamical laws are (or may be) temporally symmetric many non-fundamental and special science laws are temporally asymmetric. Most prominently, the second law of thermodynamics says, roughly, that the entropy of an isolated system (the universe as a whole) never (or almost never) decreases over time. Lewis says that he thinks that there is a connection between his account of counterfactuals and the second law but that he '... does not know how to connect the several asymmetries ... and the famous asymmetry of entropy' (1986). The project of spelling out that connection has only just begun and will be continued here.[5]

11.1 Scientific and Metaphysical Background

In this section I will go over some of the metaphysical and scientific background that is needed for my discussion of counterfactuals.

In the following I assume that *physicalism* is true. I won't try to define 'physicalism' but take it that it entails that all contingent truths, including true counterfactuals and special science laws and causal relations supervene on fundamental physical truths including the laws of physics.[6] Characterizing 'fundamental *physical* property/relation/law' is a problem that has

---

[4] Loewer (2003) argues that an intrinsic temporal direction is neither necessary nor sufficient for grounding the temporal asymmetry of counterfactuals and the others.

[5] Horwich and Sklar both suggest that there is a connection between counterfactuals and the second law (and statistical mechanics) and more recently Albert (2000) has developed the connection in ways that I will build on. A somewhat different account, also connected to Albert's, is taken by Kutach (2003).

[6] Lewis (1999) defines *physicalism* this way: 'Among worlds where no natural properties alien to our world are instantiated, no two differ without differing physically.' A property is alien to a world if it is not instantiated there or at any nomologically possible world. While physicalism has many adherents it is opposed by those who think that some fundamental properties are *mental* and others who think that there are *emergent* macro properties or emergent macro laws.

led some philosophers to say that 'there is no question of physicalism'. However, in my view the project of fundamental physics is to find (or make plausible proposals concerning) the fundamental ontology and laws and I think there is reason to think that this project has met with some success. Physicists currently have an idea (or rather a number of ideas) of what the complete and correct physical ontology may be like. I will be sticking my neck out insofar as I will assume that there is a fundamental physical ontology and there are fundamental laws and that they are close enough to current proposals so as to make no difference to the line of thought that I will pursue in this chapter.

In non-relativistic theories the totality of values of fundamental quantities at a time for an isolated system is the state of the system at that time. For example, in classical mechanics the state of a system consists of the (relative) positions and momenta of all the particles at a time. Fundamental laws come in two varieties. Static laws place constraints (or assign probabilities) on (to) possible states at a time. Dynamical laws place constraints on possible histories. They specify how the state of an isolated system evolves or the objective chances of its possible evolutions over time (as long as the system remains isolated). In classical mechanics the dynamical laws (e.g. Hamilton's equations) specify deterministically how the state evolves. While the ontology of the world and its laws are not now known it is very plausible that the correct account will, in ways that are relevant to this chapter, not be too different from the current candidates. In particular, the correct account will need to account for the approximate truth of classical mechanics, statistical mechanics, and thermodynamics.

Counterfactuals and causation are closely connected so a word about how causation comes into fundamental physics is in order. Russell (1913) observed that the fundamental laws—he was thinking of the differential equations of classical mechanics but the same holds for quantum mechanics—specify how the whole state of an isolated system evolves (or the chances of possible evolutions) but don't specify which parts of the state at one time are causally connected to which parts of the states at other times.[7] He concluded 'The law of causality, I believe, like much that passes muster among philosophers, is a relic of a bygone age, surviving, like the

---

[7] Russell's point has recently been emphasized by Field (2003), Elga (ch. 5 of this volume), and Kutach (2003).

monarchy, only because it is erroneously supposed to do no harm' (Russell, 1913, p. 1). I don't think that Russell meant that we should (or could) cease to use causal locutions but rather that causation is not among the fundamental relations and taking it to be fundamental leads to philosophical confusions. Given Russell's observation one can still maintain that causal claims may be true (or approximately true) and useful but if so their truth or approximate truth supervenes on the fundamental physical laws and truths.[8] Some philosophers (e.g. Lewis) think that causation can be analyzed in terms of counterfactuals (and other fundamental facts) while others think that an analysis of counterfactuals presupposes causation. Alternatively, it may be that neither can be analyzed without reference to the other but that they can be analyzed together in terms of fundamental physical facts and laws. Or it may be that although there is no analysis of causal claims they nevertheless supervene on fundamental physical laws and truths. I don't provide an account of causation in the following but I do propose an account of conditionals that doesn't appeal to causation and is thus suitable to play a role in accounting for causation, perhaps along Lewisian lines.

As I mentioned earlier, most candidates for the fundamental dynamical laws that have been taken seriously by physicists are temporally symmetric. By 'temporally symmetric' I mean, that if a sequence of fundamental events is compatible with the dynamical laws then there is a temporally reversed sequence of fundamental events that is also compatible with the laws. For classical mechanics if one sequence of particle positions is compatible with the laws so is the reverse sequence. As far as the fundamental laws are concerned, one kind of sequence is no more common or likely than its associated temporally reversed sequence. Classical mechanics is like this and so are certain versions of quantum mechanics.[9] So, for example, just as there is a sequence of positions compatible with Newtonian laws in which a ball rolling on a flat surface eventually stops while heating up, there is also

---

[8] Not everyone agrees. Tooley(1987) thinks that there could a duplicate of a world like ours (at least if our laws are indeterministic) with respect to laws and the distribution of fundamental properties but which differ in causal relations. While it might be that our concept of causation allows us to conceive of such scenarios as possible, reflection I think shows that this is not a concept of causation that has relevance for science or is instantiated.

[9] If the physical state in classical mechanics is the positions and momenta of each particle then of course the time reverse of a sequence of such states simply makes no sense. The appropriate time reverse sequence of states reverses the velocities of the particles. Since, plausibly, velocities supervene on positions the fundamental states are the positions of the particles and for any sequences of positions compatible with classical mechanical laws the reverse sequence is also compatible with those laws.

a sequence in which a ball initially at rest spontaneously begins to move while growing cooler.[10] But, of course, we have never seen this happen. Our world is full of processes that seem to be temporally directed. And, it is not just that there are processes that are temporally directed but that the temporal direction seems to be a matter of *law*.

Chief among temporally asymmetric laws is the 'second law of thermodynamics'. The second law (in one of its formulations) says that the entropy of an energetically isolated system never decreases.[11] There are various ways of characterizing entropy. Boltzmann's account is that the entropy of a system is the number (or measure) of the microstates that realize the macro state of the system.[12] A system S's macro state is specified by its composition, volume, temperature, mass density, average radiation intensity and frequency, and perhaps other macro variables. More generally, the macro state of a system at time t is specified by partitioning the spatial region occupied by the system into small spatial regions and specifying the values of macro quantities (average temperature etc.) in each of these regions at t. The macro states (or macro histories) partition the space of physically possible micro states (histories) that behave more or less lawfully. The usual thermodynamic states are like this. Obviously, the notion of *macro state* is vague and there are many precisifications that would serve the purposes of statistical mechanics. With one addition that will be introduced in Section 11.3 it suffices for my purposes to assume that the partition partitions the space of possible micro-states into very small volumes of positive measure.

The second law is exemplified by the melting of an ice-cube in warm water, the diffusion of a gas, the spreading of waves, the clumping of matter by gravitational attraction, and so on. All of these processes are entropy increasing even though they are governed by dynamical laws that allow for entropy decreasing processes. So the question is what explains the second law if its violation is compatible with the dynamical laws? The issue is central to my concerns since many temporally asymmetric processes are connected with increasing entropy.

---

[10] Or, more vividly, run any motion picture backwards and it depicts a sequence of events that are compatible with the dynamical laws.

[11] See Albert (2000) for statements of various characterizations of entropy and versions of the second law of thermodynamics.

[12] This characterization of entropy is Boltzmann's.

Part of the problem of accounting for the second law was solved by Boltzmann. He observed that in a system whose macro state is not one of maximum entropy there are 'many more' microstates sitting on trajectories that realize non-decreasing or increasing (toward the future) entropy than realize decreasing entropy. 'Many more' is not quite right since there are just as many—continuumly many—entropy decreasing as entropy increasing trajectories. The right way to put it (as Boltzmann did) is that the measure of states on entropy increasing trajectories on the usual Lebesgue measure is approximately 1 while the measure of those on entropy decreasing trajectories is approximately 0. Boltzmann understood this measure as, or as determining, a probability over possible states. Since the dynamical laws are (assumed to be) deterministic, exactly how probability should be understood in this context is a problem. I will have a little bit to say about that later.[13] This probability assumption entails that if S is an isolated system (consisting of many particles) it is very likely (probability almost 1) that the entropy of S won't decrease in time. More specifically, it entails that it is very likely that an ice cube placed in an energetically isolated pail of warm water will melt. But while this 'solves' one problem it raises another. The measure of the set of states compatible with the ice cube that are sitting on entropy non-decreasing trajectories *towards the past* is also approximately 1. That is, there are just as 'many' states whose deterministic evolutions are entropy increasing toward the past as toward the future. This means that on Boltzmann's probability assumption it is almost certain that the cube arose spontaneously out of water of uniform temperature. Of course, this is absurd.[14] We could deal with this particular problem by specifying that the macro state of the pail at the time it became isolated contained a larger ice cube in warmer water but this would still make incorrect or no predictions about times prior to that.[15] Another reaction is not to take these probabilities so seriously and employ them only as an instrument for

[13] It is usual to understand statistical mechanical probabilities as degrees of belief (measures of ignorance).

[14] More generally, while the probabilistic assumption provides good predictions for how a system will evolve macroscopically it makes incorrect retro-dictions if employed by itself with respect to the system's past; for example, it redicts that it is more likely that a photograph looking like Bill Clinton emerged as a fluctuation out of dispersed molecules than that it was the result of someone taking a snapshot of the ex-president. Taking this seriously undermines or claims to know the very theories that predict it.

[15] Something along these lines is suggested by Reichenbach (1956). His idea is that the statistical mechanical probability assumption is applied to systems that branch off and become energetically

making predictions about the future.[16] But this begs the question of why this procedure works so well and its working is apparently lawful.

There is a realist and foundational solution to the problem that goes back to Boltzmann. The solution is to add to the dynamical laws a boundary condition specifying the macro state of the universe. This condition—which is called 'the Past Hypothesis' (PH)—characterizes the macro state at, or soon after, the Big Bang origin of the universe. Cosmology suggests that this macro state is one which has very low entropy and which is highly symmetrical (one region is very much like any other).[17] The proposal then is to add to the fundamental dynamical laws the following two claims.

(PH) a statement specifying the macro state of the universe at one boundary (which is assumed to be one with a very low-entropy condition satisfying certain further symmetry conditions).[18]

(PROB) a uniform probability distribution over the physically possible initial conditions compatible with PH; that is, the initial macro state of the universe.

I take PH and PROB to be laws and I will later provide justification for so calling them. I will write $Pr(B/A)$ for the conditional probability function $PROB(B/A\&PH\&L)$; where L are the fundamental dynamical laws (which are assumed to be deterministic). $Pr(B/A)$ is the statistical mechanical probability distribution (the 'SM distribution') of B conditional on A. It is a consequence of these assumptions that as long as the state of the universe is not in equilibrium (maximum entropy) it is enormously likely that its entropy will increase in the direction away from the PH and that its entropy decreases in the direction of the PH.[19]

Figure 11.1 is a depiction of possible evolutions of micro and macro states that should provide an idea of how all this goes. The thin lines represent micro histories, the actual micro history being the straight line

---

isolated from the rest of the universe. One problem with his approach is that it leaves unexplained where these probabilities come from.

[16] Something along these lines is suggested by Leeds (2003).

[17] David Albert (2000). This idea is present in Boltzmann and is assumed by many others (Sklar, Feynman, Lebowitz) but the status of the PH and the consequences of the assumption for matters other than the second law were seldom discussed prior to Albert's book.

[18] These further conditions characterize the initial macro state of the universe.

[19] There is a caveat. Albert (2000, ch. 5) shows that Maxwell's Demon situations are compatible with PROB and PH and so if the world happens to evolve into a Maxwell's Demon situation at t the probability that entropy will *decrease* given the macro state at t is high.

COUNTERFACTUALS AND THE SECOND LAW   301

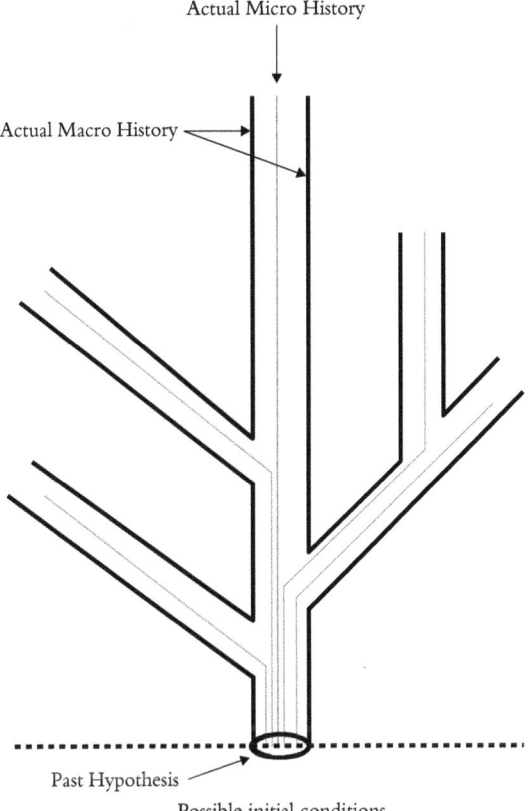

Figure 11.1. The world according to statistical mechanics

up the centre. The circle represents the special low entropy macro state postulated by PH. The circle occupies only a tiny portion of the space of nomologically possible initial conditions. The cylinders represent macro histories, the actual macro history being the cylinder that contains the actual micro history. All the micro histories that begin in the circle satisfy PH and almost all evolve towards a state of maximum entropy. However, there are a 'few' (i.e. sets of measure 0) maverick microstates whose entropy (i.e. the entropy of the macro state it realizes) does not increase, or increases for some time and then decreases, and so on. The interesting point is that although the evolutions of micro histories are governed by deterministic laws, typical macro histories appear to evolve indeterministically. That is, the macro state at t doesn't determine a unique evolution but the probability distribution specifies the probabilities of the histories that realize that macro state.[20] The branching of the cylinders represents the indeterministic evolution of macro histories. From a typical macro state in the middle of the actual macro history there will be branching in both temporal directions but there will be much more branching where the branches have substantial probability in the direction away from the time of the PH than back towards it. The overall structure is due to the fact that the macro state at t (in the middle) must end up in the direction of the boundary condition at which PH obtains (the direction we call 'the past'). There is no similar constraint at the other boundary condition (in the direction we call 'the future'). Of course, given the probability distribution and the dynamical laws it is very likely that the macro state at t will eventually evolve into a state whose entropy is maximum. But there are many many more *ways* in which the macro state at t can evolve to that condition than they can evolve from PH.

The SM probability distribution induces a probability distribution over thermodynamic propositions (propositions about the temperature, average mass distribution, etc. in small regions) and, more generally, over all propositions that supervene on microphysical histories.[21] It is plausibly a consequence of the SM distribution that an energetically isolated (or approximately isolated) system not in equilibrium (an ice cube in a pail of warm water) will (as long as it remains isolated) evolve towards equilibrium (melted ice). Because the initial entropy of the universe was so low and is likely to be increasing, and because the macro states of isolated systems are typically not correlated with entropy decreasing (increasing) micro states, it is overwhelmingly likely that a system that becomes isolated and is not at equilibrium will evolve towards equilibrium.[22]

The SM probability distribution embodies a way in which 'the future' (i.e. the temporal direction away from the time at which PH obtains) is 'open' at least insofar as *macro states* are being considered. Since all histories must satisfy the PH they are very constrained at one boundary condition

---

[20] The probability assignment over initial conditions together with the deterministic laws and the supervenience principles connecting macro to micro states determines a probability distribution over propositions characterizing macro states. So for example, it assigns a probability to the proposition that a particular ice cube in warm water will melt in the next hour.

[21] I am assuming that macro propositions are equivalent to sets of micro propositions and that the SM distribution induces a probability distribution over them.

[22] See Albert (2000).

but there is no similar constraint at other times. It is true that (almost) all histories eventually end up in an equilibrium state (there is a time at which almost all histories are in an equilibrium state) but this is not a constraint it is a consequence of the dynamics and the PH and it is not very constraining (almost all states are equilibrium states). Another feature of the SM distribution when applied to the macro state of the kind of world we find ourselves in is that the macro state of the world at any time is compatible with micro states that lead to rather different macro futures. For example, conditional on the present macro state of the world the SM probability distribution may assign substantial chances both to its raining and not raining tomorrow. On the other hand, there is typically much less branching towards the past. The reason is that the macro states that arise in our world typically contain many macroscopic signatures (i.e. macro states/events that record other macro states/events) of past events but fewer macroscopic signatures of future states/events. Newspapers are much more accurate in recording past weather than in predicting future weather. Of course these two features of the SM distribution—that histories are very constrained at one boundary condition but not at other times and that they branch much more to the future (direction away from the PH)—are related.

Albert shows how the assumption of the PH (and the consequent branching structure) allows for the production of localized macro records of past events while there are not comparable 'records' of future events. For example, a footprint on the beach at t_2 is a record of someone having walked on the beach earlier. Albert points out that the inference from the footprint to the walker assumes something about the state of the sand (e.g. that it was soft and damp) at an appropriate earlier time. Without the assumption of the state of the beach at t_1 (or something that plays that role) we are not justified in making the inference. More generally, Albert says that a localized macro state R at time t is a record of an event E at time t' if it can be inferred from the laws of physics and the macro state S at a time t'' (where t' is between t and t'') that E occurred (or probably occurred). He thinks of S as a kind of 'ready condition' that makes the production of the record possible. In the example, the sand's being damp is the ready condition. The PH plays the role of a ready state for our universe that allows the production of records. If PH were dropped then retrodictions from local (or even universal) current macro states would be wildly inaccurate.

In view of the above the SM probability distribution together with the temporally symmetric dynamical laws determines a temporal asymmetry with respect to the probabilities of macro histories of the universe. As long as the macro state of the universe is not at equilibrium it is very likely to increase in entropy in one temporal direction and decrease in entropy in the other direction. To be clear, nobody claims (and it is not true) that the PH and the dynamical laws are themselves sufficient to account for the existence of recording systems let alone the particular records that have been formed in our world. What the PH does is to remove an obstacle to there being the accurate local macroscopic traces of conditions prior to t by serving as, in Albert's words, 'the mother of all ready conditions'.

Before continuing I want to set aside one possible misunderstanding. It might be thought that the SM distribution already assumes a future/past distinction since it is characterized in terms of 'the past hypothesis'. But it doesn't. It does assume that a certain probability distribution holds over all micro states and that this probability distribution has a temporal asymmetry built into it. But this does not assume that one of the boundary conditions is temporally prior to the other. That one of these but not the other is the temporally earliest condition is something that has to be earned. It is partly earned by the fact that entropy increases in one direction. But this is only the beginning of an explanation of temporal asymmetries that ground the distinction between past and future. To the extent that the other temporal arrows can be explained in terms of the probability distribution that grounds the asymmetry of entropy increase the PH will earn the title 'the *past* hypothesis'.

What is the metaphysical status of PROB and PH? There is a great deal of controversy concerning how to understand probabilities in statistical mechanics. The usual view is that they are 'ignorance' probabilities, that is, the degrees of belief that one ought to have concerning the initial conditions of a system given that one knows that the system satisfies certain constraints (e.g. the system's temperature). There is little discussion of the status of PH but it is generally construed as a *merely contingent* (i.e. not law-like) statement. I think that both of these interpretations are mistaken. PROB and PH (together with the dynamics) are best considered to be laws and the SM probabilities to be objective. The reason is that they underwrite many of the asymmetric generalizations of the special sciences especially those in thermodynamics and these generalizations are considered to be laws. PH and PROB can bestow lawfulness on them only if they themselves are laws and if PROB is a law then the probabilities it posits must be objective; not merely degrees of belief. It is interesting that Lewis's Humean account of laws is friendly to the proposal that PH and PROB are laws. I don't have space to go into detail here but the basic idea is this. According to Lewis a law is a contingent generalization entailed by the Best System of the world. The Best System is that true theory couched in a vocabulary that includes a conditional probability function, mathematical notions, and terms that represent fundamental properties (Lewis calls them 'perfectly natural properties'). What makes it *BEST* is that it best combines simplicity, informativeness and fit. Adding PH and PROB to the dynamical laws results in a system that is only a little less simple but is vastly more informative than is the system consisting only of the dynamical laws. Lewis's official view about objective probabilities is that they are dynamical transition chances. But his account of laws and probabilities actually makes it easy to see how probabilities can also be assigned to initial conditions as PROB does.[23]

There are two further reasons that count in favor of counting PH and PROB as laws. One is that, as I will explain later, these hypotheses play a role in evaluating counterfactuals that is similar to the role that laws play. Secondly, if PH is taken as a non-lawful contingency then relative to the statistical mechanical probability distribution it is enormously improbable. This observation has led Price (2003) to think that this 'enormously improbable' initial condition requires some explanation.[24] It is easy to find this very puzzling since obviously no causal explanation is possible for an initial condition. The problem vanishes if PH is a law since it then constrains the probability distribution so that its probability is 1. PH and PROB are no more in need of explanation than is any other law.[25] My

---

[23] On Lewis's official account simplicity and informativeness are measured relative to the language whose predicates correspond to what he calls 'perfectly natural properties'. And on his official account of chance objective chances different from 1 and 0 are incompatible with determinism. Our proposal involves measuring informativeness relative to the thermodynamic language as well as the fundamental language. Once that is done it is a natural consequence that there are macro laws and objective deterministic chances. Loewer (1996) defends Lewis's account of laws, Loewer (2004) defends Lewis's account of objective chance, and Loewer (2001) argues for the claim that statistical mechanical probabilities can be understood in terms of Lewis's account of chance.'

[24] For a discussion of this issue see Price (2003) and Callender (2003).

[25] Of course if there is an explanation of PH and PROB from some dynamical or other laws that would all be to the good and completely compatible with Lewis's account of laws and our accounts of statistical mechanics and counterfactuals.

view is that the fact that Lewis's account of laws naturally counts PH and PROB as laws and his account of probability provides an objectivist account of statistical mechanical probabilities provides strong support for Lewis's accounts.[26]

What is the epistemological status of PROB and PH? As is the case for any fundamental law (or general scientific hypothesis) there is no question *of proving* PROB and PH. They are to be thought of as *conjectures*. The question then is whether so far they have yielded correct predictions and what further investigations can be carried out to 'test' them. The first point to note is that the Boltzmann probability assumption when used to make predictions about the future of thermodynamic systems has enormous support within statistical mechanics. But without PH the probability assumption is clearly false (and indeed inconsistent since it can only apply at one time). PH overcomes that problem in a way that plausibly preserves statistical mechanical predictions while avoiding the problem of higher entropy past. Further, the assumption of a low entropy and symmetrical macro state at the origin fits in well with what cosmology says about the early universe and so has independent support. I think these are strong reasons to take PH and PROB very seriously. Also, it appears that PROB provides correct objective probabilities in situations that are not obviously connected with thermodynamics. For example, it is plausible that when we conditionalize the outcome of a coin toss on the macro state (or on the macro state in the vicinity of the coin toss) at the time of the toss it is equally likely that a flipped ordinary coin will land heads as that it will land tails. The reason is that very small differences in the micro states (i.e. initial position of the coin and its initial momentum) compatible with the macro state associated with the coin toss lead to differences in the outcome and it is plausible that the SM distribution assigns equal probability to the 'heads' and 'tails' resulting states.[27] Similarly, it is plausible that PROB conditionalized on the relevant macro state provides correct probabilities for various chemical, biological and meteorological phenomena. The fact (as Albert argued in Albert (2000) and I am arguing in this chapter) that PROB and PH can ground the existence of local records, the inferences we make about the future and the past, inferences from one part of the present

---

[26] For further discussion of these points see Loewer (2004), Callender (2003), and North (2004).

[27] Strevens (2003) contains an illuminating discussion of how macro probabilities emerge from micro probabilities and dynamics.

macro state (one record) to the existence of another, and the account of counterfactuals that I will later sketch all provide strong reasons to take this account with the kind of seriousness appropriate to any very general account of the world that has so far been scientifically successful and renders coherent relations between fundamental laws and ontology and aspects of, in Sellars' expression, the 'manifest image'.

For all the reasons above I take the system of the world based on PH, PROB, and the dynamical laws very seriously. But it must be admitted that a lot more needs to be done to make it persuasive that all objective probabilities—the probabilities involved in gambling devices, natural selection, and so on—ultimately derive from the SM distribution by conditionalizing on appropriate facts. However, it must be admitted that on further investigation it might turn out that the correct probability distribution is not PROB (but one close enough so that it also grounds thermodynamics) or that there is no single objective probability distribution or any at all. But, and this is emphasized, there is as of now, as far as I know, no evidence that disconfirms PROB and PH and a lot of intriguing support for them.

Here is a summary of the physics background. The problem that confronted Boltzmann is that the fundamental dynamical laws of our world are temporally symmetric but the laws of non-equilibrium thermodynamics, and in particular the second law, are temporally asymmetric. The problem is solved in a way that avoids the reversibility objection by adding PH and PROB as laws. These hypotheses go far beyond that in that they determine an objective probability distribution over all nomologically possible micro histories (and *a fortiori* over all macro histories and all macro propositions). Even though the underlying micro dynamics is deterministic, macro-histories form a tree structure branching towards the future (away from the time at which PH holds). This suggests that the temporal asymmetry of counterfactuals and all the other temporal asymmetries connected with it are grounded in the temporal asymmetry of the statistical mechanical probability distribution.

There is another account of statistical mechanical probabilities that I want to mention briefly. It has been suggested (Albert 2000, ch. 8) that the GRW version of quantum mechanics grounds statistical mechanical probabilities. His idea is that the random GRW jumps are much more likely to move a system not at equilibrium onto an entropy increasing rather than an entropy decreasing trajectory, since the measure of the former is so much greater than the latter. Thus statistical mechanical probabilities can be reduced to quantum mechanical probabilities. If PH is added to GRW we obtain a branching tree structure that is like the tree depicted in Figure 11.1 (although, of course, the possible micro histories will also form a branching structure unlike the deterministic dynamics depicted in Figure 11.1).[28]

11.2 Lewis's Truth Conditions for Counterfactuals

Let's get back to counterfactuals. My interest in counterfactuals is not the same as that of the linguist, or philosopher of language, who aim to construct an account of the semantics and pragmatics of the natural language sentences that express counterfactuals. That project faces many daunting difficulties. English counterfactuals are expressed in a number of grammatically different ways, there are many kinds of conditionals, counterfactuals are vague, they are plausibly context relative, they have Gricean implicatures and so forth. The semantics and pragmatics of ordinary counterfactuals is a messy matter. Lewis's approach, which I will follow, is to ignore most of these difficulties. The justification for this is the assumption that there is a conditional that works more or less like core cases of counterfactuals and that is centrally related to causation, laws, and the direction of time, choice, and so forth. It is this conditional that I am aiming to articulate. Whether this approach is justified will depend on whether there is a satisfactory account of a conditional that plays this role. I will use A > B for the counterfactual conditional that I am after. On Lewis's classic account A > B is true iff either there are no possible worlds at which A is true or there are A&B worlds that are more similar to the actual world than any world at which A&−B is true. Soon after Lewis published his account Jonathan Bennett and Kit Fine objected that it yields results at odds with counterfactuals we readily accept.[29] Fine's example is this: 'If Nixon had pushed the button (say on 1 January 1973) there would have been a nuclear war' apparently is evaluated as false by the account since a world in which Nixon pushes the button but somehow there is a failure in the missile launching system

---

[28] We could also add to this account a probability distribution over initial conditions but almost any initial distribution will be soon 'washed out' and the probabilities will converge on the usual statistical mechanical ones.

[29] Fine (1975) and Bennett (1974).

COUNTERFACTUALS AND THE SECOND LAW 309

is more similar to the actual world than a world in which there is a nuclear war. Lewis responded in his paper 'Time's Arrow' that Fine and Bennett were thinking of 'similarity' as something like 'overall similarity' but that it is similarity only in certain special respects that is relevant. The question, of course, is what are these respects?

Before discussing Lewis's answer, I want to describe a more recent although less ambitious proposal due to Jonathan Bennett.[30] Bennett suggests that, when considering worlds in which A happens at time t, the worlds that are most similar to the actual world @ are ones that are very similar to, or perhaps even identical to, @ from its first moments until some time shortly before t when (assuming A(t) is false) they diverge from @ and evolve in conformity to the laws so as to make A(t) true. Bennett calls the point of divergence 'a fork' and the segment of history leading up to A(t) a 'ramp'. He says that while the fork may involve a region in which the laws of @ are violated it would not *appear* at a macroscopic level to involve anything astonishing and, in particular, would not be taken by someone who had access only to macroscopic facts to involve a violation of the laws that hold at @. The fork should occur as late as possible so as to minimize gratuitous differences between the counterfactual world and @. Bennett's view can be pictured as follows:

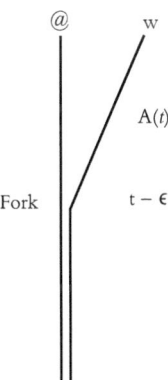

Figure 11.2

I think that Bennett's account does a pretty good job of characterizing a conditional that matches core uses of the counterfactuals that interest us.

[30] Bennett (2003).

For example, it seems to validate 'If Nixon had pushed the button there would have been a nuclear war' and, generally, it underwrites the temporal asymmetry of the kind of counterfactuals we are investigating. However, Bennett's procedure for evaluating counterfactuals *assumes* the distinction between past and future (since forks are to the future) and so it does not provide a scientific explanation of time's arrows.

Lewis's account does not build temporal asymmetry into his characterization of similarity and so holds the promise of explaining (rather than assuming) the temporal asymmetry of counterfactuals. His initial account assumes determinism. In evaluating which of two worlds is more similar to the actual world, four considerations in order of importance are relevant.

(i) The occurrence of big and widespread 'miracles' (violation of laws of the actual world).[31]
(ii) The size of the region in which the fundamental facts fail to match exactly.
(iii) The occurrence of small and very local 'miracles'.
(iv) The size and extent of dissimilarities with respect to fundamental (and other?) facts.

The last of these plays no role (Lewis expresses doubt in its relevance) in my discussion and so I will drop it. The three remaining factors determine a family of similarity relations depending on the relative weights attached to each.[32] Lewis suggests that the vagueness and context sensitivity of counterfactuals corresponds to indecision and shiftiness of the weights credited to these factors. Here is how he thinks the account applies to Fine's example. He considers four worlds at which Nixon pushes the button as candidates for most similar to the actual world @:

(1) w1: worlds which match the actual world exactly until a short time before the button pushing when there is a 'small miracle'—a small violation of @'s deterministic laws that leads to a decision to push the button and then by the laws of the actual world to the button being pushed and to a nuclear war.
(2) w2: worlds that conform perfectly to the actual laws but (since we are assuming determinism) differ from the actual world for all times and so over a large region.

---

[31] It may be that not only the size of the region but the extent of the violation is also relevant.
[32] Lewis says that the third of these is 'of little or no importance.'

(3) w3: worlds like w1 but at which a small miracle after Nixon pushes the button lead to a state that although isn't an exact match to the actual history (from a time shortly after the button pushing) is very similar to the actual world.
(4) w4: worlds like w1 but at which a miracle after Nixon pushes the button leads by the actual world's laws to exact convergence with the actual world.

Lewis claims that all the legitimate ways of weighing the three considerations for evaluating world similarity count w1 as more similar to @ than w2 since in w1 the region of match is so much greater than in w2 while the region of violation of the actual laws in w1 is small. He also claims that w3 is less similar to @ than w1 since the small differences between

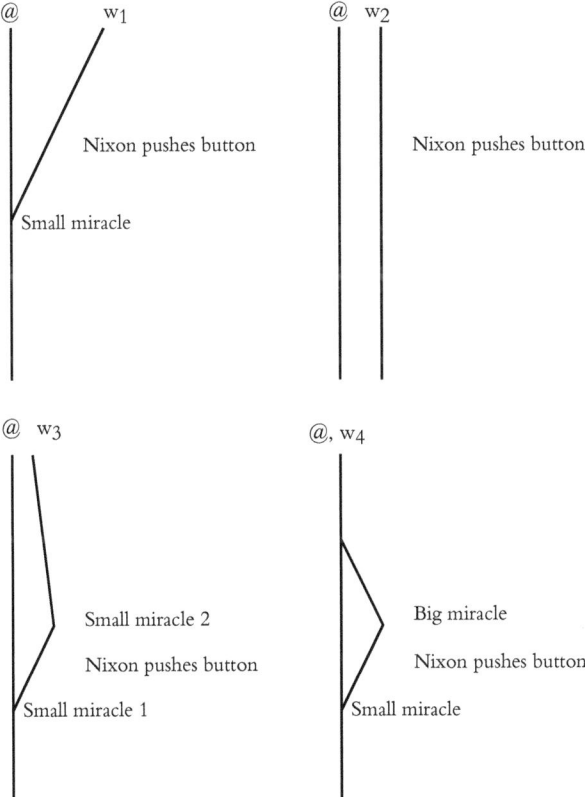

Figure 11.3

w3 and @ at times shortly after t inevitably amplify to great differences at later times. w4 would count as at least as similar to @ as w1 if the size of the miracle required to obtain exact convergence were small. But Lewis claims that certain features of @ require that the violation of the actual laws that occurs after t must be very big and widespread to bring about exact convergence. So w4 loses out to w1 in the contest for similarity. He claims that as a consequence Fine's conditional is evaluated by the account as true.[33]

Lewis thinks that the temporal asymmetry of counterfactuals is explained in terms of the asymmetry of 'miracles'; it takes only a small miracle for a world to fork from @ but a big miracle is required to achieve exact convergence to @. His reason for thinking that exact reconvergence requires a large violation of law is that he thinks that in @ there is an asymmetry of 'over determination'. He says that events at a later time over determine events at an earlier time but not *vice versa*. In the example, it is plausible that Nixon's pushing the button at t is compatible with the state of the world outside of Nixon's brain just prior to t. Once he has pushed the button there will be many dispersed traces in the form of radiation of various kinds that within a second occupy a sphere with a radius of 186,000 miles. This suggests that it takes only a 'small miracle' to result in Nixon's pushing the button but a large miracle is required to erase all the traces of that button pushing. Popper describes another vivid example. A rock is tossed into a quiet pool producing concentric waves. The rock's hitting the water is over determined by the many waves but the waves are determined only by the rock's hitting the water. Popper's idea is that an event e (or at least the kind of event that is referred to in an ordinary counterfactual) has many widely dispersed effects f1, f2, f3, ... that together

---

[33] I have heard it said that since according to Lewis the truth conditions of counterfactuals invoke possible worlds their truth-makers involve other worldly events, laws, and so on. One hears the objection 'What does the fact that a match (even if it is a counterpart) lights in some other world have to do with whether *this* match in *our* world would light if struck?' But the objection is a mistake since it is a mistake to think that on Lewis's view the truth-makers of counterfactuals are other worldly items. On his account the *truth-makers* of counterfactuals are actual fundamental physical facts and laws. The truth-maker of a (non-trivial) true counterfactual is a very complicated (and highly disjunctive) fact about the laws and fundamental physical facts. Possible worlds (and the similarity relation) are a device for characterizing those facts but are not part of the counterfactuals truth-maker. Of course Lewis does believe that there are many causally disconnected universes (his possible worlds). But someone who thinks that possible worlds are properties or sets of propositions or fictions or even someone who takes modality as primitive can still accept his account of the *truth-makers* of counterfactuals.

with the laws determine it—that is, each of the fs nomologically imply e but over determination of future by prior events is relatively rare. If this is what he means by 'determine' then (assuming determinism and laws like those of classical mechanics) he is wrong. For a localized event e(t) and time t' there will be a unique nomologically necessary and sufficient condition f(t') for e(t). Further, f(t') will typically be highly non localized; not the sort of condition we think of as an event at all. However, I think Lewis's intuition is correct when he says that events in our world typically leave many traces but have few predictors. Underlying this is the SM distribution and its role *in allowing* for the production of local macroscopic traces. So Lewis was onto something when he suggests that the asymmetry of over determination may have something to do with 'the famous asymmetry of entropy'.[34] To bring out the role of the SM distribution I will first discuss a problem for Lewis's account that has recently been highlighted by Adam Elga (2000).

Elga argues that Lewis's account fails to ground the temporal asymmetry of counterfactuals and so delivers incorrect verdicts for the counterfactuals that interest us. On his account the temporal asymmetry of counterfactuals is supposed to be grounded in the temporal asymmetry of 'miracles'. But one may wonder where this temporal asymmetry of miracles can come from when the fundamental laws are temporally symmetric and considerations of perfect match are indifferent as to whether the regions matched are in the future or the past. Bennett seems to have worries along these lines. He mentions that there may be a world u to which another world v converges by a small violation of the laws of u. 'Easy convergence' is at least a possibility. But this doesn't show that Lewis's account is mistaken since a world to which easy convergence is possible may be very unlike @. But the existence of a world in which a typical local antecedent (e.g. Nixon pushes the button) that converges to @ via a small violation of law would be devastating for Lewis's account. That there are such worlds was pointed some years ago by David Albert and has recently been nailed down beautifully by Adam Elga.[35] Elga's argument goes like this. Let the actual world @ have a first and last moment and let @* be the time reverse of

---

[34] Lewis (1986) remarks 'I regret that I do not know how to connect the several asymmetries I have discussed and the famous asymmetry of entropy' (p. 51).

[35] Albert made this point in a seminar at Princeton in 1996. Elga spells out this problem for Lewis's account in (Elga 2000).

@.³⁶ The dynamical laws of @* are the same as those of @. @* evolves in conformity with the laws until a time t^* *prior* to the time t (the time at which Nixon is supposed to push the button). At t^* there is a small violation of the laws which leads to a fork on which Nixon pushes the button at t; that is to say, there are particles that are so located so that they constitute a Nixon counterpart whose hand is on a button connected to nuclear missiles and so on. Call the resulting world w6. Now let w6* be the temporal reverse of w6. w6* has a past very different from @. It starts off in a state of high entropy. As it evolves towards time t entropy decreases as particles in it arrange themselves into the shape of Nixon pushing the button. Other particles and fields are so located that all it takes is a small miracle a time, a little after t, to get this world to match exactly the actual world; no nuclear war, no traces of Nixon's button pushing, no traces at all of w6*'s past but rather 'records' of the history of @. It then evolves just as the actual world towards higher entropy. w6* is a strange world but it conforms to the actual dynamical laws everywhere except for a small region.

If the laws are restricted to the dynamical laws (as Lewis thinks) then Lewis's account of similarity counts w6 at least as similar to @ as w1. It matches @ in all of its future a short time after t and conforms to the laws

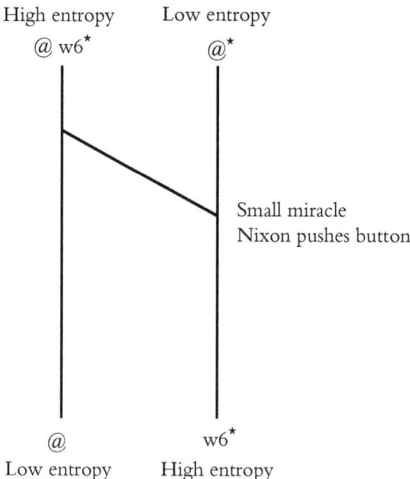

Figure 11.4

---

³⁶ In classical physics this means that the positions of particles at @*'s first moment matches @'s final particle positions and scalar field values but velocities and field vectors are reversed.

of @ except for a small region. This has disastrous consequences for Lewis's account. It means that his characterization of similarity fails, contrary to his belief, to underwrite the temporal asymmetry of counterfactuals and so fails to account for the counterfactuals connected with causation, decision, and so on. The problem is that @* is an anti-thermodynamic world. Whereas in @ the entropy of isolated systems (almost) never decreases in @* entropy (almost) never increases. w6* is like @* in that initially it is in a very high entropy state (like @*) and evolves towards lower entropy until, by virtue of a small miracle, it matches the actual world and then evolves as it does towards higher entropy.

In light of the discussion of statistical mechanics, this problem for Lewis's account should come as no surprise. The thing to do to repair Lewis's account is to add PROB and PH. The simplest way of doing that is to stipulate that conformity to PH and PROB are important in evaluating similarity to the actual world. The offending words @* and w6 will then count as less similar to @ than is w1. Adding them in this way provides another reason (in addition to their simplicity and informativeness) to regard PH and PRB as laws since they play the role of laws with respect to counterfactuals. This solution should be especially pleasing to those who favor Lewis's Humean account of laws since it counts PROB and PH as laws in virtue of their inclusion in the Best Theory of the world. It is not at all clear that alternative accounts can make sense of restrictions, especially probabilistic restrictions, on initial conditions as laws. I call Lewis's account that includes PH and PROB as laws 'the amended Lewis account'.

The amended Lewis account is not open to the Albert-Elga objection but there are still problems with it. It is plausible that there are worlds that satisfy the past hypothesis, match the actual world exactly until a short time prior to Nixon's pushing the button and then diverge by a 'small miracle' from the laws of the actual world and then reconverge by another small miracle to the actual world; that is to say, worlds like w4. Lewis is correct in thinking that almost all worlds are not like that but general statistical mechanical considerations suggest that there are worlds in which a small violation of the actual laws lead to Nixon's button pushing as in w1 but in which unlike w1 no records are produced. If so, only a little miracle is required for reconvergence. I can't prove that there are worlds like this (genuine proofs in statistical mechanics of complex systems are hard to come by) but the worry is that there are trajectories in phase space that satisfy the second law for a long time and then realize anti-thermodynamic behavior. Given PROB and PH the set of such worlds has a very tiny measure but the existence of any such worlds would be a big problem even for Lewis's account amended. If there is a single world that is like w4 then on Lewis's account the Nixon counterfactual will be evaluated as false. The remedy is to evaluate counterfactuals in a way that connects them more closely to the SM distribution.

11.3 The SM Account of Counterfactuals[37]

In this section I characterize a conditional—I call it the SM-conditional—that is temporally asymmetric, behaves in certain ways like Bennett's and Lewis's conditionals, and tracks the statistical mechanical probability distribution. The SM conditional that I will presently describe is one of many conditionals that respects the probabilistic structure induced by PBOB, PH, and the dynamical laws. It is offered here in a tentative way as part of an effort to see whether it captures that structure in ways that connect with the core counterfactuals that they are themselves connected with causation, choice, and memory and so on.[38]

Central to the account of SM conditionals are a special class of conditionals that involve decisions. These conditionals have the form 'If P were to decide to A then the probability of B would be x'. I make two assumptions about decisions. First that they are localized events (or states) in a person's brain that are smaller than macroscopic events but have positive probability and secondly that they are correlated with motions of her body. For a given macro state of a person's brain the decisions 'open' to her are those decisions that are compatible with the macro state of her brain. If determinism is true then the microstate of the world prior to my making the decision determines what decision will be made. But let's suppose (perhaps this is merely a

---

[37] A different account of conditionals and their relationship to statistical mechanics is developed in Kutach (2003). I don't compare the accounts here but will do so in a subsequent paper.

[38] At this point Albert and I slightly part company. He (as have I) insists that any procedure for evaluating counterfactuals will need to respect and reflect PROB and PH. But he has been steadfastly agnostic in print about exactly what that procedure might look like. In talks and seminars, Albert has sometimes been willing to entertain counterfactuals with determinate consequents, but as we will see the account presented here is restricted to counterfactuals with probabilistic consequents. There are a few other places where my account sticks its neck out in ways that go beyond what Albert and Albert and I have written and said in talks.

fiction) that even if determinism is true, a person can directly control which decisions she makes. Further, I will suppose that given the macro state M of the world (including the agent's brain) the various decisions that are available to her are all equally likely.[39] Decisions are thus indeterministic relative to the macro state of the brain and environment prior to, and at the moment of, making the decision. This *indeterminacy* captures the idea that which decision one makes is 'open' prior to making the decision.

Now suppose I am choosing between alternative decisions $D_1(t')$ and $D_2(t')$. If I knew the statistical mechanical probabilities $Pr(B/M(t)\&D_1)$ and $Pr(B/M(t)\&D_2)$ for various Bs, I would be in a position to know the *objective* probabilities of D_1 and D_2 each leading to B. SM conditionals express information about these objective probabilities. Here is how.

Consider future subjunctive decision conditionals of the following form:

($) If at t I were to decide D_1 then the probability of B would be x.

My proposal is that ($) is true if $Pr(B/M(t)\&D(t)) = x$.[40] For example, the conditional 'if I decide to bet on the coin's landing heads then the chance I would win is 0.5' is true at t iff the statistical mechanical probability of winning given the t macro state and my decision is 0.5.[41]

Decision conditionals are temporally asymmetric. They inherit the temporal asymmetry of the SM distribution. Alternative decisions that can be made at time t typically can make a big difference to the probabilities of events after t (i.e. events further away from the time at which PH holds) but make no difference to the probabilities of macro events prior to t. The reason for this is the predominance of local macro signatures of the past (but not of the future). It also depends, of course, on our biological structure on which very small differences in the brain get *magnified* into differences in bodily movements and these, in some cases (e.g. the Nixon example) get magnified into vast differences in the world. So it is enormously plausible that decision conditionals are temporally asymmetric. Whether

---

[39] The assumption that each possible decision is equally likely is certainly false but I don't think this simplification affects the following account.

[40] The reason we are interested in evaluating the conditional in terms of Pr(B/M&D) is that the *macro state* is a natural limit on the extent of accessible information and for typical conditionals this probability will be approximately equal to Pr(B/D&K) where K is a very small part of the *macro state*. In our example, for a typical coin flip everything but the nature of the flip is irrelevant to the probability.

[41] Gambling devices are designed so that generally the macro state external to the gambling device itself is irrelevant to the probabilities of outcomes. But, of course this isn't always correct.

I conditionalize on D1 or D2 then the probabilities of macroscopic past events *insofar as they are recorded* in the present will be unaffected but the probabilities of the macroscopic future may very well be very different.

Here is a worry. Suppose that K is a statement that an event k occurred prior to time t and there is no macroscopic signature of K at t so $Pr(K/M(t))$ is very small and $Pr(-K/M(t))$ is high. For example, k might be the event of the destruction of Atlantis (supposing it to have occurred and left no macroscopic traces). Then $Pr(K/M(t)\&D)$ will be very small for all decisions (since these neither destroy nor create records) as well. This might lead one to think that the account says you could eradicate Atlantis by deciding to lift a finger. But of course this is not so. By lifting your finger you do not alter the probability of K at all. On the other hand you, or rather the president, by lifting his finger at *t* might well destroy the world at a time after *t*.

Here is another worry. It is a consequence of this account that the probability of the past micro state is correlated with present alternative decisions. But this does not mean that a person can affect the past in the sense of having control over past micro events. Control by decision requires that there be a probabilistic correlation between the event of deciding that p be so and p being so and one's knowing (or believing with reason) that the correlation obtains. But it is immensely implausible that there is any past micro state *m* that fulfills the first part of this condition let alone both parts. So while it is true on the account that if Nixon had pushed the button the probability that the past would have been different in some micro respects is 1. But since we have no idea what those respects are we have no control over past micro conditions.

Note that the temporal asymmetry of decision conditionals is derived from the asymmetry of the SM probability distribution since the asymmetry of local macro signatures is derived from the SM probability distribution. It is not 'put in by hand' as it is in Bennett's account. The asymmetry of these conditionals is part of the explanation of why our decisions can make a difference to the future but not the past and that partly explains our feeling that the future is 'open'.

Decision counterfactuals are a very special kind of conditional. What about conditionals whose antecedents are about a non-actual past decision and conditionals whose antecedents are not decisions? Consider, for example, 'If I had decided to flip this coin at noon yesterday the probability it would have landed heads is 0.5.' The obvious suggestion is that to evaluate it we look at the macro state M(noon). If the probability of the coin's landing heads conditional on M(noon) and my deciding to flip it is 0.5 then the conditional is true. It is plausible that the probability that Kennedy served out his term given the macro state when Oswald decided to pull the trigger is pretty high. It remains high conditionalizing on the micro-event of Oswald's deciding at the last moment not to pull the trigger. If so the SM conditional 'if Oswald had decided not to pull the trigger it is likely Kennedy would have served out his term' is true.

So far I have discussed conditionals whose antecedents are decisions. This enabled us to keep the macro state fixed while altering the microstate so as to realize the alternative decision. Because the macro state can be preserved these alternatives don't 'back track' macroscopically.[42] But what about conditionals whose antecedents are macro events? Let's start with conditionals whose antecedents are bodily motions or actions; for example, 'if Nixon had pushed the button at t there would have been a nuclear war'. The natural way to extend the account to these conditionals is to find the latest time t^* prior to t at which it is open to Nixon to decide to push the button and which he immediately makes this decision and evaluate the conditional in terms of the probability of there being a nuclear war conditional on Nixon deciding at t^* to push the button and pushing it at t. This procedure will evaluate 'if Nixon had pushed the button at t there very likely would have been a nuclear war' as true given the usual assumptions about the circumstances of the button pushing.[43] This counterfactual antecedent does backtrack a bit. If Nixon had pushed the button at t the probability that he would have decided to do so at t^* is near 1. But this doesn't undermine the asymmetry of choice since it is only via decisions that we can affect bodily movements.

What about antecedents that are not actions? Consider for example, 'if a fire had started in the forest at noon it would have very likely destroyed all the

---

[42] This isn't quite correct. There may be past macroscopic events that leave no records in the present macro state and are very unlikely relative to the present macro state. For example, suppose that there was an Atlantis that left no traces. The probability of there having been Atlantis conditional on the current macro state would then be very tiny. So on our prescription if I had decided to lift my left hand (instead of my right) it is unlikely that there would have been an Atlantis. This is a strange consequence but not really unsettling since it doesn't entail that we can have control over past events.

[43] Of course, if the button was not connected or if Haig had arranged for Nixon's brain to be monitored so that should he decide to push the button it becomes disconnected we would obtain a very different result.

houses by the next day'. The natural extension of the account is to find a time t^* prior to t at which there are micro histories which diverge from the actual macro-history, and lead to a fire starting in the forest at noon. If Pr(houses destroyed the next day/M(t^*)& fire started at noon) is high, the counterfactual is true. We can see why we might be interested in this probability and the corresponding counterfactual. If Pr(houses destroyed/M(t^*)& fire) is high, while Pr(houses destroyed/M(t^*)& no fire) is low, then whether or not the houses are destroyed depends on whether or not there is a fire. So whether or not there is a fire provides (or would have provided) a kind of 'handle' on whether or not the houses are destroyed. We are interested in fire histories that diverge from the actual macro history because the divergence point (what Bennett calls 'the fork') is the size of a decision and it is decisions over which we have direct control. As a 'first stab' at an account I propose:

($\$\star$) 'If A($t$) had been true then the chance of B would have been x' is true iff t' is the latest time at which a divergence from the actual macro history similar in probability to a decision event can occur and Pr(B/A(t)&M(t')) = x.[44]

The fact that my proposal for evaluating non-decision conditionals makes use of the distinction between past and future doesn't undermine my claim that the asymmetry of counterfactuals is grounded in the SM probability distribution. The temporal asymmetry emerges from decision conditionals and depends on the SM distribution. So we have an explanation in terms of physics—a *scientific* explanation—of why we can affect the future but not the past. The proposal for evaluating non-decision conditionals is parasitic on decision conditionals.

There are problems with ($\$\star$). In the example, there may be many ways in which fires can be started by 'small' events occurring a bit before noon some of which lead to the houses being destroyed and some not. The trouble is that the antecedent is under specified. Different additions to the antecedent—the fire starts in the north corner and so on—may lead to different conditional probabilities. One way to handle this would be to

---

[44] Requiring that the divergence event is similar in size (i.e. in probability) to a decision rules out enormously unlikely 'fluctuations' that can occur at later times.

average over the ways in which the fire could start using the probabilities of the small events that lead to fires at t'. Another difficulty is that there may be times prior to the latest time at which it is much more likely that a fire starts. Perhaps, a camper threw away a lighted cigarette a bit before t' that didn't start a fire but came close. In that case I am inclined to think that if a fire had started it would have likely been at this earlier time in the vicinity of the thrown cigarette. Perhaps some sort of averaging over an interval prior to t' would help here.

There is also a worry about backtracking. If Nixon had pushed the button at t then the probability is high that at some time t' prior to t a decision size event would have occurred that evolved the state by the laws so that Nixon pushed the button at t. There is similar backtracking on Lewis's account. I don't see this as especially worrying, since it doesn't at all undermine the claim that we cannot control the past. However, certain other backtrackers are more troubling; for example, 'if there were a nuclear war at t then the button would have been pushed at t'. The trouble is that this suggests backward causation. Perhaps this can be handled by characterizing causal connection in terms of probabilistic correlations of the right sort and causal priority in terms of the temporal direction of control by decisions. But I leave the job of characterizing causal connection and causal priority for another occasion.

Even though I have not presented a complete account of counterfactuals it may be worth comparing it with Lewis's account.

(1) The SM account is not a similarity account like Lewis's but rather it is based on probabilities provided by the laws. However, in those cases to which both accounts apply the SM account plausibly gives very nearly the same result that Lewis thought his account gives and that Bennett's account does give. On the SM account it is plausible that 'If Nixon had pushed the button at t it is very likely that there would have subsequently been a nuclear war' comes out as true. Same for 'If Nixon had pushed the button at t it would have still been likely that...' where... is filled in by any macro past event.

(2) The SM account avoids the problem of 'converging' histories that Albert and Elga pointed out. Lewis can take a step towards avoiding this

problem as well by requiring that similar worlds are ones at which PH obtains. But it is only a step since Lewis's account still allows that there may be some worlds like w4 but in which it only takes a small miracle to yield convergence to the actual world. Such worlds contain very, very unlikely fluctuations but it is not clear that the dynamical laws rule them out. If there are such worlds, then as I pointed out Lewis's account is sunk. The cause of the trouble is the fact that perfect match counts as one of the criteria of similarity. The SM account doesn't work like that and so doesn't have this problem.

(3) The SM account goes some distance towards explaining the intuitive appeal of the idea that in evaluating counterfactuals we look at alternatives that are very much like the actual history until a short time prior to the time of the antecedent. But where Bennett simply declared this and Lewis tried and failed to account for it in terms of his similarity criteria the SM account *explains* how the laws of physics and the centrality of decisions underlie this way of evaluation.

(4) The SM account explains how the temporal asymmetry of counterfactuals is connected to the second laws of thermodynamics. Both are grounded in PROB and PH.

On the other hand, Lewis's similarity account is more general than the SM account. His account applies to counterfactuals that have both probabilistic and non-probabilistic consequents where the SM account applies only to the latter. It must be granted that ordinary language counterfactuals generally do not have probabilistic consequents. There is work yet to do to connect the SM account to ordinary language counterfactuals. Also, the SM account depends on the particular laws (and specifically PROB and PH) that obtain in our world. Lewis's account doesn't depend on the particular laws in the actual world. Perhaps the best way of looking at the relationship between the accounts is that while Lewis (and Bennett) were offering something close to an *analysis* of counterfactuals that is supposed to validate our intuitive judgments the SM account proposes a conditional that tracks the SM distribution in an especially interesting way. I say a bit more about this in the next section.

11.4 Why the SM Conditional?

Commenting on the criteria of similarity in Lewis's account, Paul Horwich says:

> Now these criteria of similarity may well engender the right result in each case. However, it seems to me problematic that they have no pre-theoretical plausibility and are derived solely from the need to make certain conditionals come out true and others false. For it is quite mysterious why we should have evolved such a baroque notion of counterfactual dependence. Why did we not, for example, base our concept of counterfactual dependence on our ordinary notion of overall similarity? As long as we lack answers to these questions, it will seem extraordinary that we should have any use for the idea of counterfactual dependence, given Lewis's description of it; and so that account of our conception of the counterfactual conditional must seem psychologically unrealistic. (Horwich 1987, p. 172)

Horwich's point is that there are infinitely many kinds of conditionals satisfying similarity semantics corresponding to different similarity relations. His question is: 'What is so special about the similarity relation that underlies the conditionals that express counterfactuals?' Neither Lewis nor Bennett addresses this question. I think that the question can be answered for the SM conditional and insofar as Lewis's account (or rather the account repaired to deal with the Albert-Elga objection) approximates the SM conditional account it can be answered for that account too. The answer is, roughly, that the information expressed by SM-counterfactuals is important for us because it tracks the statistical mechanical probability distribution in ways that are important for the consequences of our decisions. Knowledge or partial knowledge of this distribution is relevant to successful decision-making. People whose degrees of belief approximate the statistical mechanical probability distribution are objectively more likely to succeed in satisfying their desires (assuming they are otherwise rational) than people whose degrees of belief diverge from this distribution. So if I know that if I were to strike the match now it is likely to light, then I know that Pr(light/strike&M(now)) is close to 1. If I want to start a fire this knowledge is very useful.

The obvious objection to this proposal is that facts about the statistical mechanical probability distribution are too arcane to be common knowledge. Most people don't know statistical mechanics—the dynamical laws and PROB—and don't know the full macro state. But if we don't know these, how important can the SM-conditional be? The answer is that we often do know (or believe) many special science generalizations and laws (typically qualified *ceteris paribus*) that enable us to approximate statistical mechanical probabilities. Most of us don't know that this is what we are approximating. So, for example, we know that it is highly likely that the newly fallen snow will melt by evening. On the SM account what we know is made true by the fact that, conditional on the current macro state, the statistical mechanical probability that the newly fallen snow will melt is high. Of course we know only a little bit of the entire macro state—for example that the ambient temperature is above freezing—but most of the rest of the macro state is irrelevant to the melting of snow. The fact that it is irrelevant is, of course, itself a feature of the SM distribution. That our world contains many 'almost isolated' processes is a deep and enormously significant feature of the dynamical laws and the statistical mechanical laws.[45] But because there are such processes and because we know a lot about them we can often know the truth values of decision SM-conditionals.

11.5 Conclusion: Future Projects

In this chapter I have continued the project associated with Boltzmann, Reichenbach, Lewis, and most recently Albert of attempting to ground the various arrows of time in statistical mechanics. The main contribution here was to spell out a conditional that approximates the conditional that Bennett and Lewis were attempting to characterize and which we all agree is intimately connected with decision, memory, causation, and so on. The account succeeds in capturing the temporal asymmetry of that conditional where Lewis's account fails. Where Bennett puts in the temporal asymmetry by hand this account obtains it from the statistical mechanical distribution; that is, it is earned—not stipulated. Because of the connection between the conditional and the objective statistical mechanical distribution we have the beginnings of an answer to Horwich's question of why conditionals

---

[45] See Elga (ch. 5 in this volume) for the beginnings of an explanation of why this is so.

evaluated in a particular way are especially important. But there is a great deal that is left undone. In the first place the grand scheme of the world composed of PH, PROB and the dynamical laws requires much more defense and development than has been given here. It needs to be investigated exactly how it looks in the context of current accounts of space time and quantum mechanics. Secondly Horwich's question deserves a much fuller answer than I gave it here. The SM-conditional is one among many conditionals that could be defined that respect and reflect the probability distribution derived from PROB and PH. More investigation is needed to determine if there is reason to choose this one or any one of these and whether one is in some way 'better' for the job of accounting for the other temporally asymmetric notions. Thirdly, the SM-conditional is restricted to conditionals with probabilistic consequents. It remains to be seen if it, or something close to it, can be extended to conditionals whose consequents are not probabilistic since these are closer to ordinary usage. Fourthly, I assumed in the paper that the dynamics are deterministic. It remains to be seen if the SM-conditional (or a natural extension of it) is appropriate in the context of indeterministic dynamics. Finally, the account will prove its mettle if it can be connected to an account of causation. There is some hope of characterizing causation (or one concept of causation) in terms of counterfactual dependence and there is reason to think that the SM-conditional might play the role in a counterfactual analysis along Lewisian lines.

References

Albert, D. (2000). *Time and Chance*. Cambridge Mass: Harvard University Press.

Beebee, H. (2000). 'The Non-Governing Conception of Laws of Nature', *Philosophy and Phenomenological Research*, Nov. 2000: 571–94.

Bennett, J. (1974). 'Counterfactuals and Possible Worlds', *Canadian Journal of Philosophy*, 4: 381–402.

Bennett, J. (2003). *A Philosophical Guide to Conditionals*. Oxford: Oxford University Press.

Callender, C. (2004). 'Measures, Explanation and the Past: Should "Special" Initial Conditions Be Explained?', *British Journal for the Philosophy of Science*, 55: 195–217.

—— (2003). 'Is There a Puzzle about the Low Entropy Past?', Hitchcock, C. (ed.), *Contemporary Debates in the Philosophy of Science* (ch. 12), Malden, Mass: Blackwell.

Elga, A. (2000). 'Statistical Mechanics and the Asymmetry of Counterfactual Dependence,' *Philosophy of Science* suppl. 68: 313–24.
Field, H. (2003). 'Causation in a Physical World', in M. Loux and D. Zimmerman (eds), *Oxford Handbook of Metaphysics*. Oxford: Oxford University Press.
Feynman, R. (1965). *The Character of Physical Law*. Cambridge, Mass: MIT Press.
Fine, K. (1975). 'Review of Lewis's Counterfactuals,' *Mind* 84: 451–8.
Horwich, P. (1987). *Asymmetries in Time*. Cambridge, Mass: MIT Press.
Kutach, D. (2003). 'The Entropy Theory of Counterfactuals,' *Philosophy of Science*, 69: 82–104.
Lange, M. 2005 'Laws and Their Stability', *Synthese*, 144: 415–32.
Leeds, S. (2003). 'Foundations of Statistical Mechanics,' *Philosophy of Science*, 70: 126–44.
Lewis, D. (1986). 'Time's Arrow', in *Philosophical Papers*, vol. II. Oxford: Oxford University Press.
—— (1999). 'New Work for a Theory of Universals', in *Philosophical Papers*, vol. III. Oxford: Oxford University Press.
Loewer, B. (1996). 'Humean Supervenience', *Philosophical Topics*, 24: 101–27.
—— (2001). 'Determinism and Chance', *Studies in the History of Modern Physics*, 32: 609–20.
—— (2003). 'Time and Law' lecture at the Einstein Forum 2003, Berlin.
—— (2004). 'David Lewis's Account of Objective Chance', *Philosophy of Science*, 71: 1115–25.
Maudlin, T. (2005). 'Remarks on the Passing of Time', *Proceedings of the Aristotelian Society*, volume CII (part 3): 237–52.
North, J. (2004). Time and Probability in an Asymmetric Universe. Dissertation; Rutgers University.
Price, H. (2003). 'There is Still a Problem About the Low Entropy Past' in C. Hitchcock (ed) *Contemporary Debates in Philosophy of Science* Malden, Mass: Blackwell.
Reichenbach, H. (1956). *The Direction of Time*. Berkeley: University of California Press.
Russell, B. (1913). 'On the Notion of Cause.' *Proceedings of the Aristotelian Society* 13: 1–26.
Sklar, L. (1993). *Physics and Chance*. Cambridge: Cambridge University Press.
Strevens, M. (2003). *Bigger than Chaos*. Cambridge, Mass: Harvard University Press.
Tooley, M. (1987). *Causation: A Realist Approach*. Oxford: Clarendon Press.

12

The Physical Foundations of Causation

DOUGLAS KUTACH

The prominence of causation in twentieth-century metaphysics is curious considering it partly stems from a classical theory of interaction by mechanical contact that has long been superseded by more sophisticated physics. Specifically, the idea that a central organizing principle of nature is a causal relation between events is not motivated by a serious examination of fundamental physics. What we do find in our best fundamental theories are equations expressing relationships between physical quantities at different times and places, equations that have no obvious connection with the concept of causation. Bertrand Russell in 'On the Notion of Cause' (1913) took this observation as evidence for his argument that there is no law of cause and effect, and that causation is dispensable. While there is some justice to Russell's claim, the utility of causal notions demands an explanation. If there is just physics, why do the ideas of cause and effect serve us so well? And if there *is* a physical explanation for the usefulness of causal notions, would that not arise by demonstrating how causation reduces to the physical?

The answer to these two questions, I suggest, is that different aspects of the physics justify different principles about causation, and together these elements suffice to explain the utility of our notion of cause. Yet, the physics as a whole does not support a reduction of the robust notion of cause that philosophers usually care about, the kind of causation for example that applies to ordinary events and is useful for assigning causal responsibility and matches important pre-theoretical intuitions about causal interaction among ordinary objects.

At bottom, causation is a result of two quite different aspects of our fundamental physics: boundary conditions and dynamical laws. The dynamics vindicate our thinking of the cause as somehow necessitating the effect, and the boundary conditions vindicate our thinking of the cause as happening before the effect. Yet these two components do not cohabit peacefully. On the one hand, the necessitation relation in the physics applies only to detailed physical microstates, not to coarse-grained events. On the other hand, the causal asymmetry is grounded in the physics only insofar as one is concerned with relations among coarse-grained events, not among the detailed microstates. There is no single level of description for the relata where all the constitutive properties of causation apply. Thus, our concept of causation is a kind of arranged marriage, with the bride of necessitation and the groom of asymmetry being ill-matched but wed nevertheless for social utility.

Yet, if we allow ourselves to mix our reasoning about both microscopic and macroscopic (coarse-grained) descriptions of events and processes, we can justify a causal-like relationship that explains key facts about causation *simpliciter*. It explains why recourse to the idea of cause and effect is so fruitful in realistic situations. This account of causation fails to underwrite popular philosophical conceptions of causation primarily by not fully justifying intuitions about how salient causes should be distinguished. In the stock example where Billy and Suzy throw rocks at the bottle, and Suzy's rock breaks the bottle significantly before Billy's rock arrives, everyone is supposed to agree that Suzy's throw was a cause of the bottle's breaking but not Billy's. In the account presented here, both throws are counted loosely speaking as causes, but for pragmatic reasons (that are well-grounded by the physics) we attach a lot more importance to Suzy's cause than Billy's. Standard robust accounts incorporate such pragmatic features into the deep structure of causation, claiming that Billy's throw is not a cause, period. Yet the necessarily capricious selection of which events count as salient combined with the robust theorists' oversimplification by insisting that each event counts as either cause or non-cause, turns out to be the source of the most serious problems for robust accounts, for example, late pre-emption, double prevention, and causation by omission.

12.1 Causation and Physical Necessity

The 'cement of the universe' is David Hume's famous phrase describing our conviction that causes necessitate their effects. Setting aside indeterminism momentarily, the empirical basis of this necessitation aspect of causation is captured by the principle 'same cause, same effect'. If we have two situations identical with respect to the precise cause and relevant environment, they both have the effect occurring. It embodies the pragmatic upshot of causal theorizing, as it allows us to draw straightforward inferences from empirical observations to general rules or laws about causation that we can then use to achieve a desired effect by creating the appropriate cause in a suitable environment. Yet, the 'same cause, same effect' principle by itself is nearly vacuous because it doesn't indicate the relevant factors for judging whether two given causal situations are similar enough to count as effectively the same cause.

It turns out fortunately that there exist structures in our fundamental physical theories that we can use to clarify these conditions: dynamical laws establishing nomological determination between microstates at different times. With these laws in hand we can interpret 'same exact cause' to mean 'same local microstate', and 'same relevant environment' to mean 'same exact microstate'. In this sense, deterministic dynamical laws vindicate our thinking of the world as obeying the 'same cause, same effect' principle. Crucially, the microstructural relations allow us to determine how much difference in the effect exists between two causal situations that have only approximately the same cause.

The folk concept of causation applies not to microstates, but to macroscopic objects, events, and processes. To make the connection between these entities and the nomological determination relations, we group together some of the local physical facts that reasonably fall under some convenient description like 'striking a match' or 'neuron firing'. In many circumstances, such an underdescription of the physics serves well as a proxy for the microscopic physics that in practice is epistemologically inaccessible and too complicated for making inferences. For example, we say 'neuron A's firing causes neuron B's firing' to describe physical situations that could be more accurately described as 'the A-microstate nomologically determines the B-microstate'. The causal terminology omits reference to physics outside the neuron even though the micro-facts in one's liver are needed to necessitate B's firing just as much as the micro-facts in A. This omission is often excusable because usually A-microstates with minor differences in the liver nomologically determine microstates that also include B's firing. That is, the vague language is good enough when the physics in the area around A and B is sufficiently insensitive to the kinds of physical facts one typically finds further away in the liver. But because sensitivity to the external physics is a matter of degree, no matter how you individuate these coarse-grained events, there are always going to be borderline cases where the causal terminology substantially misrepresents the microphysics. This is a key source of intractable difficulties for robust theories of causation, which try to defend much more of the folk theory of causation.

The justification for counting 'neuron A causes neuron B to fire' as a good approximation of the real physics is that the physics of neuron B is more sensitive to the physics of neuron A than it is to the liver. In order to measure sensitivity objectively we need to compare various microscopic modifications to the initial state, which requires that we have some objective measure over the possible A-microstates and their nomological determinants. Fortunately, we have objective probability measures in the theory of statistical mechanics that can arguably quantify sensitivity by allowing us to compare counterfactual macrostates by quantifying what proportion of their microstates lead to what effects. We justify the focus on the A-microstate as the cause by noting that the fraction of alterations to the neuron-A part of the microstate leading to B's not firing greatly exceeds the fraction of alterations to the liver part leading to B's not firing.

Once we have statistical mechanical probabilities in our theory of causation, it is easy to add fundamental dynamical probabilities, the kind that exist in stochastic theories where nature makes random jumps according to probabilistic rules. When we have this kind of indeterminism in our theory, typically we no longer have a 'same cause, same effect' principle, but instead a 'same cause, same probability of effect' principle, which works well enough in many circumstances.

Incorporating statistical mechanical probability is also good because it allows us, even in a deterministic environment, to associate causal processes with probability-raising processes, widely recognized as a decent first-order approximation to the folk theory of causation. Accounts of causation as probability-raising are also known to have serious problems like pre-emption and fizzling (Schaffer 2003), but such problems, even if insoluble, need not count decisively against the theory being partially developed here because it already admits that the underlying physical principles may not perfectly capture folk intuitions. What will eventually be needed is an explanation of why other grounding principles of causation interfere in cases like pre-emption and fizzling to override the rule of thumb that causation involves probability-raising.

We are now in a position to explain why it is often fruitful to conceive of causation as transitive. Whenever the A-microstate nomologically determines the B-microstate and the B-microstate nomologically determines the C-microstate, it follows that the A-microstate nomologically determines the C-microstate. That is, we have transitivity at the level of microscopic determination. Putative counterexamples to causation involve the coarse-graining of events that prevents the straightforward application of transitivity. For example, Schaffer (2003) describes A, a boulder rolling down the hill towards a hiker's head, which causes B, the hiker to duck, which in turn causes C, the hiker's survival. If causation is transitive, we are forced to accept the apparently counterintuitive claim that the boulder rolling towards the hiker caused him to survive. It's easy to see why it is counterintuitive at the level of coarse-grained description: A makes C less likely in the sense that there is a bigger proportion of boulder-rolling-microstates that determine the hiker's death than the proportion of boulder-stays-still-microstates that determine the hiker's death. This probability lowering comes despite the fact that A raises the probability of B, and B raises the probability of C. The coarse-grained description focuses our attention on the fact that there is a rolling boulder, which lowers the probability of survival and hence counts as at least some kind of reason for denying that A causes C. Nevertheless the particular boulder that rolled was the most significant part of a larger microstate that determined that the hiker survived, which gives us a reason to say A did cause C. The fact that such purported counterexamples to transitivity are seen not to be counterexamples when we describe the same facts at a more fine-grained level, strongly suggests that the justification for the causal transitivity ultimately lies in the nomological determination part of causation and is sometimes obscured under a more coarse-grained description of reality.

12.2 Causal Asymmetry

The other prominent aspect of causation is the causal asymmetry, the fact that causes temporally precede their effects. It is very tempting to conceive this asymmetry as somehow embedded in the local physics—that there is something in the striking of the match that made it burn but nothing in the burnt match that made it previously struck. However, a careful look at the underlying physics does not support the idea that the causal asymmetry is localized. Surprisingly, the apparent asymmetry between the striking and the burning is grounded instead by way of special boundary conditions of the early universe.

12.2.1 Fundamental Physical Asymmetries

To seek the physical ground of the causal asymmetry, we need first to examine the kinds of temporal asymmetries existing in plausible fundamental physical theories. There are two possible types worth consideration here. The first is a temporal orientation, which locally defines one direction in time as dynamically different from the other. The most common version of this temporal orientation is implicit in the use of a stochastic dynamics. In a theory with stochastic dynamics, one has laws of nature specifying that chance processes sometimes occur and that any time a chance process occurs, the physical state on one temporal side of the chance process depends in a fundamentally probabilistic way on the physics of the other temporal side. The independent side is what we call 'the past' and the dependent side, 'the future'.

Indeterministic and stochastic dynamics of the kind that have been proposed as serious fundamental physical theories look superficially like they might give a plausible explanation of the causal asymmetry. Because a stochastic dynamics has rules for calculating the probability of future microstates from the current microstate, there is a determinate probability for what would have happened had a given cause not occurred. Any full microstate without the cause has some nomologically determined probability for the effect occurring. Thus, there will be facts of the matter about what local chunks of matter raise the probability of an effect.

However, even though stochastic rules make the future chancy, all the kinds of stochastic laws that appear in realistic fundamental theories—like laws about quantum mechanical wave collapse—fail to make the past determinate. In fact, they don't constrain the past at all, and so restrict the past even less than they do the future. This has the consequence that they cannot, by themselves or together with other deterministic dynamical laws, justify an objective probability measure over the past states compatible with some hypothetical present state. The probabilistic rules that go only from present to future cannot be applied in going from present to past unless we have some independent grasp on the initial probability distribution, which is not given by the dynamics. So the temporal orientation determined by the stochastic dynamics, far from explicating causal asymmetry, makes the problem worse by making the past less fixed than the future.

The other kind of physical asymmetry is an asymmetry in boundary conditions. For example, we know there is a smooth, bunched up distribution of matter and energy at the temporal end of our universe we call the past, and another clumpy, spread out distribution a good distance into the other temporal direction we call the future. A long history of investigation into the foundations of statistical mechanics indicates this difference cannot be explained in terms of the dynamics. If we take for granted all the gross features of the physical state in the early universe, the dynamics does tell us that matter at the other end of time will be spread out and clumped. However, if we take for granted the macroscopic features of the physical state at some future time, we cannot infer the existence of the smoothly concentrated matter of the early universe. This is true even though a deterministic dynamics entails that the exact future state determines every feature of the early universe. The asymmetry exists at the coarse level of description where we have less than a full microscopic state to use for inferences.

There are many kinds of physical asymmetries that cannot be explained by dynamical asymmetries but only by boundary conditions. For example, we have thermodynamic asymmetries in the dispersal of gases and the flow of heat. These thermodynamic asymmetries are grouped together theoretically in that they all can be summarized by the rule that the entropy of an isolated system virtually never decreases (as we go forward in time). The dynamics cannot explain thermodynamic asymmetries because from the point of view of the microphysics, the kinds of dynamical behavior one needs to repeatedly drive a wide variety of systems towards lower entropy are far too variegated. One can see this by way of standard examples in the literature on statistical mechanical explanations of entropy increase. Consider a gas inside an isolated tank, where we idealize the gas as molecules interacting only by elastic collision, and we take the gas at $t = 0$ to be uniformly distributed only in a small volume V of the tank. Because the gas is uniformly distributed, there will almost certainly be many gas molecules on the edge of V that happen to have their velocities pointed towards the empty space in the rest of the tank. These molecules will spread out over the first few seconds and after a short time the gas will be uniformly distributed throughout the tank and will stay there at equilibrium. Let S label the microstate of the gas and tank at $t = 1$ min. We know that there is a physically possible microstate S^* which is just S with all the particle velocities reversed. The classical dynamics makes S^* evolve in a way that is macroscopically the reverse of S, so that S^* will sit at equilibrium for almost a minute, and then the particles will hit each other in just the right 'improbable' combination to make the gas collapse to the small volume that S started in.

The question we are interested in is whether dynamical laws acting at a local level could explain the behavior of gases macroscopically like S^*. While one could cook up a dynamics that has certain chance moments where dispersed gases collapse, such dynamics would fail to reproduce the relevant behavior of S^*'s evolution. To see this, augment the example by having V be the inside of a small canister inside the tank that is opened to let the gas escape. For an embellishment also imagine that there are other canisters in the tank with small leaks making them useless for holding gas. Picturing the evolution of S^*, we have the uniformly distributed gas doing nothing for almost a minute and then collapsing into the one functioning container and being sealed in by an apparently spontaneously shutting lid. In this example, the dynamical law, even if it had spontaneous gas collapses, would not only have to collapse the gas but would have to collapse it into the one leak-free container. It would need to be responsive in a reliable way to facts that are very hard to describe in any way other than as macroscopic facts about will happen in the future. The particle motions would need to conspire to be in just the right microscopic condition to instantiate future states that are overwhelmingly unlikely but satisfy a compact macroscopic description. This kind of responsiveness, local dynamical laws do not have.

The correct explanation of thermodynamic asymmetries comes by way of an asymmetric boundary condition. Specifically, we posit low entropy in the past and no similar constraint in the future. This entropy constraint on the macroscopic state of the entire universe often goes by the name 'the past hypothesis' (Albert 2000). It follows from the past hypothesis in addition to the dynamics and standard probability measure we already have, that the universe will be likely to exhibit thermodynamic asymmetries of the kind we regularly see. It is exactly the structure of such boundary constraints that makes dynamical evolution of a macrostate towards the past exhibit the seemingly conspiratorial motion that we in fact see happen in reverse. And it is this feature that explains the seemingly local character of the causal asymmetry, as we will soon see.

12.2.2 Counterfactual Asymmetry

In order to apply causal terminology to coarse-grained events in a way that justifies our selection of some events as salient, we needed a structure that made possible a measure of the causal sensitivity of various chunks of physics. We found this measure in the theory of statistical mechanics. We can equivalently treat our comparison of various microstates as a comparison among counterfactual possibilities, the possible worlds that possess the microstates in question.

The recourse to counterfactuals in elucidating causation is familiar, and suggests the possibility that the causal asymmetry can be fully explained as a counterfactual asymmetry. For example, the asymmetry of causal dependence between the striking and burning of the match can be expressed as, 'It's true that had the match not been struck, there would have been no flame, and it's false that had there been no flame, it would not have been struck.' In making this idea precise, it's worth noting that our offhand judgments about counterfactuals do not universally justify such an asymmetry. In a significant number of cases we judge that if the effect had not happened, its cause wouldn't have happened either. For example, we get counterfactual dependence of the past on the present when we imagine the cause and effect as part of a fixed system. A ring falls into the sink, bounces around randomly and slips past the drain cover, causing it to land in the plumbing. It is reasonable that if the ring were not in the plumbing, it would be because it didn't slip past the drain cover. We also get backwards dependence in more theoretically-minded evaluations such as when we focus on the underlying deterministic microphysics. When an A-microstate nomologically determines a preceding B-microstate, it is reasonable to claim that B would have been true if A had obtained in the exact way specified by the A-microstate.

Because our evaluation of counterfactuals can often in ordinary circumstances fail to possess the asymmetry needed for causation, to explain the causal asymmetry by the counterfactual asymmetry we need either a theoretical refinement of counterfactual reasoning or else a way to understand the causal asymmetry as only loosely tied to the counterfactual asymmetry. I will first consider two different justifications for seeking a causation-friendly refinement of our ordinary counterfactual reasoning and dismiss them as insufficient. Then, I will go on in Section 12.2.3 to elaborate a theory that treats the counterfactual asymmetry as only a rough approximation, allowing the so-called backtracking reasoning associated with counterfactual dependence of the past on the present. Nevertheless, even without a strict counterfactual asymmetry, we will have enough of a counterfactual asymmetry to justify the causal asymmetry for all practical purposes.

The most famous attempt at the refinement strategy is David Lewis's 'Counterfactual Dependence and Time's Arrow' (1979). Lewis recognizes that some contexts allow the counterfactual dependence of the past on the present, but distinguishes what he calls the 'standard resolution'. The standard resolution of counterfactuals is stipulated to disallow the undesired counterfactual dependence of the past on the present. To justify the appeal to a standard resolution, Lewis presents a theory of counterfactual evaluation that tries to clarify the criteria we effectively have in mind (or should have in mind) when we think about counterfactuals. His theory tells us which respects of similarity to use in comparing possible worlds and is designed so that the most similar worlds with the antecedent A true (assuming A is actually false) will turn out to be worlds with the exact same microscopic past up until a short time before the events mentioned in A happen and afterwards a different but lawful future. A legitimate review of Lewis's theory deserves more attention than I can give here, so I will set it aside with the suggestion that his theory, if it were patched to avoid all the known counterexamples (e.g. Edgington, (1995), for a review and Schaffer, (2004), for discussion of some more recent examples), would be too baroque a system to count as a justification of the standard resolution. Furthermore, Lewis's theory comes with no explanation of why our local physical environment satisfies the 'overdetermination asymmetry' he posits as the explanation of counterfactual asymmetry. In Section 12.2.3, I discuss why the past hypothesis and dynamical laws often establish something like Lewis's overdetermination asymmetry, and with this physical explanation in hand, Lewis's recourse to miracles and the associated priority rankings is made superfluous.

A second potential justification for concerning oneself only with the standard resolution is that decision-making situations seem to require it—that when deciding among alternatives, one shouldn't use backtracking reasoning. The hypothesis here is that the counterfactual asymmetry needs only to be justified for the kind of situations where we humans are able to influence the world by way of decisions. Then, we project this asymmetrical aspect of our agency to other objects in nature, making causation seem asymmetrical when it is really only decisions and our perspective that are asymmetric.

In one example, (Downing 1959, Lewis 1979), Jack and Jim have a fight and the next day Jim considers asking a favor of Jack, but decides not to. There are two reasonable ways to understand what would have happened had Jim had asked the favor. On one reading, Jack would refuse because he is still angry. On another reading, we fix on Jim's pride as a salient characteristic to keep constant and infer that if he were to ask Jack for a favor, it would have been because there was no fight the day before. Then we reverse direction and infer that if there hadn't been a fight, Jack wouldn't be angry and so would likely grant the favor.

The reason decisions appear counterfactually special is that it is pragmatically irrational to use backtracking reasoning when making decisions. It is exceedingly unwise for Jim, who remembers the fight quite well, to use the backtracking reasoning that leaves subject to his volition whether yesterday's fight occurred. He shouldn't think (indicatively) that if he asks Jack, Jack will grant the favor because there wasn't a fight. Due to the prevalence of situations like these, it is arguably an empirical fact that people who use such backtracking reasoning will typically be less successful in achieving their aims than those who use the standard resolution, *ceteris paribus*.

Regardless of the plausibility of justifying the standard resolution of counterfactuals in this way, the idea that we are projecting the counterfactual asymmetry onto the world is undermined for two reasons. First, counterfactuals involving decisions can sometimes be naturally interpreted in backtracking-friendly ways, like decisions that are psychologically difficult or decisions to accomplish tasks that only make sense in other contexts. For example, if I had decided yesterday to convert to Asatru, a religion I know nothing about, that would have been in part because

I had earlier learned something about Asatru. If I had decided yesterday to descend Denali, it likely would have been that I had previously ascended. The reason decision-counterfactuals don't usually involve significant backtracking is that we usually imagine decisions where a person is already restricting the options to those that are possible given what he or she knows.

Second, limiting the scope of our explanation to decisions misses the point. The goal is to explain why *the world* is friendly to a folk physics that invokes causation. It is not enough to explain why the counterfactual or causal asymmetry applies merely to decisions. We need the explanation to apply to the wider environment, so that we can explain why it does us no good to try to cause events by way of processes that take place after the fact. The asymmetry demanding explanation is the temporal asymmetry in the kinds of causal chains we can hope to create.

12.2.3 *Counterfactual Asymmetry via Boundary Condition Asymmetry*

Consider the following explanation of a counterfactual asymmetry that permits backtracking and counterfactual dependence of the past on the present while still giving us a kind of causal asymmetry. The basic idea is to evaluate counterfactuals dealing with physical events by way of the objective chance that the consequent would obtain among the relevant possible worlds where the antecedent obtains. The counterfactual conditionals of primary concern are ones where the antecedent A describes facts that can plausibly be thought of as some coarse-grained events localized in some spatial region R at time t. For these conditionals, take the microstate of the actual world at t, consider all the localized microscopic modifications needed to make A obtain, and then let the objective probability distribution and the dynamics tell us how likely it is for the consequent C to obtain. One can think of this objective conditional probability $prob(C/A)$ as a measure of the degree of assertibility (or acceptability) we ought to have in the conditional if we knew all the actual facts and laws, or one can use the standard semantics for counterfactuals and claim that $A \boxright$ the objective chance of C is $prob(C/A)$. This evaluation procedure extends to more vague counterfactuals like 'If people ate more vegetables, there would be fewer cases of diabetes', by thinking of them as generalizations about more specific counterfactuals, that is what would happen if particular people ate particular vegetables at particular times.

In this theory, what it means for the past to be fixed is for it to turn out (for almost all actual coarse-grained past facts, F, and almost all localized antecedents, A) that '$A \,\square\!\!\rightarrow$ the objective chance of F is very high'. The explanation for why the past is fixed is that when we counterfactually suppose A, we include two important (but defeasible) background assumptions in addition to the dynamical laws. We assume that the past hypothesis PH is true and that microscopic facts M outside of R are held fixed (merely because we are restricting consideration to localized counterfactual alterations). For ordinary antecedents A, making explicit the tacit hypotheses means $prob(C/A)$ should be understood as $prob(C/A\&PH\&M)$. The effect of having the constraints M and PH is plausibly that lots of macroscopic past facts are highly probable given these two constraints. It is important to realize that whether this is true for any particular antecedent depends on the actual dynamics and the antecedent itself, and unfortunately reality is far too complicated for us to test whether particular counterfactual assumptions really do hold the past fixed. So we can proceed only by making intelligent guesses about how the dynamics will work under the given constraints using the most reasonable rules of thumb available.

A quick sketch of an example should convey how PH and M team up to fix the past. For major historical facts, like Napoleon's rule of France, we have existing macroscopic evidence spread over a large spatial region. If we try to imagine what is the most likely way for a world starting out like ours to evolve into a world with all this widespread evidence, it is reasonable to guess it would be largely through worlds where Napoleon did really rule France and not through worlds with accidental accretions of misleading evidence. Minor historical facts lack macroscopic traces, but if we look at the present detailed microstate, it is plausible that we have lots of microscopic bits of evidence like images streaming away from Earth, and so forth. These microscopic traces could plausibly conjoin to make highly likely the historical events they seem to jointly imply. The presence of micro-traces also adds credibility to inferences based on the existence of macroscopic traces. While macroscopic evidence is sometimes misleading by way of hoaxes, cover-ups, accidents, and so on it is difficult to hide all the macro-traces without leaving at least microscopic traces of the cover up. Insofar as we have at least microscopic traces (at t) of a previous macroscopic fact F, holding fixed the macrostate of the early universe and the present microstate outside R makes F very likely to occur in the worlds where the antecedent is true.

An important caveat is that there is no fact of the matter in general about how large R should be. To some extent our ordinary reasoning about counterfactuals involves inferring from facts inside R at t to previous times and then backtracking to t in a way that might require us to adjust the microstate outside R. For example, we might start out thinking that in order to make Jim ask the favor, we need only adjust the physics of Jim's head and leave untouched the microphysics elsewhere. But if we keep fixed Jim's pride, then it is likely there would have been no fight yesterday, and if there had been no fight yesterday, there would be microtraces of this fact in the present, so we are postulating microscopic changes outside R after all. This shows that at best that the evaluation of ordinary counterfactuals involves delicately balancing the consequences of what we have in mind when we consider antecedents that can backtrack. The details of the full theory are covered elsewhere (Kutach 2001, Kutach 2002).

The picture this theory is supposed to vindicate is that the counterfactual A-worlds are mostly worlds that start off microscopically very nearly identical to the actual world, with the motion of every particle in the counterfactual universes so nearly like the actual microstate that there is no noticeable difference until some reasonably short time before t. During that time, the microscopic differences very, very slowly build until some (usually brief) time before t, the differences become big enough to set off dynamical changes that quickly become macroscopic, eventuating in A being true. The counterfactual differences come about through normal lawful evolution of physics, and do so in ways weighted probabilistically by the physics, with no miraculous funny business. This picture supports our intuition that there is some limited counterfactual dependence of the past on the present, dependence insofar as the past had to be different in order for the present to arise from the past by way of the actual dynamical laws. Yet, the two constraints make counterfactual differences in the past mostly microscopic. The probabilistic feature also supports our intuitions that usually forward-directed counterfactual processes, like actual processes, can be understood through the usual procedures of conditionalizing on known facts. For example, if we have a device with a chancy mechanism that 15 percent of the time rings a bell via causal process P, 45 percent of the time rings the same bell via a different causal process Q, and 40 percent of the time doesn't ring the bell, then we can conclude the following: If the bell doesn't actually ring, then the chance that P would have occurred if the bell had rung is 25 percent.

The past hypothesis plays a critical role because if we don't keep it as a constraint, the most likely worlds making A true and the microstate outside R fixed would be worlds with an anti-thermodynamic past. One can see this from the example in Section 12.2.1, where S^* is evolving towards a condensed volume V. Due to the dynamical fragility of the evolution towards low entropy, it only takes a small deviation in the motion of a single particle to disrupt an evolution towards low entropy into an evolution towards high entropy. In the real world, there are lots of ordinary forces that can spread the effects of motion, so any localized disturbance will quickly spread to other parts of the universe. Thus, without the past hypothesis, the past would be far more sensitive to the present than the future would be.

This account of counterfactuals supports something like David Albert's (2000) explanation of the asymmetry of knowledge and causation. Although a long tradition exists where thermodynamic asymmetries are explained by the past hypothesis, attempts have failed to explain other asymmetries—like the fact that our knowledge of the past is more precise or more abundant than our knowledge of the future—by connecting them to entropy increase. Albert's key insight is that epistemological and causal asymmetries are not explained by the past hypothesis in virtue of the past hypothesis making highly likely the entropy increase of mental structures or causal sequences but instead are explained directly in terms of the past hypothesis itself, regardless of whether the entropy of mental structures or causal sequences is even definable.

In giving an account of causal asymmetry, Albert first defends an epistemological claim, that what we know about the world (assuming a classical statistical mechanical world-view) can in principle come by way of a certain inference procedure where we derive probabilities of facts from the conjunction of the laws, the past hypothesis, the objective probability distribution, and the current 'directly surveyable condition of the world', which he characterizes as the world's current macrostate plus possibly a few microscopic features one might be able to introspect (p. 96). The resulting epistemological asymmetry is connected to the counterfactual and causal asymmetry because when we consider alternative possibilities, we are usually concerned with worlds like our own, meaning worlds where this inference procedure is effective. Thus, we can use the inference procedure to make inferences about what counterfactually would have happened under certain postulated circumstances. Albert then argues that this justifies our believing that among all our causal handles, that is, things over which we have immediate control, there are a 'far wider variety' that affect the future.

In my specification of $prob(C/A\&PH\&M)$, the past hypothesis, laws, and probability distribution play essentially the same role as in Albert's argument, although M is characterized in a way that fixes the past more. The two most important differences in the accounts are that my truth (or assertibility) conditions for counterfactuals (1) are explicitly independent of any epistemological claims, and (2) apply to counterfactuals beyond those involving volition and decisions. Regardless of differences, the following theory of causation can be interpreted as one way of making precise the sense in which our causal handles on the world are temporally asymmetric.

12.3 Causation

An event c is a *contributing cause* of event e if and only if c is an essential part of some microstate that nomologically determines an objective probability for e. For c to be an essential part means that the microstate doesn't determine the probability for e if we exclude the fact that c occurred. From there, we can go on to make finer distinctions designed to determine how important any cause is compared to other contributing causes of e. Our folk notion of causation at least roughly tracks the notion of important cause.

One feature that usually serves as a good sign of c being an important contributing cause to e is that replacing c with some alternate physical state leads to E being significantly less likely to obtain, where E is some coarse-grained description of e. To make this more precise, consider all the possible microstates constructed by taking the actual microstate S at the time t when c occurs and modifying it by replacing the spatial region occupied by c with some physics that involves C not occurring, where C is a coarse-grained description of c. There are, in general, many different ways to make C not occur, depending partially on how narrowly C is characterized and partially on what other background assumptions we make about what kinds of non-occurrence are contextually relevant. Supposing there is some reasonable space of such microstates, let $C^\dagger$ be the proposition corresponding to the possible worlds that are nomologically compatible with these microstates. $C^\dagger$ captures the relevant sense in which c does not occur. A good (but defeasible and not fully objective) measure of whether c is an important cause of e is whether the probability of E is boosted more by C than by other things. We can measure the degree of sensitivity of E on C or the degree to which c is a probability-raiser for e, by way of $prob(E/C^\dagger)$, the objective probability that E occurs among the relevant worlds where C doesn't occur: If $prob(E/C^\dagger)$ is significantly lower than $prob(E/C)$ or is relatively low compared to other contributing causes, we can typically say that c is an *important cause* of e.

An example of simple causation is when Billy throws a rock at a bottle, breaking it one second later. Suppose that the event c_1 of Billy throwing the rock occurs at time t, as part of a microstate S that extends across a sphere of radius one light-second. Assuming some dynamics compatible with relativistic locality, the contributing causes are all the various chunks of S, including innocuous events like c_2, an ant crawling on the top of the Taj Mahal. Billy's throw is underwritten as a relatively important cause because the bottle's breaking, E, depends more on c_1 than on other contributing causes like c_2. The dependence of E on c_1 is the objective probability that the bottle will break in worlds that are just like actuality except with Billy not throwing the bottle. This is low by the presumption that nothing else in the environment makes the probability of E high, that is that there are no backups. E is not dependent on c_2 because virtually any reasonable way of making the ant absent from S will have negligible dynamical consequences for the bottle.

This defeasible marker for causal importance can also be expressed using counterfactual conditionals. We can say event e under the description E counterfactually depends on c if and only if had c not occurred, E would be likely not to have obtained. This way of expressing the dependence relies on a special interpretation of the counterfactual: that we interpret c's not having occurred in terms of an objective probability distribution over the $C^\dagger$ worlds, resolving the vagueness of 'had c not occurred...' in a way that excludes backtracking reasoning.

Another important physical feature that plays a role in determining causal importance is the presence of intermediate determining microstates. In all important fundamental theories that physics has so far uncovered, when one state A nomologically determines a later state B, it does so while also determining any temporally intermediate states to be such that they also determine B, except possibly for esoteric cases that are irrelevant to ordinary causation. In common cases of causation, not only is $prob(E/C^\dagger)$ significantly lower than $prob(E/C)$, but in all the intervening states, there are events that are related to previous and later states by this same probabilistic connection in a chain or continuum. In cases where an intermediate determining state contains no events that connect to previous and later causes with this probabilistic relation, we think of this as a case of a broken causal link and thus have a good reason to reckon c as unimportant. Important causes, the intuition goes, deliver their importance through a causal process. (See also, Schaffer 2001)

These considerations guiding our evaluation of causal importance are imperfect and far from exhaustive. But since my aim here is not to give a full account of causation, I will just suggest that evaluating whether c causes e by measuring *importance* offers some benefits over more robust analyses. First, because conflicting kinds of importance might bear on causal judgments, causation could be treated as sometimes involving interest-relative features. This may prove useful in acquiring a better fit between the theory of causation and judgments of culpability. Second, different kinds of causal importance might have irreconcilable differences, and if we can identify the source of conflict, we might resign ourselves to the presence of seemingly persuasive counterexamples that we cannot accommodate. Third, by shifting much more of the pragmatics of causation away from the physics, we can allow greater discrepancies between our theory of causation as it is in the world and causation as it seems to us. This lets us violate ordinary convictions about causation in order to achieve peace with physics while nevertheless accounting for the usefulness of problematic folk convictions. An instructive example is the case of causal asymmetry.

12.4 Causal Asymmetry Revisited

The conditions for a contributing cause involve only facts about nomological determination and involve no temporal asymmetry that makes the past fixed. Yet we can find a causal asymmetry in the criterion for important contributing causes. It arises from the use of the past hypothesis as a background condition implicit in $C^\dagger$. Assuming the arguments in Section 2.3 are good, the past hypothesis will fix most macroscopic facts about the past under localized counterfactual suppositions about the present. Supposing that we have a fully (bidirectional) deterministic dynamics, my wiggling of my finger, c, constitutes a contributing cause (on the theory so far) of Napoleon's invasion of Russia, E, but because $prob(E/C^\dagger)$ is nearly one, c doesn't count as an important cause. Indeed, present contributing causes of the past are so often unimportant, it is convenient and almost always permissible to ignore this kind of backward causation and just imagine causation to have an asymmetry where only earlier events can be contributing causes to later events.

However, even though the past hypothesis by and large fixes the past, there exist cases where the past counterfactually depends on the present, and these cases threaten to count as instances of backward causation. Nevertheless, it turns out that even where the past depends on the present, we are unable to exploit this dependence to accomplish anything practical. Hence, we are for practical purposes justified in assuming a uniform causal asymmetry and explaining away the limited counterfactual dependence of the past on the present in ways that are causally innocuous. I consider three cases below.

12.4.1 Common Causes

The treatment of counterfactuals in Section 12.2.3 permitted backtracking reasoning in the evaluation of what would have happened had c not occurred. This seems to license a mutual counterfactual dependence of two events on each other, and thus mutual causation, in cases where we would ordinarily say that the two events are merely effects of a common cause. For example, let v be the infection of a person with some specific virus, r be the rash that appears one day later made highly likely by v (in the sense that $prob(R/V^\dagger)$ is low), and let f be the fever that appears two days later made highly likely by v in the same way. Imagine there are no other likely mechanisms by which r or f can occur, for example, the patient is isolated, the rash has the signature color of this virus, and so forth. In this case, it is perfectly reasonable to claim counterfactual dependence of f on r, that if the patient hadn't gotten the rash, she wouldn't have gotten the fever on the grounds that there is no other plausible way to get the rash other than by having the virus. Nevertheless, counterfactual dependence is not a sufficient condition for one event being an important cause of the other, even when conjoined with nomological determination.

The importance of r as a cause of f is measured by $prob(F/R^\dagger)$, which is the likelihood that the fever occurs, given that the patient is in a state just like actuality but with no rash. Hypothetically removing merely the rash from the patient at some time t does not necessarily involve removing the virus as well. The microstates that determine $R^\dagger$ are all those molecular configurations where the blood vessels near the skin are smaller, but factors in the blood are left as is. Under the presumptions in the example, the virus would still be there in the body to make F likely, and so $prob(F/R^\dagger)$ is be high, and so the rash doesn't count as an important cause of the fever. If one argues that for some reason the virus needs to count as part of the rash, then the low value for $prob(F/R^\dagger)$ will signal that the rash-virus combination is an important cause of the fever, but this is no surprise since the rash-virus *is* an important cause of the fever in virtue of its viral part.

12.4.2 Faking Traces

Because traces in the future make highly likely the events of which they are traces, one might think we can affect the world by way of affecting what traces exist, and thereby backwardly cause an effect by way of creating the traces that occur afterwards. Indeed we *can* do this, but it always turns out to be merely a disguised case of what we have heretofore understood as ordinary forward causation.

We can safely suppose an absurdly generous upper limit on our power—that we have no more control over the world right now than we would have if we could freely instantiate any microstate of our choosing, compatible with the dynamical laws and the past hypothesis. Suppose I try to use this power to instantiate a state right now that will cause traces ten seconds from now (in the future) of my having thrown a stone into a pond five seconds from now, and thereby make it likely that the stone entered the pond. The existence of traces, the concentric outward traveling ripples on a smooth pond surface together with a stone bearing my fingerprint at the pond's bottom, and so on, do indeed make it likely that the stone entered the pond, but we can distinguish two distinct classes of dynamical evolution that start with my choice of microstate and eventuate in the traces. The first class includes all those evolutions that include the stone not entering the pond, that is worlds where the traces are misleading. Instantiating a microstate that causes *misleading* traces does nothing to make the stone enter the pond, because the stone does not in fact enter the pond. The second class includes all those evolutions that include the stone entering the pond. These are worlds where the dynamical evolution flows from my initial choice of microstate through the stone entering the pond and later to the ripples, and so on. All these worlds are just worlds we interpret as having the ordinary causal order, where I have caused the traces of the stone entering the pond by way of throwing the stone. Since I am unable to affect events usefully in a backwards way even when granted the extraordinary power to instantiate the microstate of my choice, I am equally unable in my more humble actual circumstances. And since the world cannot be controlled by way of such backwards causation, we can safely ignore it.

12.4.3 Single Trace Causation

The overall counterfactual fixity of the past arises through the joint effect of the past hypothesis and the existence of traces in the present, but the *recent* local past is still usually counterfactually dependent on the present. This results from the normal, lawful evolution of the world into the counterfactually postulated state. For example, there are no apples nearby, but if I were to see an apple right now in front of me, it would be that the apple existed in front of me a millisecond ago. This comports with the principle that current situations evolve naturally out of previous circumstances, with nothing miraculously teleporting into view or springing into existence *ab nihil*. The question is whether this short-range counterfactual dependence should be understood as a kind of backward causation, and the answer is that technically, according to the rule for measuring which contributing causes count as important, it is a case of backward causation. Nevertheless, it cannot be exploited to do anything practical, and so we can safely ignore it.

Suppose my control over the world comes by way of some limited control over what is going on in my head, what you might call an act of will, w, falling under a coarse-grained description W. The choices I could have made but didn't are $W^\dagger$, the worlds possessing microstates where (1) I choose something other than W, (2) the facts outside my head are just as they actually are, and (3) the past hypothesis holds. By the criterion for important contributing causes, the kind of past event over which I can have control will satisfy some event type E, such that $prob(E/W^\dagger)$ is low. Circumstances where $prob(E/W^\dagger)$ is low are situations where there are insufficient traces in the present to force the high probability of E.

For simplicity, we can imagine this situation as one where some event e has a single trace W in my mind with all other potential traces of e being shielded out by a physical barrier that blocks traces.

Assuming present traces really do counterfactually fix most macroscopic facts about the past, the macro-facts outside the shielded region, including macro-facts prior to the shielding's existence, cannot imply or make probable e's occurrence. Whether e occurs must as a consequence depend crucially on the microscopic details of the prior physical situation, just as macroscopic facts about who wins the lottery depend critically on microscopic facts about particular ping pong balls. This implies that the only kind of *macroscopic* counterfactual connection that e has to other parts of the universe come by way of w. There are different ways to interpret this situation. On one way of looking at it, w does backwardly cause e to occur, but since e has no causal connection to anything else at the macroscopic level, we are unable to gain knowledge of its effects or use the causal connection to accomplish aims other than causing events inside the shielded region. Hence, knowledge of this kind of causal connection is useless. On another way of looking at it, e dynamically arises from the chaotic microstructure of the universe, and any traces in my head that e depends on can be understood as merely a reliable detection of e. A reliable detector's state, after all, depends counterfactually on the state of the detected phenomena, but this in itself does not imply that the detector *causes* the phenomena to be what it is. Given the lack of practical utility for conceiving of the counterfactual dependence as backward causation, it is more convenient to think of the causation as unidirectional and interpret the dependence as a mundane instance of detecting e's occurrence.

12.5 Conclusion

The objective structure underlying causation, that is what the physics appears to tell us exists, is a local nomological determination relation among physical states at different times, possibly with additional stochastic relations, neither of which contains the kind of time asymmetry able to explain the kinds of asymmetries that occur in ordinary cases of causation among salient events. There is another feature of the physics that can explain the kinds of asymmetric phenomena in causation, features about the universe's boundary conditions. The boundary conditions do not support a counterfactual asymmetry at the microscopic level, but at the level of salient events, there is an important (to us) difference between the way past events and future events typically counterfactually depend on the present. Given a very strong constraint on the physics at the end of time we call the past, salient events in the past, tend to depend less on variations of the present state than do similar future events. More precisely, past events are usually resilient under counterfactual changes to the present that involve only small, localized modifications, or changes that seem as though they can easily arise from small modifications in their recent past. Furthermore, the boundary condition implies that the evolution of the physical state towards the past is highly conspiratorial. As a consequence, the kinds of actions we are capable of performing either have no effect on the past or they have an effect on the past that is unpredictable and uncontrollable. This being the case, whatever backwards causation does exist is of no practical value, and is safely ignored by our ordinary concept of causation. Because the only kind of manipulable causation is forward causation, we are pragmatically justified in projecting the rough counterfactual asymmetry into the world, treating it as a universal asymmetry localized in material processes. Causation is ultimately our amalgamation of the nomic determination structure in the fundamental physics and our misconceiving the manipulability asymmetry as somehow built into the local physics.

References

Albert, D. (2000). *Time and Chance*. Cambridge: Harvard University Press.
Downing, P.B. (1959). 'Subjunctive Conditionals, Time Order, and Causation', *Proceedings of The Aristotelian Society*, 59: 125–40.
Edgington, D. (1995). 'On Conditionals,' *Mind*, 104 (414): 235–329.
Hall, N. (2000). 'Causation and the Price of Transitivity', *Journal of Philosophy*, 97: 198–222.
Kutach, D. (2001). *Entropy and Counterfactual Asymmetry*, Ph.D. Dissertation, Rutgers University.
Kutach, D. (2002). 'The Entropy Theory of Counterfactuals', *Philosophy of Science*, 69:1.

Lewis, D. (1979). 'Counterfactual Dependence and Time's Arrow', *Noûs*, 13: 455–76. Reprinted in *Philosophical Papers, Volume 2* (1986). Oxford: Oxford University Press.

Russell, B. (1913). 'On The Notion of Cause', *Proceedings of the Aristotelian Society*, 13: 1–26.

Schaffer, J. (2001). 'Causation, Influence, and Effluence', *Analysis*, 61:11.

—— (2003). 'The Metaphysics of Causation', *The Stanford Encyclopedia of Philosophy*, Edward N. Zalta (ed.), http://plato.stanford.edu/archives/spr2003/entries/causation-metaphysics/.

Schaffer, J. (2004). 'Counterfactuals, Causal Independence and Conceptual Circularity', *Analysis*, 64: 284.

13

Causation, Counterfactuals, and Entropy

MATHIAS FRISCH

13.1 Introduction

Bertrand Russell famously argued that the notion of cause has no place in modern fundamental physics, where it has been replaced by the concept of functional dependency (Russell 1918). Fundamental dynamical laws specify how one thing follows after another, but since these laws are time-symmetric, they do not support an asymmetric distinction between cause and effect of the kind that appears to be part of our commonsense notion of cause. In a recent paper (Field 2004), Hartry Field has endorsed Russell's conclusion but has pointed to a problem resulting from Russell's thesis. Even if we were convinced by Russell's thesis, we cannot simply excise any 'weighty' asymmetric notion of cause from our conception of the world, since, as Nancy Cartwright has argued (Cartwright 1979), just such a notion seems to be essential for the distinction between effective and ineffective strategies: In deliberating which actions will further our goals we need to appeal to a robust distinction between causes and their effects, for, intuitively, we can influence the occurrence of an event by affecting the occurrence of its causes but not by influencing its effects. In fact, Field maintains that trying to reconcile the apparent need for causation in a theory of effective strategies with Russell's thesis is 'the central problem

---

I would like to thank audiences in Sydney, at Reed College, the University of Munich, and the University of Konstanz for their comments and criticisms on parts of various earlier incarnations of this paper. I want to thank especially Adam Elga, Gerhard Ernst, Doug Kutach, and Huw Price, for helpful discussions and Barry Loewer for extremely detailed comments and criticisms on an earlier draft of this paper.

in the metaphysics of causation'. What, then, is the source of the causal asymmetry? How do we locate causation within a fundamental universal physics with time-symmetric laws? And if invoking the concept of cause in fundamental physics did indeed prove to be a 'relic of a bygone age', how is it that we have come to understand the world in asymmetric causal terms?

In this chapter, I want to examine one recent attempt at providing answers to these questions: David Albert's entropy-account of causal influence (Albert 2000, ch. 6).[1] Albert appears to agree with Russell's conclusion that the fundamental dynamical laws of physics neither support nor require a 'weighty' concept of cause, yet he argues that the fact that the universe had an extremely low-entropy past can account for our possession of such a concept. That is, according to Albert, the causal asymmetry can be explained in terms of an asymmetry of physical initial or boundary conditions. Yet Albert denies that there is a completely general asymmetry of causation or causal influence. Rather, he claims that the causal asymmetry is grounded in a counterfactual asymmetry that exists for small, yet macroscopic hypothetical interventions of the kind for which we humans could in principle be responsible. Thus, if Albert were right, there is no tension between fundamental physics and the demands of a theory of effective strategies despite the fact that the micro-dynamical laws *alone* do not support an asymmetric notion of cause. For the causal asymmetry would turn out to be due to an asymmetry of initial conditions and not due to an asymmetry of the dynamical laws; and the asymmetry would arise precisely for those kinds of possible macro-interventions that play a role in a theory of effective strategies. Field's problem would be solved.

Similar to David Lewis (Lewis 1986a, 1986b), Albert advocates a counterfactual analysis of causation. Thus, at the heart of his account is an argument for an asymmetry of counterfactuals appealing to entropy considerations, which in turn is meant to explain the asymmetry of causation. Yet Albert's account has the advantage over Lewis's that it does not invoke the dubious notion of miracles that can be ranked with respect to their sizes. Moreover, while Albert adopts Lewis's core idea that facts about the present allow us to

---

[1] Interestingly, Field himself also suggests that the solution to the problem may lie in recognizing the importance of statistical regularities to the concept of causation, similar to those that arguably can account for the concepts of entropy and temperature. As Field points out, such a solution to the metaphysical problem has the consequence that the causal asymmetry is absent on the micro-physical level.

make inferences about the past in a way different from inferences about the future—Lewis's notion of *postdeterminants* is echoed in Albert's notion of *records*—his account does not rely on the problematic (and in fact provably false!) thesis of an asymmetry of overdetermination between the past and the future.[2] Thus, Albert's account might be understood as offering a defense of Lewis's overall project that avoids some of the latter's deep problems.

In this chapter I want to argue, however, that Albert's thermodynamic account of the causal and counterfactual asymmetries is problematic as well. Section 13.2 will be devoted to introductory remarks in which I will draw some distinctions that will be important in the subsequent discussion. In Section 13.3, I will summarize what I take Albert's argument for the time-asymmetry of counterfactuals to be. I will criticize the account in Section 13.4, where I will argue that, as it stands, Albert does not offer compelling reasons to accept his account of the causal and counterfactual asymmetries and that, in fact, there are good reasons to reject at least part of the account. In Section 13.5, I will briefly discuss a version of an entropy account, inspired by Albert's account, that Barry Loewer defends in this volume. I will end with a brief conclusion.

13.2 Counterfactuals and Causes

Albert, like Lewis before him, argues for a time-asymmetry of counterfactuals: In certain standard contexts the future counterfactually depends on the past, but the past does not counterfactually depend on the future. Yet one might appeal to considerations similar to those advanced by Russell to argue that all scientifically respectable counterfactuals are time-symmetric. Take a physical theory with time-symmetric dynamical laws that pose a well-defined initial value-problem. Then we can both predict and retrodict the evolution of a system governed by that theory, given the state of the system at a certain time. Moreover, the laws do not only allow us to derive the evolution of an actual system, but also allow us to determine how the evolution would have been different had the system's 'initial' state been different, where it makes no difference whether the 'initial' state occurs before or after the state in which we are interested. Thus, the laws seem

---

[2] For criticisms of Lewis's account see (Elga 2001) and (Frisch 2005).

to support both forward-looking and backtracking counterfactuals equally: If the state of the system at the initial time were different, both its past and its future would have to be different. Just as the future counterfactually depends on the present initial state, so apparently does the past.

Now, both Lewis and Albert believe that at least in worlds as complex as ours we do not evaluate the truth of a counterfactual by simply letting the relevant counterfactual present state of the world evolve in accord with the dynamical laws. Lewis appeals to a complicated similarity metric between worlds and maintains that the closest counterfactual worlds to ours are 'miracle worlds' diverging from the actual world, in which the laws of the actual world are not exceptionless truths; while Albert argues, as we shall see in more detail shortly, that our inferences about the past are also constrained by the hypothesis of a low-entropy past. Yet there clearly are standard scientific contexts in which we draw inferences about states of a system at different times but in which our reasoning does not presuppose a rich and complex world and not one with thermodynamic features. In those contexts we draw inferences based on special, highly idealized circumstances in which the system in question can be represented as a relatively simple, perhaps purely macroscopic system—for example, as a mechanical or electromagnetic system. And, *pace* Lewis and Albert, the a appropriate procedure for drawing counterfactual inferences in such cases can simply be to investigate the evolution of possible states of the system, given certain initial or final values and the relevant dynamical laws. For example, in the context of examining possible trajectories of balls on a billiard table, it might be correct to assert the backtracking counterfactual that if a certain ball had gone into the corner pocket, then it would have to have been struck differently from the way it actually has been struck; just as it might be the correct thing to say that if the ball were struck differently, then it would roll into the corner pocket.

Thus, there certainly appear to be contexts for evaluating counterfactuals in which forward-looking counterfactuals are not privileged. Nevertheless, I think that Lewis and Albert are correct in claiming that there *also* is a sense in which we think that the future but not the past depends on the present and, hence, that there also are contexts in which counterfactuals are time-asymmetric.

We appear to take the past to be counterfactually independent of the present in contexts which we intuitively think of as causal. On this point,

I think, there is relatively widespread agreement, even though accounts of the precise relation between causal and counterfactual claims differ widely. One way to spell out the connection between causation and asymmetric counterfactuals is in terms of the notion of hypothetical interventions: Interventions into a cause influence the occurrence of its effects, but intervening into a putative effect cannot influence the occurrence of its causes (see Pearl 2000, Woodward 2003). One might then try to distinguish between asymmetric, intuitively causal contexts, on the one hand, and symmetric contexts, on the other, by invoking a distinction between *closed* and *partially open* systems. Counterfactuals associated with closed systems appear to be symmetric: Each set of possible initial conditions at a time defines a different closed system whose past and future evolution is given by the relevant dynamical laws. Systems with different initial states will in general have both different futures and different pasts. By contrast, counterfactuals associated with interventions from the outside into an otherwise closed system might be thought to be asymmetric, since interventions may be taken to affect only the future evolution of the system but not its past.[3]

Two aspects of this scheme are worth being made explicit. First, the scheme is most naturally spelled out as not advancing an account of the causal asymmetry. *Intervention* arguably is itself a causal notion and the account simply stipulates that interventions influence the causal 'future' of a system and not its causal 'past'. Second, it appears to be crucial to this way of thinking about causation that causal systems have an environment or 'outside' from which interventions can occur. This may suggest the Russellian claim that there is indeed no room for the notion of cause in a *universal* physics that aims to have models of the universe as a whole among its class of models.[4]

In sharp contrast with this scheme, Albert proposes an account of 'intervention' that applies to closed systems as well, thereby promising to provide a place for causal notions even within a universal physics. Hypothetical interventions, on Albert's account, are treated by postulating counterfactual initial states, where both future and past evolutions of the

---

[3] The notion of hypothetical interventions functions in some ways similar to Lewisian miracles, with the advantage that an interaction of the system with its environment takes the place of counterlegal time-evolutions.

[4] See, for example (Hausman 1998).

system in question are then determined by what Albert calls our *normal procedures of inference* (NPI), which include use of the fundamental dynamics, but do not invoke Lewis-style miraculous violations of the laws. Instead of appealing to a difference between open and closed systems, Albert's account suggests that the difference between symmetric and asymmetric counterfactuals is due to a difference between thermodynamic systems and non-thermodynamic systems. In the former case, Albert argues, a counterfactual present state that differs only locally from the actual present will be overwhelmingly likely to have evolved from a past identical to the actual past.

A central *explanandum* for Albert is, as he puts it, our 'fundamental conviction [...] that the future *depends on what happens now*—that the future depends on what we *do* now—in a way that the past does not' (Albert 2000, p. 125).[5] Thus, what he wants to 'get to the bottom of' (*ibid.*) is how it is that we have come to have a certain conception of the world. He takes for granted that our conception includes the notion of an asymmetric causal dependence of the future on the present and his account is meant to offer an explanation of this aspect of our conception.

What would it be for an account to be successful in getting to the bottom of why we conceive of the world in time-asymmetric causal terms? One possible explanation of our belief in an asymmetry of causal dependence would be to give a genetic account of our conviction. The aim of such an account could *either* be to show how humans in general have actually developed a certain conception of the world *or* to offer a developmental story of how individual human beings come to acquire such a conception during childhood. Another kind of explanation of our conviction would be to propose a philosophical reconstruction that does not trace the actual mechanisms that have produced our conception but instead simply argues that being able to draw certain distinctions is advantageous to beings like us. If one could show that conceiving of the world in certain ways was to our advantage, that might go some way towards getting to the bottom of our having a certain conviction, since this might be part of an evolutionary account of why it is that we might have developed a certain conception. Finally, our concern might be epistemological and we might be interested in showing that, *contra* Russell, at least part of our causal conception of the

---

[5] The italics in this and in all subsequent quotes are in the original, unless otherwise noted.

world is justified or legitimate in that it can be grounded in considerations arising from fundamental physics.

Which of these different projects is Albert engaged in? Unfortunately Albert is not entirely clear on this issue. Partly his concern seems to be epistemological in that he appears to be interested in delineating to what extent our conviction in an asymmetry of dependence is justified. Understood in a certain sense, our conviction is clearly false, Albert would say: If we assume a deterministic physics with time-symmetric laws, then the past is determined by the present just as the future is. But Albert then argues that there also is a sense in which our conviction is supported by a kind of counterfactual reasoning we engage in, which in turn is grounded in the entropy-asymmetry. Yet there are also central aspects of Albert's account that are quite explicitly and unambiguously aimed at explaining how we actually go about making certain inferences—inferences involving records of the past—that are at the heart of our belief in an asymmetry of dependence. I will address the question to what extent Albert might succeed with these different projects in Section 13.4.

What exactly *is* our fundamental conviction? Despite the ease in which Albert moves from the main sentence in the above quotation to the sub-clause set off by dashes, they in fact express two quite distinct convictions. On the one hand, we believe in a general causal dependence: Events have causes and these causes are in an event's past, or at least not in its future—quite independently of whether or not we take the events in question to be under our control. On the other hand, there is our conviction that our own actions can influence the future but not the past. Different accounts of causation will differ on the proper relation between these two convictions. On many views, agency is itself a causal notion and the asymmetry characterizing our actions is merely a sub-class of a more general causal asymmetry. That is, on such a view an account that got to the bottom of our convictions would have to explain our belief in a general asymmetry of dependence and the belief in an asymmetry of agency would simply follow as a special case. By contrast, so-called *agency accounts* of causation, such as the ones defended by Menzies and Price (1993) and Ahmed (ch. 6 in this volume), argue that the notion of agency is primary and that our general causal convictions are the result of our tendency to project the asymmetry of agency onto the objects, in a manner similar to what is often held to occur in the case of secondary qualities.

Even though Albert himself does not raise this issue explicitly, it is obvious that his account squarely falls into the first category. For he argues that in worlds like ours 'more or less *any present feature of the world you can think of*' can amount to a '*causal handle*' on 'more or less any future one' (Albert 2000, p. 128), simply because the relevant counterfactuals that ground our concept of causal dependence come out true. That is, on Albert's account there is a broad class of counterfactuals—those with antecedents that postulate small macroscopic changes to the present—that are time-asymmetric because what he calls 'our normal procedures of inference' for evaluating such counterfactuals yield time-asymmetric results. Our procedures of inference presumably involve calculating probabilities of past or future macro states conditional on present macro states. This asymmetry accounts for the time-asymmetry of causal dependence and, in particular, the asymmetry of influence.

By contrast, the entropy account defended by Loewer (ch. 11 in this volume) privileges counterfactuals involving possible human interventions, similarly to agency accounts of causation. Loewer argues that statistical mechanical considerations entail in the first instance that so-called 'decision counterfactuals' are time-asymmetric. This asymmetry of decision counterfactuals is then 'spread over the objects': Other counterfactuals, not postulating human decisions, are time-asymmetric, on this account, in virtue of the fact that we take their antecedents to be the result of small micro events that are analogous in size to human decisions.

On Albert's account, the counterfactual 'If the billiard ball were 20 cm to the right of its actual location, then the past evolution of the balls on the table would have to have been different from the actual evolution' comes out false, because of the result of applying our normal statistical procedures of inference to the counterfactual antecedent state. By contrast, on Loewer's account, the counterfactual comes out false, since we treat the change in the ball's present location as if it were the result of a human decision (and intervention). In this chapter I will focus mainly on Albert's entropy account of causation. I discuss Loewer's account briefly in Section 13.5 and more fully in Frisch (forthcoming).

13.3 Albert's Argument

Instead of using the more familiar locution of 'the cause of an event', Albert introduces the notion of one event being a 'causal handle' for another. In light of the remarks in the preceding section it may come as somewhat of a surprise, however, that he suggests that one can introduce this notion even in the absence of any thermodynamic considerations. Albert points out that if we constrain the *remote* past of any physical system, then only very special alterations of the present can lead to a different *recent* past, while many alterations of the present may lead to a different future. To illustrate this point he asks us to consider a collection of billiard balls on a frictionless plane such that ball 5 is currently stationary with the additional constraint that ball 5 was moving 10 seconds ago. Given the additional constraint, that ball 5 has been involved in a collision in the past 10 seconds is determined by facts about the present state of ball 5 *alone*. That is, alterations to the present state of the balls *not* involving changes in the state of ball 5 cannot change the fact that ball 5 was involved in a collision during the last 10 seconds. (In fact, there will be many alterations to the present not involving ball 5 that will result in a present state inconsistent with the additional constraint on the past.) Yet there are many changes to the state of the balls not involving ball 5 that will result in a different future evolution of ball 5. From this Albert concludes that there are a far wider variety of 'what we might call *causal handles* on the future of the ball in question here, under these circumstances, than there are on its past' (Albert 2000, p. 128).

An obvious objection at this point is that evaluating counterfactual situations compatible with an apparently *ad hoc* time-asymmetric constraint on the past, but not the future, tells us nothing about the causal structure of the case and, in particular, cannot license any conclusions about an asymmetry of causal influence. Why are 'these circumstances' the right circumstances for assessing the causal structure? The asymmetry obviously is the result of imposing an asymmetric constraint on possible alterations. If instead we were only interested in possible changes to the present state of the system of billiard balls that are compatible with an additional constraint on the *future* evolution of the system without in any way constraining the past, then many more backtracking than forward directed counterfactuals would presumably come out true. Why then should we impose a constraint only on the past? One answer might be that the past is fixed while the future is open. Alternatively, we could maintain that we ought to keep the causal history of the event fixed, but not its future effects. But obviously these answers would beg the question, if our aim is to give a counterfactual analysis of the notion of causal influence.

Ultimately, however, Albert is not interested in non-thermodynamic systems such as the idealized billiard balls. Rather, his aim is to argue that the counterfactual asymmetry arises in systems that are complex enough to exhibit thermodynamic features. Thus, Albert himself would probably agree that introducing the notion of a causal handle simply by reflecting on the consequences of imposing a time-asymmetric constraint on the motion of the billiard balls is somewhat misleading. The new ingredient in the case of thermodynamic systems is that there is a special asymmetric constraint on the past—the hypothesis of an extremely low-entropy initial state. This condition, which Albert calls 'the *past-hypothesis*', is a central assumption in standard accounts of the thermodynamic asymmetry that the entropy of a closed system never decreases. There it is needed to avoid the *reversibility objection* against the most straightforward attempt of accounting for the increase in entropy. While one can argue that the entropy of a given low-entropy macro-state increases, assuming an intuitively plausible probability distribution over micro-states and a Newtonian time-symmetric micro-dynamics, the same type of argument also allows us to conclude wrongly that the present macro-state is at a local entropy minimum and evolved from a higher entropy past. The undesirable retrodiction that entropy decreased in the past can be blocked, if the distribution of micro-states is conditionalized not only on the present macro-state but also on a low-entropy past and, ultimately, a low-entropy initial state of the universe.

Now, why should we keep the past-hypothesis fixed in assessing counterfactual changes to the present? How is this hypothesis different from the apparently question-begging assumption to hold the past state of billiard ball 5 fixed? We may, as Albert claims, have good inductive reasons to believe that the past-hypothesis is satisfied in the actual world. Yet there is much that we know inductively about the future in the actual world that we do not keep fixed in assessing counterfactual changes to the present. Albert's answer is that in assessing the truth of counterfactuals we need to consider other worlds that are in important ways like ours. In particular, the counterfactual worlds to which we appeal in assessing the results of counterfactual changes to the present have to license the same normal procedures of inference as the actual world does. And Albert argues that these procedures rely crucially on assuming the truth of the past-hypothesis. If all counterfactual reasoning must rely on our normal procedures of inference from the state of the world at one time to the state at other times and these procedures presuppose the past-hypothesis, then all counterfactual reasoning must presuppose the past-hypothesis.

In somewhat more detail, Albert argues the following. In order to assess the truth of a counterfactual, where the antecedent involves some small yet macroscopic hypothetical alteration to the actual world, we need to look at counterfactual worlds that are like the actual world except for the small change and then use our normal procedures of inference to determine the past and future evolutions of such worlds. In the case of forward-looking counterfactuals, these procedures amount to taking the present macro-state of the world, assuming an equi-probability distribution over micro-states compatible with that macro-state and evolving the state forward in accord with the dynamical laws. Thus, counterfactuals such as 'If the light switch were flipped, then the light would go on' come out true (assuming that in the actual world the light is and remains off and I don't flip the switch), if most micro-states compatible with the initial macro-state evolve into micro-states corresponding to a macro-state such that the light is on.[6]

However, backtracking counterfactuals such as 'If the light switch were flipped, then the light would have to have been on prior to that' are not supported by our normal procedures of inference.[7] If we simply evolved the counterfactual present state backward in accord with the macroscopic

---

[6] Note that given his appeal to procedures of inference, one might think that, unlike Lewis, Albert is offering *assertibility* conditions of counterfactuals and not *truth* conditions. Yet Albert concludes his discussion of counterfactuals by saying 'And it follows—if all this is right—that the future does indeed counterfactually depend on what we do now, and the past [...] does not' (*TC*: 130). This suggests that Albert is committed to the view that whatever follows from our normal procedures of inference is true.

[7] This example is not Albert's own, who also follows Lewis in trying to stack the deck through his choice of examples. Albert considers the forward-looking counterfactual 'If the president pushed the button, there would be a nuclear explosion' and contrasts it with the backtracking counterfactual 'If the president pushed the button, then there would have been an explosion'. He says that there 'are (for example) no worlds *at all*, even *remotely* like our own, in which the [normal procedures of inference] translate small hypothetical present differences in the present position of anybody's finger into differences between a certain thermonuclear device's exploding or not exploding two minutes *ago*' (*TC*: 129–30). While it might well be that this particular backtracking counterfactuals comes out false, this of course does not show that backtracking counterfactuals are false in general. The passage continues as follows: 'And that (as I said before) is precisely because there is a past-hypothesis and not a future one. That (to put it another way) is because there are—vis-à-vis such things as the *past* explosion

regularities with which we are familiar, then the light's being off after the light switch was flipped presumably would have to have been preceded by the light's being on earlier. For generally, flipping the switch is accompanied by a change in the light's state from on to off or vice versa. And this fact would seem to allow us to draw inferences in a time-symmetric manner. We can, it seems, infer the *past* state of the light from its present state and the fact that the switch was flipped, just as we can infer its *future* state from its present state and the fact that the switch was flipped. But Albert argues that this inference would violate the presupposition that there are *records* of the past. In our example, records of the light's being off in the actual world might include the fact that there is no light that has recently escaped through the window and that the light bulb is relatively cold. Albert's notion of record plays a role analogous to that of Lewis's postdeterminants: A record is a relatively localized fact about the present from which we can infer the occurrence of some event in the past. If the present contains a record of the light's having been off, then we can infer that the light was in fact off. Like Lewis, Albert believes that we take many of the local facts we do hold fixed to be sufficient for the occurrence of certain (relatively localized) events in the past. Yet, unlike Lewis, Albert is well aware that there are no local facts about the present which *alone* are sufficient for the occurrence of some past event. Rather, inferences appealing to records, Albert holds, are always inferences from facts at two different times to a fact at a third time in between, since facts cannot function as records of the past unless we assume something about the more remote past that functions as a 'ready condition'.

Recall Albert's example of the collection of billiard balls. Albert points out that the fact that ball 5 is currently stationary functions as a record of the fact that the ball underwent a collision in the last ten seconds *given* the additional constraint that ball 5 was moving ten seconds ago. In other words, that the ball was moving 10 seconds ago functions as ready condition allowing us to record the ball's collisions in terms of its present state of motion. Without the ready condition we cannot infer whether ball 5 underwent a collision from the current state of ball 5 alone but would need to know the present state of the entire collection of balls.

---

of thermonuclear devices (or the lack of them)—such things as *records*, as *memories*' (*ibid.*) I will criticize equating the truth of the Past Hypothesis with the existence of records later.

What does all this have to do with the past-hypothesis? Ultimately, Albert claims, the single assumption that on its own can ensure that we can treat facts about the present as records of the past is the past-hypothesis. In treating a fact as record we need to presuppose that some ready condition in the more remote past was satisfied. But how do we know the latter fact? Again, we need some record of the ready condition's being satisfied, which in turn requires an even earlier ready condition. According to Albert, this regress ends with the past hypothesis, which functions as a first 'mother of all ready conditions'. That is, Albert's amended version of Lewis's thesis of overdetermination of the past by the present is that *given the past-hypothesis*, localized facts about the present are records of the occurrence of certain localized facts in the past. Thus, while the light bulb's being cold alone is not sufficient for the light's having been out, it is sufficient, or at least is overwhelmingly probable, on Albert's view, in conjunction with the assumption that our universe had an extremely low entropy past.

The structure of Albert's argument, then, is this. Albert argues for two claims:

(*i*) The past-hypothesis is true, if and only if there are records of the past.

(*ii*) If there are records of the past, then there is no counterfactual dependence of the past on the present.

From (*i*) and (*ii*) Albert's first conclusion follows:

(*iii*) If the past-hypothesis is true, then there is no counterfactual dependence of the past on the present.

If we add to this a counterfactual account of causation, we arrive at the final result that if the past hypothesis is true, then the future, but not the past, causally depends on the present—that is, in Albert's own terminology the present contains *causal handles* on the future but not on the past.

At the heart of the account is the idea that the past counterfactually depends on the present, if there is an actual present event c and actual past event e such that our normal procedures of inference license us to accept the counterfactual 'If c had not occurred e would not have occurred.' This raises the question as to how exactly according to our normal procedures of inference we evaluate the truth of counterfactuals. Yet given the question's central importance to Albert's account, it is surprisingly difficult to find a precise statement of Albert's answer to this question. The most plausible proposal (which has been implicit in some of what I said above) appears to be that our normal procedures of inference involve calculating conditional probabilities. On this proposal the truth of the counterfactual 'If c had not occurred, e would not have occurred' is assessed, on Albert's account, by looking at the class of worlds that satisfy the past-hypothesis and whose present macro-states match the macro-state of the actual world as much as possible, given that c does not occur in those worlds. The past and future of these worlds is then determined by the micro-dynamical laws. If in most of these non-c worlds e does not occur, then the counterfactual is true, otherwise it is false.[8]

Thus, on what I take to be the most plausible reconstruction of Albert's account, the past does not counterfactually depend on the present, exactly if for all (suitably localized and small) actual present macro-events c and for all past macro-events e the following condition is satisfied: The conditional probability of es not occurring is extremely low, *given* the micro-dynamical laws, the past-hypothesis, and that the present is unchanged except for cs not occurring;[9] that is, if the past does not counterfactually depend on the present, then

(1) $\Pr(\sim e / \sim c \& S_a \& PH) \approx 0$

for all actual events c and e such that e is in the past of c, where PH is the past-hypothesis. Here I am taking events to be the goings-on in some region of space at a particular time. The event c is the complete actual present macro-state in some suitably small region of space. S_a is the remainder of the present macro-state of the world. Thus, $c \& S_a$ is the complete present macro-state of the world. Also I am assuming that $\sim c \& S_a$ is nomically possible—that is, that the occurrence of c is not implied by S_a together with synchronic constraints imposed by the laws.

---

[8] A similar scheme for assessing assertibility conditions for counterfactuals is advocated in (Kutach 2002).

[9] One might worry that since conditional probabilities come in degrees it is not clear how this can result in conditions for the truth or falsity of backtracking counterfactuals. But since the relevant thermodynamic probabilities are usually either absurdly small or very, very close to one, this might perhaps license Albert's conclusion that 'the future does indeed counterfactually depend on what we do now, and the past [...] does not'.

13.4 Criticism

13.4.1 Records and Ready Conditions: The Puzzling Regress

Let us assume for the moment that Albert can indeed establish claims (*i*) and (*ii*). Would that be enough to explain our fundamental conviction that the future depends on the present, but the past does not? Our conviction, on Albert's account, is due to how we reason counterfactually. According to (*ii*) there is a connection between counterfactual reasoning, on the one hand, and memories or records, on the other. Quite plausibly, such a connection can indeed explain why there are certain standard contexts in which we take the past to be counterfactually independent of the present. We have memories and records of the past but not the future; and this fact might very will be the reason for why we take the past to be counterfactually independent of the present. For arguably, we keep the past, but not the future, fixed in assessing the results of counterfactual changes to the present, precisely because we have memories of the past.

Now, even the connection between the existence of memories and the fixity of the past is not completely obvious and straightforward. For it does not strictly follow from the fact that we keep all those events in the past fixed of which we have memories or records that we should keep the *entire* past fixed when reasoning counterfactually. Hypothetical changes to the present do not affect what we *know* to have happened, we might think, but what about past events of which there are no records or traces? Or past events of which we do not know there to be traces? Or past events whose traces are unknown to us? It is conceivable that these cases could have been treated differently in our reasoning practices, and an account that promises truly to get to the bottom of our fundamental conviction ought to offer an explanation for why our conviction apparently does not distinguish between them. Why do we believe that the entire past is independent of what we do now and not just those events of which we have memories? But perhaps the fact that we take the entire past to be counterfactually independent of the present can simply be explained by a tendency of ours to generalize from those events of which we have records or memories—and which we therefore keep fixed—to the past as a whole.

More problematic than Albert's account of the role of records in counterfactual reasoning is his appeal to a low-entropy past to explain our inferential practices involving records. According to Albert, 'the reason there can be records of the past and not of the future is nothing other than that it seems to us that our experience is confirmatory of a past-hypothesis and not a future one' (Albert 2000, p. 118). And indeed, that the past-hypothesis holds is, for Albert, strictly equivalent to the fact that there are records and memories of the past; the second claim is just the first when we 'put it another way' (Albert 2000, p. 130). Thus, Albert's account might be intended to offer at least a partial explanation of why it is that we engage in reasoning based on records and in causal reasoning. First, such reasoning would be impossible in a world without a past-hypothesis; and second, the fact that our reasoning based on records can be reconstructed as arguments that assume the past-hypothesis as premise shows why it is advantageous to possess certain conceptual distinctions or to have the ability to draw certain inferences. Albert appears to argue that our ordinary reasoning practices appealing to records happen to latch on to an important feature of the world—a feature that becomes evident once our inferential procedures are reconstructed as involving an appeal to the past-hypothesis. This purported insight might then be cited as part of an evolutionary explanation for why beings like us might have developed certain reasoning practices.[10]

But this does not yet fully capture what Albert says that the role of the past-hypothesis in our reasoning is. As we have seen, Albert argues that every inference based on records is, as a matter of fact, an inference from two times—from a record and an earlier ready condition—to a time in between: 'The sort of inference one makes from a recording is [...] from *two* times to a *third* which lies *in between them*' (Albert 2000, p. 117). Thus, anyone using a fact about the present as record of the past needs to possess some information about the more distant past as well. The puzzle for Albert then is 'how it is that we can ever manage to *come by* such information' (Albert 2000, p. 118). This additional information needed to draw inferences based on records cannot be obtained by means of retrodiction, Albert points out, because otherwise such inferences would ultimately be reducible to standard retrodictions from 'initial data' and the laws. Thus, he argues, the fact that some ready condition obtained itself has to be known through a further record—a record which in turn requires knowledge of a second

---

[10] Loewer (ch. 11, in this volume) is much more explicit about the fact that he wants to give an account of our concept of cause that shows its evolutionary advantages.

ready condition. This, according to him, leads to a regress which can be blocked *only if* there is 'something we can be in a position to *assume* about some other time', where this other time 'must be *prior in time* to everything of which we can potentially ever *have* a record, which is to say that it [that is, the 'something' we have to be in a position to assume] can be nothing other than the initial macrocondition of the universe as a whole' (Albert 2000, p. 118). Thus, according to Albert there is a puzzle concerning how we '*come by*' information regarding ready conditions, or how ready conditions are 'established' and this puzzle results in a regress that can only be blocked by positing the past-hypothesis as 'mother of all ready conditions'.

But what exactly is the puzzle? I think the only way to interpret what Albert actually says without doing too much violence to the text is to read him as raising a puzzle concerning how we *actually* reason. If inferences based on records are in fact inferences from two times to a third time in between, then nobody can actually draw inferences from putative records without having information concerning the relevant ready conditions and Albert's account is meant to explain how we arrive at this additional information. Albert introduces the role of ready conditions by discussing the billiard ball example that we encountered above: the fact that ball 5 is currently stationary is *by itself* not a record of anything and knowing this fact alone does not put us in a position to infer anything about the ball's past. The present state of ball 5 only becomes a record of past collisions when it is conjoined with the additional fact that the ball was moving ten seconds ago. Thus, we simply cannot draw any inferences about the past (beyond mere guessing) based solely on the 'record state' of ball 5 and without having some belief or making some assumption about the ready state of our 'recording device'. The example illustrates Albert's view that the record relation is a function from two variables—a record variable and a ready state variable—to a variable representing the recorded state. To compute the value of the variable representing the recorded state we need the values of *both* other variables as inputs. If we are concerned, as Albert is, with the question as to how we can make inferences about times other than the present based on what we know about the present, then it might indeed seem quite puzzling how we, as a matter of fact, come to possess information about the value of variables representing earlier ready conditions.

How else might we understand Albert's puzzle? Perhaps an alternative interpretation, of the puzzle is as a puzzle about justification. On this interpretation, Albert's worry is not how we as a matter of fact establish or come by certain ready conditions but how we *justify* our assumptions about ready conditions and, therefore, our inferences based on records. Admittedly, this second interpretation does some violence to the text. The questions as to how we ever *manage to come by the information* that ball 5 was moving ten seconds ago and how a ready condition is *established* seem straightforwardly to be concerned with the question as to how we as a matter of fact can obtain certain information about the past rather than with how we justify our beliefs about past ready conditions. Nevertheless, I want to explore this second interpretive option in what follows as well.

Whatever the correct interpretation of Albert's puzzle is, it is clear that the role the past-hypothesis is meant to play in its solution is one *in addition* to being a necessary condition for the possibility of records and in addition to explaining why reasoning based on records can be valuable in virtue of its tracking certain features of the physical world. For whatever the role of the past-hypothesis as 'mother of all ready condition' is in solving his puzzle, it is a role that Albert worries we may instead, albeit mistakenly, be tempted to assign to the ready state of our brains during normal sense perception (see Albert 2000, p. 118, footnote 6). But the claim that it is a necessary condition for successful reasoning based on records that our brain be in some appropriate ready state does not conflict with the claim that the past-hypothesis is a necessary condition for the possibility of records. And if Albert's aim was (merely) to establish that, as Loewer puts it in Chapter 11 (p. 323) of this volume, 'the information expressed by SM-counterfactuals [and by records] is important for us because it tracks the statistical mechanical probability distribution' (Loewer this volume), then an appeal to the ready state of the brain as mother of all ready conditions, as alternative to the past-hypothesis would not only be mistaken, as Albert claims, but would simply be nonsense. By contrast, an assumption about the condition of our brains might at least with some initial plausibility be offered as an alternative answer both to the question as to how we as a matter of fact go about to establish ready conditions and to the question as to how we might justify our inferences based on records: we as a matter of fact come by information about the past by consulting our memories and we justify beliefs about the past by appealing to the veridicality of our memories. Albert's footnote appears to be directed against such a view.

Albert's paradigm example of a record, as we have seen, is the present state of motion of a billiard ball, which can function as a record of past collisions, given the ball's earlier state of motion as 'ready condition'. It is obvious in this example that any inference to the recorded state has to be an inference from both the record state and an earlier ready state. Yet this is less obvious for a different kind of record—our memories of past events. In the former case, what the content of the record is—what the record state is a record *of*—is not given by the record state alone but only in conjunction with the ready state. By contrast, memories are intentional states and it is part of the content of a memory what it is a memory of. Thus, there is an important difference between the state of the billiard ball as record and our memories: in the former case but not in the latter a ready state is necessary to fix the content of the record.

One might think that, nevertheless, even inferences based on memories are inferences from two times to a third in between. Even in the case of memories, one might hold, an appeal to a ready condition is implicit, since we only infer p from a memory that p when we believe the memory to be veridical; and we believe a memory to be veridical only if we believe that our brain was in some appropriate ready state when we perceived that p. Yet even if this line of reasoning were correct, the appeal to ready conditions would fulfill two quite distinct roles in the two cases. In the cases of memories its only job is to support the claim that our memories are veridical, while in the billiard case the ready condition is needed to fix the content of the record. I doubt, however, that we generally do reason from our memories along these lines. Generally, it seems to me, we simply assume that our memories are veridical and are, as a matter of fact, neither concerned with establishing that our brains were in an appropriate ready condition nor interested in coming by the relevant information about the appropriate ready conditions. Only in situations where *doubts* about the veridicality of our memories arise do we appeal to information that suggests that our brains were in appropriate ready states at the time when we perceived whatever it is we take ourselves to remember. That is, unless there are doubts about the veridicality of our memories, the inferences in which we actually engage make no appeal to ready conditions and simply have the form 'I have a memory that p; therefore, p.'

At the very least, then, how we think of the puzzle concerning ready conditions has to be qualified in important ways. Yet one might think that my considerations leave Albert's main claim intact: there still appears to be a puzzle how we come by relevant information about earlier ready conditions. In the case of the billiard balls the information is needed even to fix the content of the record state and to be able to draw any inference at all from that state; and in the case of memories it is needed at least when we are asked to justify our belief in the veridicality of a particular memory.

What then is Albert's solution to the puzzle? First, Albert's negative claim is that there is simply *no* solution to the puzzle at the first level, as it were. Each ready condition itself needs to be established by means of a further record, which requires yet another ready condition. This leads to a regress, which, as Albert says, 'obviously' would go on ad infinitum. It is impossible to stop the regress for any particular record short of positing a 'mother of all ready conditions' at the time of the early universe: 'this mother', he says, '*must* be *prior in time* to everything of which we can potentially ever have a record' (Albert 2000, p. 118, first italics are mine). Thus, Albert's second, positive claim is that the puzzle can be solved by an appeal to the past hypothesis as mother of all ready conditions.

But as an account of how we actually reason Albert's account surely must be mistaken. Most people (and, I want to submit, all people most of the time) do not assume any initial macro condition of the universe when they engage in reasoning appealing to records of the past—let alone *the* actual initial macro condition. And most people are not at all in a position to assume what that initial condition might be. According to Albert, reasoning appealing to records is possible precisely since 'it seems to us that our experience is confirmatory of a past-hypothesis.' (Albert 2000, p. 118) But to most people it does not seem that this is so, for the simple reason that most people do not possess the relevant concepts. For something to seem to us *that* it is thus and so, we must be able to form a coherent thought concerning it being thus and so.[11] But quite plausibly most people do not grasp the content of the past-hypothesis or of the second law of thermodynamics, and in fact the content of their experience may seem to them to be

---

[11] What, on Albert's account is the relation between our being in a position to assume the past-hypothesis, our in fact assuming the past-hypothesis, and it seeming to us that our experience is confirmatory of the past-hypothesis? Albert does not tell.

confirmatory of a quite different past than that postulated by modern cosmology. Nevertheless, most people arguably engage in reasoning appealing to records.[12]

If Albert's account fails to provide an answer to the question as to how we obtain information about ready conditions in cases like that of the billiard balls fails, what then is the answer to the puzzle? It seems to me that if all records were like the position of the billiard balls there would be no solution to the puzzle and reasoning from records would simply be impossible. In the case of the billiard balls, Albert's puzzle seems inevitably to result in a regress. How else could we possibly come by the necessary specific information about the ready state, if not either by means of retrodiction or through yet another record? But there are records that differ from this case in crucial respects. Consider, first, record states with intentional content, such as memories. For such records we need to be in a position only to assume that the ready state *whatever it may have been* was a state of a kind that normally results in veridical records. It seems less plausible that such a very general assumption about ready states needs to be established by means of a further record. More plausible, I want to suggest, is that the general veracity of our memories is one of our most fundamental convictions that underwrites much of our reasoning about the world and that, when challenged, we support our belief that a certain memory is veridical by appealing to how this memory *coheres* both with other memories we have and our current perceptions. These other memories might be of earlier states of affairs, but often they will concern other events simultaneous with the event remembered.

Moreover, the regress can be blocked even for many record states without intentional content. Often, it seems to me, when we take certain facts about the present to function as records of the past, we assume that the recording system in question had earlier been in its 'normal' or 'typical' state, which

---

[12] Given how obvious it is that an appeal to the past-hypothesis cannot be part of our actual reasoning practices involving records, one may wonder if I have misinterpreted Albert. And, indeed, in personal discussions Albert has insisted that he is not proposing his theory as an account of our actual reasoning practices. Yet it is difficult for me to see how else one might interpret the passages cited above. The puzzle concerning ready conditions raised by the billiard ball example and expressed by the question as to how we 'come by' information about ready conditions appears unequivocally to be a puzzle about how we do as a matter of fact obtain information relevant to drawing inferences based on records.

functions as ready condition. And what we assume the normal state of the system to be is given by our past experiences with systems of the kind at issue. Thus, when we observe diverging ripples on a pond, we take this to be a record of the fact that some wave source, like a stone, had just broken the surface of the water. Here we implicitly assume that there were no coherent concentrically converging waves present on the pond prior to the source; and we make this assumption not necessarily because we have a record of the pond's having been still, but because our past experience with wave media suggests that there are no coherent waves in the absence of sources.

One worry about this sketch is that in appealing to past experiences of what constitutes a 'normal' ready state of a system, the account relies on our already having information about the past, yet how this is possible is precisely what is at issue. How can we, in the present, have information about the 'normal' states of systems in the past? Don't we already have to rely on records or memories of past experiences to assume certain 'normal' past states as ready conditions? But I want to suggest that (at least in principle) we can think of information about past normal states as being inferred from the totality of our present experience. Certain assumptions about the past are part of the best explanation of the totality of our present experience (including our memories) and of the success of our predictions about how future observations will turn out. It seems that at least sometimes we come by information about the past by positing past states that best account for our present beliefs, experiences and memories.

Moreover, we can use the same considerations to *justify* beliefs about the past. That is, our beliefs in normal ready states (such as the belief that there are no coherent waves on ponds prior to the presence of wave sources) are supported by their 'conspicuous success [...] in making *predictions* about how *future* particular observations are likely to *come out*, [...] and because [they manage] to render various of our *other* most fundamental convictions (about the veracity of memories [...]) compatible with one another'. Here I have simply adopted Albert's explanation of how our belief in the past-hypothesis is grounded in an inference to the best explanation (Albert 2000, p. 119).[13] Thus, I want to propose that we can use the very

---

[13] Just before the passage quoted is, by the way, the only place in the section on pp. 117–19 where Albert speaks of the question of the justification of our beliefs. Albert here asserts both that what *justifies* our belief in the past-hypothesis is an inference to the best explanation and that *we believe* in the hypothesis for these reasons.

same inferential procedures that might be used to support a belief in the past-hypothesis to support beliefs in much more recent ready states: In both cases our assumptions about the past are supported by the fact that they make the best sense of aspects of our present experience and are predictively successful.

Now, some of what Albert says suggests that we can use this inference procedure *only* to justify our belief in the past-hypothesis and that it is strictly impossible to infer anything about more recent ready states in this manner. For, as we have seen, he holds that in order to establish *any* ready condition there '*must* be something we can' assume, which '*must* be *prior in time* to everything' of which we can have a record (Albert 2000, p. 118, the first two emphases are mine). But it is difficult to see why this inference technique supposedly works 'globally' and for the totality of our experience, but not more 'locally' for certain aspects of our present experience. Why should we be able to infer a low-entropy state of the early universe as best explanation of the totality of our experience, yet be unable to infer that ponds are 'normally' still as best explanation of our present experiences (including putative memories and records) as far as they relate to small bodies of water?

On Albert's behalf, one might try to point to the fact that the past hypothesis is needed to guard against Loschmidt's reversibility objection: Unless we assume a low-entropy past, the most likely past evolution is a fluctuation out of equilibrium. But again, it is unclear why we cannot use more 'local' inferences to the best explanation to establish merely that the more recent past ought to have been one in which entropy behaved appropriately such that ripples on a pond could indeed function as records of past wave sources.[14] In addition, Albert might argue that the past-hypothesis is 'more fundamental' in that belief in it can provide the best, most unified explanation of the totality of our experience. For, he maintains, everything we know of the world can be deduced from what we know about the world's present macrocondition; 'the standard microstatistical rule; the dynamical equations of motion; the past-hypothesis' (Albert 2000, p. 119). But even if Albert could show that the past-hypothesis *can* function as the 'mother of all ready conditions', this does not imply that, as he claims, the

---

[14] To what extent the asymmetry of wave phenomena is related to that of thermodynamics is a difficult question. For a discussion of this issue see (Frisch 2005).

past-hypothesis *must* be assumed in any inference involving records. Thus, inference procedures similar to the one to which he appeals ought to be able to provide us with direct information (not relying on knowledge of the 'mother') about the 'mother's offspring' as well and Albert's regress argument can be blocked at its very first steps.

Thus, it appears that we do not need to appeal to the past-hypothesis in justifying our beliefs at least about the recent past. Finally, there is also a worry about appealing to the past-hypothesis as ultimate justification of our beliefs about the past. Albert presents the following skeptical problem. According to Loschmidt's reversibility objection, the most likely evolution of the present state of the world is as a fluctuation out of equilibrium. How is it then, that we could ever come to have any reason to believe in the past-hypothesis? One might think that we can know of a low-entropy past simply through records or memories of the past. Yet any putative record is itself already part of the present state and, thus, just like anything else in the present appears to be most likely the result of anti-thermodynamic fluctuations. Albert's solution to the problem is two-fold. On the one hand, he argues that we can have inductive grounds for accepting the past-hypothesis. For we can make predictions from the assumption that the universe had a low-entropy past which we can confirm. On the other hand, and as Albert says, 'more profoundly' (Albert 2000, p. 119), the past-hypothesis is supported by the fact that it renders our fundamental convictions, such as the veracity of memories, correct.

Yet it is difficult to see how the past-hypothesis could be confirmed by successful predictions if it cannot be confirmed (and in fact is apparently disconfirmed) by the sum-total of our experiences at any one moment in time. For once the predicted outcome is realized, our reasons for taking that outcome to be the result of a prior prediction are again based on records or memories. If, as Albert suggests, our experience *now* provides us with no reasons to accept the past-hypothesis, then neither does our experience a few minutes *later*, after we have conducted an experiment to confirm our predictions.[15]

Thus, we are left with Albert's second solution: our reason for believing the past-hypothesis is that it is needed to account for some of our fundamental convictions, such as the veracity of memories. This spells

---

[15] I owe this point to discussions with Dan Parker.

trouble for Albert's theory both as an account of our actual reasoning practices and as an epistemological justification of these practices. If we assume, with Albert, that our memories are veridical, then, it seems, at the very least we assume what the contents of our memories are. Thus, we take ourselves to be in a position to draw inferences about the past without actually having to first postulate the past-hypothesis as ready-condition. Moreover, the account cannot provide a justification for our inference practices either. For if our justification for positing the past-hypothesis is that it renders our memories veridical, then we cannot in turn justify the veracity of our memories by appealing to the past-hypothesis, on pain of circularity.[16]

In this section I have argued that Albert's account fails as an explanation of our fundamental conviction that the future, but not the past, depends on what happens now. *Assuming* the past-hypothesis is a necessary condition neither of our drawing inferences based on records nor of our being in a position to justify these inferences. Yet despite the fact that Albert's argument for the necessity of assuming the past-hypothesis in inferences about the past plays a prominent role in his overall discussion, his core thesis that our counterfactual and causal reasoning can in principle be recovered from positing the past-hypothesis relies rather on the claim that the past-hypothesis is sufficient for the reliability of records and, ultimately, for the asymmetry of causal dependence. It is to the sufficiency claims expressed in (*i*) and (*ii*) above to which I want to turn next.

13.4.2 *Records and the Past-Hypothesis*

As we have seen, Albert maintains that local facts about the present can function as records of the past, if we can assume certain facts about the remote past as 'ready condition', and that 'the initial macro-condition of the universe as a whole' can function as the 'mother of all ready

---

[16] Albert seems to be of two minds about how much weight to attach to the fact that we believe in the veracity of our memories. On the one hand he suggests that a belief in the veracity of our memories is among our 'most fundamental convictions' that even plays a role in justifying our belief in the past-hypothesis. On the other hand, he denies, as we have seen, that a belief in the veracity of memories can function as a 'mother of all ready conditions'. As one reason for this he offers the observation that the evidence of our senses can be overridden (see Albert 2000, p. 118, footnote 6). But the fact that it is possible for a *particular* memory to be overridden by other evidence is compatible with the claim that we generally assume our memories to be veridical and that this assumption ultimately 'grounds' all our reasoning based on records.

376 MATHIAS FRISCH

conditions'. (Albert 2000, p. 118). From this he immediately and without further argument concludes the following:

And so it turns out that *precisely* the thing that makes it the case that the second law of thermodynamics is (statistically) true throughout the entire history of the world is *also* the thing that makes it the case that we can have epistemic access to the past which is not of a predictive/retrodictive sort; the reason there can be records of the past and not the future is nothing other than that it seems to us that our experience is confirmatory of a past-hypothesis but not of a future one.   (Albert 2000, p. 118)

And a little further on he says:

[E]verything we can know of the past and present and future history of the world can be deduced, in its entirety [...] from the following four elements: what we know of the world's present macrocondition [...]; the standard microstatistical rule; the dynamical equations of motion; the past-hypothesis.   (p. 119)

Hence, Albert takes the claim that 'the initial macro-condition of the universe as a whole' functions as ready condition to imply premise (*i*) — the claim that the past-hypothesis can play the role of 'mother of all ready conditions'.

But to conclude from the claim that the initial macro-condition of the universe can function as ready condition that the past-hypothesis alone is such a ready condition is a *non-sequitur*. For the past-hypothesis provides us with significantly less than a full specification of the initial macro-state of the universe. All the past-hypothesis asserts is 'that the world first came into being in whatever particular low-entropy highly condensed big-bang sort of macro-condition it is that the normal inferential procedures of cosmology will eventually present to us' (96). And clearly whatever it is that cosmology will eventually present us with, this will fall far short of a complete account of the initial macro-state of the universe. Thus, Albert owes us an argument for why the broad constraints on the early universe posited by the past-hypothesis (as opposed to a full specification of the macro-state of the early universe) are sufficient to ensure that local facts about the present can function as records of the past.

To illustrate this point, we might imagine a slightly amended version of Albert's billiard balls. Let us assume that ball 5 is currently *moving* and was *stationary* five seconds ago. Further, let us imagine that the balls are moving on a table with very weak frictional forces. Given the ready condition that ball 5 was stationary, the fact that the ball is currently moving functions as a record of a collision in the last five seconds. The ready condition in this case exactly specifies the value of one of the system's state-space variables. But obviously it does not follow from the fact that the ball's having been *stationary* can function as ready condition that also the claim that the system of balls was in a *low-entropy* initial state can function as ready condition. From the fact that the system of balls was in a low-entropy state we can conclude that the most likely evolution of the system of balls was one that is thermodynamically normal and, hence, that the ball's currently moving is not due to random 'anti-frictional' forces exerted by the table. But without the further assumption that the ball was stationary five seconds ago, we cannot exclude the possibility that the ball has been moving without collisions for more than five seconds.

Loewer (ch. 11, in this volume) stresses that in addition to a low-entropy constraint we need to posit that the state of the early universe also satisfied certain symmetry constraints (without specifying what these constraints are). But again the billiard ball example suggests that such an additional constraint still falls short of what is needed. Even if we were told that the initial low-entropy state of the system of balls (more than five seconds ago) was the highly symmetric special macro-state when the balls are racked, we could not infer from the fact that ball 5 is currently moving that it underwent a collision in the last five seconds.

As far as I can tell, Albert does not offer any argument that in the case of the universe as a whole the assumption of a low-entropy past alone can function as ready condition. Moreover, Albert's discussion suffers from the fact that he does not distinguish clearly between the claim that the past-hypothesis is *sufficient* for the reliability of records and the claim that it is *necessary* in various senses. In the previous section I criticized Albert's claim that assuming the past-hypothesis is a necessary condition of drawing inferences based on records. In the text Albert moves from this claim without any discussion to the claim I quoted above that, 'anyway, [...] everything we can know of the past' can be deduced from the past-hypothesis in the standard way—that is, to the sufficiency of the past-hypothesis for the reliability of records. There is one additional set of considerations Albert advances, namely that there would be no reliable records in worlds that do not satisfy the past-hypothesis. In any such world, Albert says, the most probable way in which putative records originate would be as results of random fluctuations from a maximal-entropy state and, hence, they would not be correlated with the relatively low-entropy states of which they are supposed to be records. But this argument can at most show that the past hypothesis is necessary for the existence of records.

Thus, there are three distinct theses we ought to distinguish: First, that *assuming* the past-hypothesis is a necessary condition of our actually drawing inferences based on records. Second, that the fact that the past-hypothesis *holds* is a *necessary* condition for the reliability of records. And third, that the past-hypothesis is a *sufficient* condition for the reliability of records. The second thesis is arguably true; the first, I have argued, clearly false; and Albert provides no argument for the third.

Yet we can try to imagine what kind of considerations one might advance in support of the sufficiency thesis. How, we might ask, could it be that local facts about the present are associated with a past different from that of the actual world? One way to construct such a situation is to postulate some small yet macroscopic change to the actual present and then evolve the resulting state backward in time. The effect of such local changes will in general be that *other* local facts are no longer associated with the same past events with which they were associated in the actual world: such present facts constitute fake records, as it were. In the amended billiard ball example ball 5 is presently moving in the actual world, and this is associated with the ball having undergone a collision in the past five seconds. But there can be changes to the state of balls *other* than ball 5 that, if we evolve the state of the balls backward in accord with the laws, will result in a history where ball 5 did not undergo a collision in the last 5 seconds. We might say that in the corresponding counterfactual world the fact that ball 5 is presently moving constitutes a 'fake' record of its past evolution. Of course in such counterfactual worlds the ready condition that ball 5 was at rest five seconds ago is not satisfied. The crucial question is whether the past-hypothesis would likewise not be true in such a world.

Adam Elga, in a somewhat different context, has presented an argument that suggests that the past-hypothesis would indeed not be satisfied in most counterfactual worlds resulting from localized macroscopic changes to the actual world (Elga 2001).[17] Elga points out that the time-evolution of the actual world *toward the past* is thermodynamically extremely unlikely.

---

[17] Loewer (ch. 11, in this volume) stresses that the argument presented by Elga is originally due to Albert.

(This is most easily seen if we imagine that the direction of time were flipped.) Moreover, the evolution toward the past is extremely sensitive to small changes in the micro-state of the world: most small changes will result in worlds that evolve in thermodynamically normal ways toward the past—that is, worlds that violate the past-hypothesis and behave anti-thermodynamically in the normal time-sense. For example, in the case of the billiard balls it is probable that changing the position of any of the balls will, through small changes in the gravitational force, disturb the normal thermodynamic behavior of thermodynamic sub-systems in the vicinity. Further and further into the past, more and more sub-regions of such a world will be 'infected' by the anti-thermodynamic behavior with the result that the remote past of the world will have high entropy.

The upshot is that localized macroscopic changes to the present result both in an anti-thermodynamic past *and* in 'fake' records of the past. But this again is not enough to establish (*i*). One might think that the argument is simply this: If there are fake records in a world, then the past-hypothesis is false in that world. Thus, taking the contrapositive, if the past-hypothesis is true, then there are no fake records. Now, one may doubt that the Elga-Albert considerations do in fact show that in worlds with fake records the past-hypothesis does not hold (since there might be ways to construct worlds with fake records without violating the past-hypothesis). But even if we grant this step in the argument, the conclusion does not follow, for the argument is not valid in a probabilistic context.

Let us grant that the probability of the past-hypothesis given that there are fake records is extremely small. That is, formally:

(2) $\Pr(PH/R\&{\sim}e) = \varepsilon$,

where R is a record of some event e in its past in the actual world. (2) says that the probability of the past-hypothesis is extremely small given that the record R is present without the event e of which it is a putative record having occurred. From this we would like to conclude (*i*)—that is, that the past-hypothesis ensures the reliability of records; or, in other words, that the probability of an event e not occurring, given the past-hypothesis and record R of its occurrence, is extremely small:

(3) $\Pr({\sim}e/R\&PH) = \delta$.

Yet, (3) does not follow from (2). In fact, as a simple application of Bayes' theorem can show, in order to get from (2) to (3) we need to assume as additional premise that

(4) $\Pr(PH/R) \geq \Pr(\sim e/R)$.

But (4) is false. The left-hand side of (4) is the probability that a world had an extremely low-entropy past, given certain information about its current macro-state. This probability is absurdly small, as Loschmidt's reversibility objection has taught us. The right-hand side is the probability that a past event e did not occur given the presence of a putative record of e (and nothing more!). This probability will in general not be all that small. In fact, Albert's own argument relies crucially on the assumption that this probability is in general not small, for it is precisely this assumption that makes an appeal to ready conditions necessary in the first place and that ultimately is supposed to underwrite the connection between records and the past-hypothesis. If $\Pr(e/R)$ were close to one, then an inference from R to e would not need to involve an appeal to a ready condition prior to e.

For (3) to be true, it would have to be the case that the probability of PH is much, much lower, given fake records, than the probability of PH given the record state. But the Elga-Albert argument cannot establish this claim. Of course it is extremely improbable, given the present state of a world that differs from the actual world by a small macro-change, that this world satisfies the past-hypothesis. Yet that the actual world evolved from an extremely low-entropy state, given its present macro-state, is similarly improbable. This, after all, is just the reversibility objection that is circumvented by simply postulating a low-entropy past. That a small counterfactual macro-change to the present will again result in a micro-state which evolved from a low entropy macro-state is no less probable than the low-entropy past of the actual world.

12.4.3 Records in Counterfactual Worlds

The second premise of Albert's argument is the claim that if there are records of the past, then there is no counterfactual dependence of the past on the present. In discussing an example of a putative case of backward counterfactual dependence, Albert supports this claim by saying that the past could not have been different since a different past would have to have left traces or records that ought to be part of the present. Since by assumption the counterfactual present is identical to the actual present except for a small, local change, the counterfactual present contains traces of the actual past but not of any counterfactual past events. Hence, the past does not counterfactually depend on the present. This argument relies on (3)—that is, that the probability of any past event e not occurring, given its present records R and the past-hypothesis PH is very small. We have just seen that the claim that the past-hypothesis alone can ensure the reliability of records is problematic, but for present purposes I want to grant that claim and see what follows from it.

According to premise (*ii*), (3) implies (1), or equivalently

(5) $\Pr(e/\sim c \& S_a \& PH) \approx 1$.[18]

As in the case of premise (*i*), however, Albert provides no argument for (*ii*). If in (3) δ were strictly equal to zero, then (5) would indeed follow and (*ii*) would not need to be introduced as an independent premise but would simply be a consequence of the probability calculus. Since, however, macro-states are only probabilistically given in terms of the underlying micro-states and their dynamics, we need a justification for the move from (3) to (5). It may be plausible that conditionalizing on the entire state S_a of which the records R are a part does not change the probability of e, that is to say, that (3) implies

(6) $\Pr(e/S_a \& PH) = 1-\delta$.

But what is less clear is why conditionalizing on $\sim c$ as well should not significantly affect the probability of e. Again, Albert owes us an argument.

There is one particular class of events for which it is perhaps most obvious that there is indeed the need for an argument here—those events e that are complete macro-states of cross sections of the backward light cone of c in the relatively recent past of c and where there are macro-laws governing the system that are near-deterministic. In such cases, e determines the occurrence of c with 'thermodynamic certainty,' as it were:

(7) $\Pr(c/e \& PH) \approx 1-\varepsilon$, with $\varepsilon \approx 0$.

---

[18] Intuitively, the difference between (3) and (5) is this. According to (3), in the vast majority of worlds in which R and PH occur, R is a reliable record of e. According to (5), R is a reliable record of e even in the majority of those counterfactual worlds in which c does not occur.

Cleary there are systems like this—systems that appear to behave deterministically and non-chaotically on the macro-level. There are, of course, also macro-systems which behave chaotically or in which the macro-dynamics is probabilistic (such as coin tosses); and there arguably are systems for which we cannot write down any macro-dynamics. Here I want to focus on systems which we model in terms of a deterministic macro-dynamics.[19] As a concrete example, we can once more think of a version of Albert's billiard ball example: As event c we can pick the current velocity and position of ball 5 and as e we can choose the macro state of the world five seconds ago in a sphere centered on the current location of ball 5 with a diameter of five lightseconds.[20] In particular, e includes the state of the entire collection of billiard balls five seconds ago. Records of aspects of e might be, among other things, the positions and velocities of balls other than ball 5 and light waves that were reflected by the collection of balls five seconds ago. Our macro-dynamics gives us *ceteris paribus* laws, according to which in modeling the billiard balls we can ignore everything in the past light-cone aside from what happens on the billiard table itself.

Since the probability of c is completely determined by events in its backward light cone, e will screen off c from any event outside of the light cone of c. In particular,

(8) $\Pr(c/e \& PH) = \Pr(c/e \& S_a \& PH)$.

Moreover, for many systems for which (7) holds, $\Pr(\sim e/\sim c \& PH) \approx 0$ ought to hold as well. For example, if counterfactually ball 5 were at the other end of the table now, then it seems dynamically unlikely that it would have been at its actual past location a short while ago. However, it follows from (5) and the definition of conditional probability that

(9) $\Pr(e \& \sim c/S_a \& PH) \gg \Pr(\sim e \& \sim c/S_a \& PH)$.

That is, it is much more probable that the actual past event e occurs without the actual present than that neither the past nor the present are those of

---

[19] My aim here is not to show that Albert's normal procedures of inference fail in *all* cases, only that there are standard cases where it looks as if the procedures do not yield the result Albert needs.

[20] Is it okay to consider events of that size, given that Albert's focus is on small macro-changes for which we might take ourselves to be responsible? The answer is: Yes, since the hypothetical change is small—a change to the state of ball 5. And we are simply asking whether changes to the occurrence c could have as a consequence changes to the occurrence of e, just as Nixon's pushing the bottom would have large consequences in the future.

the actual world. And this is so, even though the occurrence of *e* makes it dynamically extremely improbable that *c* does not occur and there are no purely dynamic constraints that make $\sim e \& \sim c$ improbable.

More intuitively, we can put this result this way. Premise (*ii*) assumes that we treat inferences based on records as much more reliable than predictions and retrodictions based on the dynamics. The dynamics alone would predict that a locally different present macro-state would in general have resulted from a different past macro-state. According to (*ii*), however, any such retrodiction is overridden, as it were, by the assumption that records are reliable. Yet one might worry that this gets things backwards: No matter how reliable our records are, they never can be more reliable—and will in general be far less reliable—than any inferences we can draw based on a complete macro-state, the past-hypothesis, and the dynamics. Of course the conjunction of (5), (7) and (8) does not contain a contradiction. Yet it is far from obvious (and requires an argument) why the record condition (3) ought to commit us to (5), and hence (9).[21]

Think about the collection of billiard balls. Intuitively, it seems that if the position or velocity of ball 5 were different, then the overall state of the system of billiard balls would have to have been different five seconds ago. Contrary to this, Albert's account would claim that since there are records of the state of the balls five seconds ago, including the present state of the remaining billiard balls, it is overwhelmingly likely that the macro state of the system of balls would have been exactly the same as the actual state. This claim strikes me as highly counterintuitive. Once again, we might test our intuitions by taking the racked up billiard balls as stand-in for the past-hypothesis. Let us assume that after the break the billiard balls move for ten seconds. And let us then ask what our dynamics predicts for the state of the balls just five seconds ago, given that we assume that ball 5 came to rest somewhere else from where it actually did, the positions of the other balls remain unaltered, and the 'past-hypothesis' that the balls ten seconds ago were racked up. I take it that Albert's intuition is that something thermodynamically 'odd' must have happened to the balls in order for ball 5 to end up at a macroscopically distinct present location and quite plausibly this intuition is correct. But what is difficult to see is why the most plausible

---

[21] Of course Albert maintains that the micro-dynamics and the statistical postulate *alone* make all the wrong retrodictions. But what is at issue here is whether the dynamics and statistics further constrained by the past-hypothesis still are less reliable than our records of the past.

past evolution of the counterfactual system of billiard balls is supposed to be one that *exactly* matches the actual macro-evolution until immediately before the present and only then diverges in some thermodynamically unexpected way. It might seem more plausible that our dynamics will tell us that when there is an apparent 'tension' between our records, as given by the location of all the billiard balls except for ball 5, and the present state of ball 5 itself, then there will be some kind of 'trade-off' where the relatively not-too-distant past of the counterfactual system will be different from that of the actual system in ways not entirely compatible with the reliability of records.

Now, ultimately the question whether or not it is a consequence of the dynamics, the statistical postulate, and the past-hypothesis that worlds that differ from the actual world locally and macroscopically are overwhelmingly likely to have had exactly the same macroscopic past as the actual world should not be a question that is settled through a battle of intuitions. Whether Albert's thesis is right is a question for the relevant physics. My aim in arguing that Albert's thesis is counterintuitive is only to urge that there is indeed still the need for a physical argument. That is, my aim is to combat the impression that once we realize the importance of the past-hypothesis as additional constraint on the dynamics, then the counterfactual independence of the past from the present follows almost immediately and that there really is no need for a detailed physical argument for why records of the past, in a sense, 'trump' what might be suggested by a macro-dynamics.

13.4.4 *Possible Replies: Transition Periods and Degrees of Counterfactual Dependence*

In the last section I have suggested that a more careful examination of the relevant physics might show, at the very least, that local changes to the present would have to have been preceded by a transition period during which the counterfactual past differed macroscopically from that of the actual world. Could Albert not simply concede this point without abandoning his account? Perhaps Albert need not show that, as far as small macroscopic changes to the present are concerned, there is absolutely no counterfactual dependence of the past on the present. A weaker claim might be sufficient, namely that whatever counterfactual dependence of the past on the present there is, it is much less and dies off much faster than any dependence of the future on the present. That is, one might try to argue that neither the record condition (5) nor the counterfactual independence claims (1) hold for *all* past events and that events in the very recent past of c will exhibit some counterfactual dependence on c, but that nevertheless there is a significant asymmetry in the degrees of counterfactual dependence that is sufficient to account for our asymmetric notion of causal dependence.[22]

However, this defense faces serious problems of its own. First, nothing in my criticism relies on the fact that e is an event in the very recent past. As long as e is recent enough for there to be a macro-dynamics that is near-deterministic linking e and c, the arguments go through.

Second, the defense proposes to replace what appears to be a sharp and precise distinction with a qualitative and gradual difference. According to our common sense notion of causation we think that our actions can influence the future but have *no influence at all* on the past. What a defender of an entropy-account would have to explain is how we have come to believe in this sharp distinction, despite the fact that, according to the account, there is some counterfactual dependence of the past on small interventions into the present. In particular, it is not sufficient simply to argue that there might be some 'transition' period that is necessary for a world with a past that perfectly agrees with the macroevolution of the actual world to evolve in a nomically possible—even if thermodynamically abnormal—way into a counterfactual present state. For it is of crucial importance *how long* such a transition period needs to be. If the transition periods were on the order of fractions of a second, then perhaps it would be somewhat plausible to suggest that such a limited counterfactual dependence of the past on the present could somehow give rise to our fundamental conviction of a strict independence of the past from the present.

Yet the suggestion seems to be that during a transition period an extremely rare anti-thermodynamic fluctuation takes place that carries a past state that is macroscopically indistinguishable from the actual past into a counterfactual present state and it is not clear that such fluctuations could occur fast enough to render the transition period negligible. How long, for example, would it take for anti-frictional forces on a billiard table to move a ball that came to rest at one end of the table to the other end? In order

---

[22] This reply was suggested to me by Adam Elga and Doug Kutach.

for a ball that was at rest at one end of the table to end up at rest at the other end, the ball has to start rolling across the table and eventually slow down again to come to rest. The first part of the ball's trip has to be the result of anti-frictional behavior, while the second part will be in accord with normal, frictional behavior. Since anti-frictional behavior is simply the time-reverse of normal frictional behavior, the ball will take just as long for the first part of its trip as for the second part. Hence the total time it will take for a ball to travel across the table due to a combination of anti-frictional and frictional forces will be roughly twice the time it takes for a ball to roll for half the length of the table and then come to a stop. On a realistic billiard table this time is probably on the order of a few seconds—too long to be negligible. For if the transition period were that long, there ought to be, according to Albert's account, 'causal handles' on the relatively recent past of the kind that we should take ourselves to be able to exploit: just as we think that our actions now have effects even a few seconds into the future, we ought to think that our actions have effects at least a few seconds into the past. Moreover, most likely a few seconds will not turn out to be the upper limit on the time of such anti-thermodynamic transitions.

If this is right, then an appeal to anti-thermodynamic transition periods can be of no help for Albert's account. Even if a careful examination of the underlying physics revealed that the most probable history of a counterfactual world was one in which the hypothetical change to the present was due to a thermodynamically abnormal fluctuation immediately preceding the present state, the periods of mismatch in the past would probably be too long to support our fundamental conviction that the entire past is counterfactually independent of any of our decisions in the present.

13.5 Counterfactuals and Decisions

In this volume, Barry Loewer also defends an entropy account of counterfactuals. While Loewer's account is closely related to Albert's account, there are also important differences between the two accounts. One such difference concerns the question how we evaluate the truth of counterfactuals. According to Albert's account there is a single procedure for evaluating counterfactuals given by our 'normal procedures of inference'. Loewer, by contrast, proposes two distinct kinds of truth conditions—one kind for what he calls 'decision conditionals' and another for counterfactuals positing small macro changes to the world. The truth conditions for the latter are different from—and in fact incompatible with—the truth conditions suggested by Albert's 'normal procedures of inference'. I want to end this paper with a few remarks concerning Loewer's account, which I discuss more fully in Frisch (forthcoming).

Loewer apparently endorses Albert's account of the role(s) the past-hypothesis plays in the production of records. Similarly to Albert, Loewer holds that the past-hypothesis is *necessary* for the production of records—the PH 'allows for the production of localized macro records' in that it removes an obstacle to there being such records (this volume, p. 303). In addition, the past-hypothesis 'plays the role of a ready state for our universe'—that is, it is what Albert calls the 'mother of all ready condition'. Since Loewer further agrees with Albert that without positing the relevant ready states 'we are not justified in making' inferences based on records and that a ready state together with the dynamical laws and the record state entails the recorded state, he presumably also agrees with Albert that the past-hypothesis as first ready-state is necessary for *justifying* inferences based on records and that the past-hypothesis is a *sufficient* condition for the reliability of records.[23]

Loewer's account makes explicit an intermediate step in the argument from statistical mechanics to a counterfactual asymmetry—a step that Albert does not discuss but may well be implicit in Albert's account. Loewer argues that a statistical mechanical account implies a tree-structure for possible macro histories, according to which there are many small differences in the micro conditions at a time *t* not involving changes in the macro state at *t* that result in different macro futures, but there are only very few such differences that result in different macro pasts. Here is how Loewer characterizes this tree-structure:

The SM probability distribution embodies a way in which 'the future' (i.e. the temporal direction away from the time at which PH obtains) is 'open' at least insofar as *macro states* are being considered. Since all histories must satisfy the PH they are very constrained at one boundary condition but there is no similar

---

[23] Loewer stresses that the past-hypothesis and the dynamical laws are not 'sufficient to account for the existence of recording systems let alone the particular records that have been formed in our world' (this volume, p.304). But this is true of any ready condition: no ready condition is sufficient for the existence of a particular record. Rather, what Albert's account of records claims is that a ready condition and the laws in conjunction with a record are sufficient for the recorded state.

constraint at other times. It is true that (almost) all histories eventually end up in an equilibrium state (there is a time at which almost all histories are in an equilibrium state) but this is not a constraint it is a consequence of the dynamics and the PH and it is not very constraining (almost all states are equilibrium states). Another feature of the SM distribution when applied to the macro state of the kind of world we find ourselves in is that the macro state of the world at any time is compatible with micro states that lead to rather different macro futures. For example, conditional on the present macro state of the world the SM probability distribution may assign substantial chances both to its raining and not raining tomorrow. On the other hand, there is typically much less branching towards the past. The reason is that the macro states that arise in our world typically contain many macroscopic signatures (i.e. macro states/events that record other macro states/events) of past events but fewer macroscopic signatures of future states/events.   (Loewer this volume, pp.302–3)

This passage raises several questions. First, the asymmetry embodied by the tree structure is not strict; Loewer says only that there is 'typically *much less* branching towards the past'. Yet the asymmetry he ultimately wants to derive is strict: according to our folk notion of cause, there is absolutely no causal dependence of the past on the present and there are no true backtracking counterfactuals in causal contexts. One question, then, is whether Loewer can recover these strict asymmetries from the merely quantitative differences between the amount of future and past branching. One way in which this might be achieved is if all (or at least the overwhelming majority of) branches toward the past have negligible probabilities.

A second question is how exactly we are supposed to understand the claim that possible macro histories exhibit a tree structure. The passage above suggests that the branching structure is a *consequence* of the existence of records: 'the reason', Loewer says, for the relative absence of past branches is the existence of macro records. Yet immediately after this passage Loewer continues by saying that 'Albert shows how the assumption of the PH (and the consequent branching structure) *allows* for the production of localized macro records of past events' [my italics] (this volume, p.303), which suggests that the tree structure *entails* the possibility of records.[24]

No matter how precisely the tree structure and its relation to the possibility of records are ultimately to be understood, it is not immediately

---

[24] I discuss prospects and problems for both these readings in detail in Frisch (forthcoming).

obvious how any such asymmetry follows from the three assumptions that Loewer, following Albert, identifies as constituting the micro statistical account—that is, the dynamical laws, the past-hypothesis, and the assumption of an equi-probability distribution over all micro states compatible with the Big Bang initial state of the universe. Indeed one might worry that if anything the Second Law implies an upside-down tree structure. Since states closer to equilibrium occupy vastly larger regions of phase space than macro states very far from equilibrium, it follows from Liouville's theorem that there will be many different non-equilibrium states very far from equilibrium that evolve into the same state closer to equilibrium in the future. Thus, there appear to be many more changes to the micro state of a system close to equilibrium that are associated with different pasts further away from equilibrium than there are changes to the micro state of a system far from equilibrium that are associated with different futures closer to equilibrium. Loewer points out that the *PH* constrains possible past evolutions, while there is no similar constraint on future evolutions. But constraining the past to low-entropy states entails that there will be a vastly *larger* number of possible past states than of possible future states, which have higher entropies and occupy vastly larger regions of phase space.

Assuming that possible macro histories did indeed exhibit the kind of tree structure postulated by Loewer, can that underwrite a counterfactual asymmetry? It seems to follow immediately from the tree structure that Loewer's decision counterfactuals (this volume, Section 11.3) are time-asymmetric. Yet it is less clear that the asymmetry can be extended to non-decision counterfactuals. Loewer proposes the following truth conditions for counterfactuals positing small macro changes:

'If $A(t)$ had been true then the chance of B would have been x' is true iff t' is the latest time at which a divergence from the actual macro history similar in probability to a decision event can occur and $Pr(B/A(t)\&M(t')) = x$. (Loewer this volume, p. 320)

Here $M(t')$ is the complete macro state at t'. The difference between these truth conditions and those suggested by Albert's account is that the probability of the consequent of the counterfactual is conditionalized on the macro state at some time *earlier* than the antecedent $A(t)$ rather than on the macro state at t. Generally the conditional probability of an event will change, if we conditionalize it on the complete macro state at different times and, hence, it will not in general be the case that $\Pr(B/A(t) \& M(t')) = \Pr(B/A(t) \& S_a(t))$, for $t' < t$. According to Loewer's proposal the truth conditions for counterfactuals postulating small macro changes to the world, thus, are not given by Albert's 'normal procedures of inference'.

One problem for Loewer's proposal is that for macro systems that are *not* extremely sensitive to changes in the system's micro state—and there are many such systems that appear to evolve deterministically on the macro level—there will be backtracking counterfactuals and counterfactuals relating what intuitively are joint effects of a common cause which come out true according to the truth conditions. Consider the following example. In the actual world I am standing alone with an axe in a forest, with no other axe-wielding person in the vicinity. I do not swing my axe and no tree falls. Let the counterfactual event B be that I swing the axe at a tree and the event A be that the tree falls. Does my swinging of the axe counterfactually depend on the falling of the tree? Arguably there is no relatively probable decision-sized event D compatible with the actual macro history later than my possible decision to swing the axe that would result in the tree falling.[25] Hence the time t' of the relevant macro state on which we ought to conditionalize is the time of my contemplating whether to swing the axe. And the right probability to consider in evaluating the truth of the counterfactual 'If the tree were to fall, then the event that I would have swung the axe has probability p' is $\Pr(B/A$ & *the macro state of the world at the time when I contemplate*). But this probability arguably is quite large! For given that the tree falls and that at the time I was contemplating whether to swing the axe no other putative causes of the tree's falling were nearby, it is highly probable that I did indeed swing the axe. But that my swinging of the axe caused the tree to fall and not the other way around is the kind of judgment that is paradigmatically causal and any account of causation would have to get examples like this one right. How, then, do we arrive at the causal asymmetry, given that the relevant backtracking counterfactual is true?

The worry can perhaps be brought into sharper focus by contrasting Albert and Loewer's account with Lewis's theory. On Lewis's account, unlike on Loewer's, the counterfactual 'If the tree fell, I would have to

---

[25] I want to emphasize that this example depends in no way on the fact that it involves a human action that ultimately is due to a decision. One can readily construct structurally identical examples not involving human actions.

have swung the axe' comes out false. It comes out false, since according to Lewis's prescription the possible world in which the tree falls that is closest to the actual world is one that diverges from the actual world *immediately before* the tree falls and *after* my not swinging the axe. That is, on Lewis's account, when we evaluate backtracking counterfactuals, the miracle needs to be introduced temporally *between* the occurrence of the antecedent event and its putative earlier effect. Yet if a system is macroscopically relatively stable, then, on Loewer's proposal, we may have to go back in time quite far until we reach a time at which a relatively probable micro 'miracle' is sufficient to alter the system's macro evolution. In all such cases—that is, in any of the vast number of causal claims concerning relatively stable macro systems—there will be backtracking counterfactuals with high probabilities on Loewer's account. To be sure, Loewer does not offer a worked out account of how precisely counterfactual claims are related to causal claims—even though he does maintain, echoing Albert's notion of a causal handle, that in cases where the probabilities in the consequent vary significantly with changes in the antecedent, the events in the antecedent provide 'a kind of "handle"' on the consequent (this volume, p. 320). Yet to the extent that Loewer's counterfactuals are meant to match Lewis's the account appears to fail.

There is a closely related worry concerning situations that we intuitively take to involve two events that have a common cause. Consider the counterfactual 'If the sound associated with an axe striking a tree occurred, then the event that the tree will fall has probability p.' Arguably, the latest decision-sized micro event that could result in the sound might again be my decision to strike the tree and the probability p might be quite large. Thus, in the case of counterfactuals linking two effects of a common cause, Loewer's counterfactuals also do not in general match Lewis's, since on Lewis's account the counterfactual 'If there were an axe-striking sound, then the tree would fall' is false, because the account involves introducing a miracle *between* the occurrence of the sound and the striking of the axe. Loewer suggests that the problem of backtracking counterfactuals might be solved by characterizing 'causal priority in terms of the temporal direction of control by decisions' (Loewer this volume, p.321). But this proposal would be of no help with the problem of counterfactual dependence between effects of a common cause.

13.6 Conclusion

I want to sum up what I have argued. It is one of our fundamental convictions that the future depends on the present in ways in which the past does not. Albert's account is meant to account for that conviction by delineating an inference procedure according to which the future counterfactually depends on small local changes to the present macro state but the past does not. This inference procedure involves the past-hypothesis, the micro-dynamical laws, and a statistical postulate. One of my aims in this paper was simply to get clearer on the question as to what exactly it could be to get to the bottom of this conviction of ours. There are at least three distinct projects in which Albert might be engaged. The first project is that of trying to account for how we actually proceed in making certain inferences. As we have seen, Albert appears to argue that in making inferences from records, we make inferences from two times to a time in between; and in order to come by the information about the past involved in making such inferences we must assume the initial state of the universe and it must seem to us that our experience is confirmatory of the past-hypothesis. I have argued that as an account of our actual inferential practices Albert's account is inadequate. If it indeed were the case that inferences based on records only were possible if we assumed the past-hypothesis, then most of us are not in a position to reason from records. The second project is epistemological and is an attempt to provide a justification for our fundamental conviction. However, to the extent that Albert's account is meant to justify the veracity of our memories—an assumption that on his account underlies all our counterfactual inferences—it is circular, since it is our belief in the veracity of our memories itself that, according to Albert, justifies our belief in the past-hypothesis.

The third project Albert might be engaged in is one of offering a philosophical reconstruction of our counterfactual reasoning practices. Albert, on this reading might only be claiming that postulating the past-hypothesis can ensure that records are reliable and that the past is counterfactually independent of the present, even though we need not assume the past-hypothesis when we actually reason about the past. To be sure, on this third reading Albert's worry how we actually come by the information about the past needed in reasoning from records remains unanswered. Nevertheless, the fact that our reasoning practices could be reconstructed in the way suggested by Albert would obviously be extremely interesting and might play a role in an evolutionary explanation of our concept of cause. Yet I have argued that on this third reading the project is problematic as well.

First, it is highly doubtful that, as Albert claims, the past-hypothesis alone can in principle ensure that records are reliable. Even if standard arguments from statistical physics were able to show that if records were unreliable, then the past-hypothesis would most likely be violated, it does not follow that if the past-hypothesis is true, then records are probably reliable. At the very least, there is a *lacuna* in the argument here. And I have suggested that when we try to gauge the prospects of Albert's claim by thinking about relatively small, isolated thermodynamic systems as toy-examples, then it seems highly unlikely that this *lacuna* will ever be filled.

Second, I have criticized Albert's claim that, given the past-hypothesis, the existence of records ensures that the past is counterfactually independent of small macro changes to the present. As I have shown, Albert's account presupposes that, in the case of systems that are governed by a nearly-deterministic macro dynamics, the reliability of records trumps, as it were, what should be expected in light of the dynamics plus the past-hypothesis. Again, Albert owes us an argument for why we should accept his claim that it is a consequence of our physics that records strictly take precedence here. More plausible, I have suggested, might be that there is some trade-off between the reliability of records and our macro dynamics, which will have the effect that at least the relatively recent past of counterfactual systems will macroscopically differ from that of the actual system. In order to settle this question a more detailed argument appealing to the relevant physics is desperately needed.

I have argued, moreover, that Albert could not simply reply to my worries by acknowledging the existence of an anti-thermodynamic transition period during which a past identical to the actual past evolves into a counterfactual present, without thinking carefully about the length of such transition periods. For the existence of such a transition period constitutes a counterargument to entropy accounts unless it can be shown that the transition period is very, very short. One sometimes gets the impression from reading defenders of entropy accounts of causation that they assume that thermodynamic fluctuations can very quickly and within time periods that are of negligible duration ensure that a past identical to the actual past evolves into a present that differs from the actual present locally yet macroscopically. Yet if thermodynamic fluctuations are simply the time-reverse of normal thermodynamic behavior, then this assumption is unwarranted. The relaxation times of thermodynamic systems can be very long. And just as it will take a very long time for Napoleon's boots to fully decompose (to cite one of Albert's own examples) so it would take a very long time for the boots to form anti-thermodynamically out of primordial 'goo'.

Finally, I have made some brief remarks suggesting that Loewer's recent entropy account of counterfactuals does not, as it stands, fill in the argumentative gaps in Albert's entropy account of causal influence. Unfortunately, then, Field's puzzle concerning the role of causation in fundamental physics still awaits a solution.

References

Ahmed, A. (this volume). 'Agency and Causation'.
Albert, D. (2000). *Time and Chance*. Cambridge, Mass.: Harvard University Press.
Cartwright, N. (1979). 'Causal Laws and Effective Strategies', *Noûs* 13: 419–38.
Elga, A. (2001). 'Statistical Mechanics and the Asymmetry of Counterfactual Dependence', *Philosophy of Science* 68 (Proceedings): s313–s324.
Field, H. (2004). 'Causation in a Physical World', in M. Loux and D. Zimmerman (eds), *Oxford Handbook of Metaphysics*. Oxford: Oxford University Press.
Frisch, M. (2005). *Inconsistency, Asymmetry and Non-Locality: Philosophical Issues in Classical Electrodynamics*. New York: Oxford University Press.
—— (forthcoming). 'Causal Asymmetry, Counterfactual Decisions and Entropy', *Philosophy of Science*.
Hausman, D. (1998). *Causal Asymmetries*. Cambridge, UK; New York: Cambridge University Press.
Kutach, D. (2002). 'The Entropy Theory of Counterfactuals', *Philosophy of Science* 69: 82–104.
Lewis, D. (1986a). 'Causation', in *Philosophical Papers*. Oxford: Oxford University Press.
—— (1986b). 'Counterfactual Dependence and Time's Arrow', in *Philosophical Papers*. Oxford: Oxford University Press. Original edition, Noûs, 13 (1979).
Loewer, B. (this volume). 'Counterfactuals and the Second Law'.
Menzies, P and Huw Price (1993). 'Causation as a Secondary Quality', *British Journal for the Philosophy of Science* 44 (2): 187–203.

Pearl, J. (2000). *Causality*. Cambridge: Cambridge University Press.
Russell, B. (1918). 'On the Notion of Cause', in *Mysticism and Logic and other Essays*. New York: Longmans, Green and Co.
Woodward, J. (2003). *Making Things Happen: A Theory of Causal Explanation*. Oxford: Oxford University Press.

Index

agency theory 76, 120–54
  agent-probability 134–49
  circularity of 120, 121–2, 123–9, 147–8, 150, 153–4
  concept-possession 129–34
  objections to 120, 121–2, 123–9, 147–8, 149–54
  varieties of 120–9
agent-probability 130–1, 134–49, 152
  and causal judgments 140–6
  first argument 134–40
  second argument 135, 140–6
  third argument 135, 146–9
agents/agency 183–4, 186, 274–5, 279, 281–3
  and causation 76, 284, 357
  and ignorance 282, 284
Albert, D. 303, 304, 307, 313, 352–8
  and beliefs 372–3, 374
  on causal asymmetries 341–2, 352
  causal influence, entropy-account of 352, 365–86
  on counterfactuals 316 n. 38, 353, 358, 359–64
  on hypothetical interventions 355–6
  normal procedures of inference (NPI) 356
  and past hypothesis 360–1, 366, 373–4, 376
  and ready conditions 363, 366–8, 369, 370
  on records 366, 369
ambiguity 198–199
anti-fundamentalism 21–2
approximation 173–4
Aristotle 15–16, 218–19
Arntzenius, F. 89–90
association, laws of 286
astronomy 14, 45, 48, 52, 53, 56, 57
asymmetries 36, 164, 278–9
  spatial 255–62
  thermodynamic 334–5, 360
  see also causal asymmetries; temporal asymmetries

backtracking counterfactuals 354, 361, 390, 391
backward causation 265, 277, 347
Baldwin 49, 50
behaviour
  default 110–11
  propositional 229
  of sensitive systems 111–17
  of stable objects 107–10
beliefs 238–40
  causal 140–6, 238
  justification of 372–5
  reason and 239–40
Bennett, J. 308–10, 313, 322, 324
Bergson, H. 47
Berkeley, G. 121, 123
Best Theory of the world 305, 315
bi-deterministic physics 157–8
big bang cosmology 40
biology 42
Blackburn, S. 224–5, 227, 229
blobs and arrows diagrams 35–8
Boltzmann, L. 270–1
  on entropy 298–9
  probability assumption 299–300, 306
boundary conditions 69, 81, 83, 302, 303
  and asymmetries 304, 333, 338–42
  low entropy 270, 272, 300
  see also past hypothesis
brain imaging experiments 70
Brownian amplifiers 113–15, 116, 117
Bunge, M. 19, 22

Campbell, N. R. 22
Carnot, S. 29
Cartwright, N. 161, 170 n. 14, 351
  on causal laws 284–5
  on causal realism 288
  on causation, notion of 51–2
causal asymmetries 2, 352
  and bi-deterministic physics 157–8
  boundary condition asymmetries 338–42

causal asymmetries (cont.)
  common causes 345–6
  counterfactual asymmetries 335–42
  fundamental physical asymmetries 332–5
  past hypothesis and 344–5, 347
  and perspectivalism 261–5
  single trace causation 347–8
  traces, future 346–7
causal notions 66–7, 68, 70
causality 36, 46, 47–8
  definition of 49, 50
  failure of 39
causation 18, 36, 50, 162 n. 3
  associative mechanism of 232, 233, 247–8
  in contemporary philosophy 1–2
  context dependency of 176
  counterfactual model of 165–72, 173
  error theory about 193
  indispensability of 177–87
  judgements of 140–6
  notion of 51–2, 55–6
  by omission 82 n. 17
  reality of 30–1
  and spurious evidential correlations 123
  transitivity of 331
causation proper 199
causes
  dispensability of 14
  existence/non-existence of 46–7
  misleading nature of word 46, 48–9, 61–2
  notion of 45, 46, 49–51
chance 284
chemistry 42
choice 276–7, 282–3
coarse-grained systems of causation 41, 80–90
coarse-graining of events 171–2
Collingwood, R. G. 121, 123
Collins, J. 195
colour judgements 124–5, 129–30, 133–4, 229
concept-possession 129–34, 151–2
  and circularity 131–4
  and concept-introduction 129–31
conditionals
  causal 210–18
  decision 317–18
  SM 316–24
connectionless beings 242, 243, 245, 246–8

connections
  causal 230
  necessary 242–8
contextualism 193, 194–8
correlations 97–8, 147–8
cosmology 40, 300
counterfactual asymmetries 335–42
  boundary condition asymmetries 338–42
counterfactual dependency 206, 210, 384–5, 391
  and transition periods 385–6, 393
counterfactual worlds 380–4
counterfactuals 53, 58, 152–3
  backtracking 354, 361, 390, 391
  and causation 165–72, 173, 297
  and closed systems 355
  and decisions 318, 337–8, 386–91
  and deliberation 183–5
  forward looking 354, 361
  interventionist 91–3
  and partial determination 163–5
  records and 365–6, 380–4
  and second law of thermodynamics 293–325
  and SM conditionals 316–24
  standard resolution of 336–7
  temporal asymmetries of 294–5, 310, 312, 313, 353
  and thermodynamic systems 359–64
  truth conditions for 215–16, 218, 308–16, 360–1, 363–4, 389–90
Craig, E. 231–2

Davidson, D. 82 n. 16
De Angelis, E. 16
decision conditionals 317–18
decision counterfactuals 318, 337–8, 358
decisions
  and backtracking reasoning 337–8
  and counterfactuals 318, 337–8, 386–91
  determinism and 316–17
  indeterminacy of 317
default worlds 216–22
deliberation 160–1, 273, 288
  architecture of 274–9, 80
  causation and 274, 282–3, 285
  counterfactuals and 183–5
  fixed past principle (FPP) 277–8
descriptions 176, 177

determination
  causation and 156–9, 161, 176
  conditionals in 163
  and extreme locality 107–10
  folk model 106–10
  local 158–60
  partial 163–5
determinism 16–18, 48, 54
  and decisions 316–17
difference-making 163, 200–9
  default causal model 207–8
  SE framework 202–7, 209–15
dissipative physical systems 40–1
Dowe, P. 192
Dummett, M. 143 n. 6
dynamical laws 296
  and entropy 298–9
  temporal symmetry of 297–8

econometrics 70, 202
economics 70
Eells, E. 192
effectual beliefs 142
efficient causes 15, 16
Einstein, A. 56
Elga, A. 313–14, 378
eliminativism 22, 177–8
entropy 90, 270–4, 278, 302, 314–15, 341, 358
  asymmetries of 295, 298–9, 300, 301, 313
  and causal influence 352, 365–86
  and dynamical laws 298–9
  and equilibrium 304
  GRW and 307–8
  increases in 90, 264, 270, 334
  and temporal asymmetries 264–5
equilibrium 300, 302–3, 304, 388, 389
Evans, G. 128
exclusion problem 172–7
  response to 185–7
explanations
  causal 176, 177, 178, 184–6, 199–200
  fundamental 186
  physical 178
  statistical 112–13

Field, H. 83, 84, 106, 108, 351–2 & n.
  on causation, notion of 118 n.
  on language of fundamental science 178 n.

on variables 166 n. 7
final causes 15–16, 37–8, 40–1
Fine, K. 308–9
fine-grained systems of causation 84, 85 n., 86–9, 172
first causes 37, 38, 39–40
fixed past principle (FPP) 277–8
folk model of physics 106–18
  default behaviour of 110–11
  extreme locality and 107–10
  and sensitive systems 111–17
folk theories
  of causation 12–13, 31–40, 42–3, 329–30
  and perspective 250–1
FPP (fixed past principle) 277–8
free agency 122, 124, 125–7
free will 47, 49, 54
Friedman, M. 95 n.
fundamental dynamical laws 112–14, 296–9, 351
  and PH (past hypothesis) 300–7, 315–16
  PROB 300, 304–7, 315–16
fundamental dynamical probabilities 330
fundamental explanations 186
fundamental laws 70, 296–7
fundamental physical asymmetries 332–5
fundamental physics 66–7
  causation and 70, 157, 174–5, 296–7
fundamentalism 3, 14–22, 70

general relativity 13, 30
  and determinism 18
  and earlier theories 28–9
  spacetimes in 70 n. 6
generalizations 48, 52–3, 94, 97–8
  and coarse-grained systems of causation 80–90
  Hooke's law 76–9
Gold, T. 271–2
Gold universe 272–4, 277–8
gravitational astronomy 14, 45, 48, 52, 53, 56, 57
gravity 13, 16

Hall, N. 331
Hardy-Weinberg law 222
Hart, H. L. A. 194
  and causes as difference-makers 201–2, 207
Hausman, D. 70 n. 5

400 INDEX

heat 29, 31
Hesslow, G. 196
Hitchcock, C. 172, 202
  on ambiguity of context-sensitivity 198
  and causes as difference makers 204, 206–7, 208–9
  on contextualism 197, 198
Hoefer, C. 284 n.
Honoré, A. 194
  and causes as difference-makers 201–2, 207
Hooke's law 76–9
Horwich, P. 323
Hume, D.
  and beliefs 238–9
  and causal reasoning 231–5
  on causation 21, 150
  and ethical projectivism 236–41
  and inductive inference 246
  and naive regularity theory 224
  and necessary connections 242–8
  on necessity 239
  and projectivism 225–30, 232–3, 236–41
  on reason 238, 239–40
  and sceptical realism 224
  on taste 236, 238, 239–40, 241
Humean scepticism 21
Humphries, P. 192

ignorance 282–3, 284
Image of God doctrine 231–2
indeterminism 17–18, 157, 330
inductive reasoning 246–7
inferences 303, 306–7, 333
  causal reasoning and 232–5
  and counterfactuals 339–40, 341–2, 361–2, 363–4
  inductive 246
  necessary connection and 244–5
  NPI (Albert) 356
  and projectivism 225
  and records 366–7, 374
influence
  causal 352, 359–60, 365–86
  causation as 162 n. 3
inquiry 187
Insight Ideal 231, 232
interventions 355–6
  and causation 71–6, 82, 90–3, 279–81, 282

circularity of 170–1
in counterfactual model of causation 169–71
drug trial scenario 91–2, 93–4, 95, 96
hypothetical 355–6
and incompleteness 78
invariance and 76–80
and perspectivity 268–9
stability and 77–80
and temporal asymmetries 279–80
see also miracles
isolation 108–9
  and default behaviour 110–11
  sensitive systems and 111–12, 113

judgements
  causal 140–6, 286
  colour 124–5, 129–30, 133–4, 229

Kant, I. 11 n., 252–3, 289–90
knowledge 275, 282 & n.

language 178 n., 182
  causal 178–9, 183–5, 186
Laplace, P. 16–17
Lavoisier, A. 29
laws
  of association 286
  causal 284–5, 286, 288
  fundamental 70, 296–7
  static 296
  see also fundamental dynamical laws
Lewis, D. 49, 163, 192, 224, 295, 321–2
  on causation as influence 162 n. 3
  on contextualism 195–6
  on counterfactuals 308–16, 336, 390–1
  on laws 305–6
  and miracles 205, 354
  overdetermination asymmetries 336–7
  on physicalism 295 n. 6
  possible worlds semantics 205, 215
liar paradox 54
Liouville's theorem 389
local determination 158–60
locality 37, 38, 107–10
Loewer, B. 358, 386–7, 389–90
Loschmidt's reversibility objection 373, 374, 380

McDermott, M. 195
Mach, E. 21, 22
Mackie, J. L. 192, 236
manipulation 34, 170; *see also* intervention
Margenau, H. 19 & n.
Mass on the Dome 23–8
  first causes and 39–40
  Newton's First Law and 25–6
Maudlin, T. 219–20, 273 n., 294 n. 3
memories 365, 371–2
  veridicality of 368, 369, 370, 374–5, 392
  see also records
Menzies, P. 34, 121, 122–9, 150
Mill, J. S. 11 n., 16, 47, 48
minimalism 179–80, 181–2, 187
miracle worlds 354
miracles 53, 205, 315, 391
  temporal asymmetry of 310–12, 313, 314
modal language 182

Nagel, T. 20
naive regularity theory 224
necessary connection 226, 228, 230, 240, 242–8
necessity 238–9, 242
  physical 329–31
neurobiology 70
Newcomb's Problem 143–4, 145
Newton, I. 16
  First Law, and Mass on the Dome 25–6
  Second Law 23–4
Nietzsche, F. 53
Norton, J. D. 107
numerical language 182

overdetermination 336–7, 353

partial dependence 162–72
partial determination 163–5
past hypothesis (PH) 335, 339, 341, 360–1
  causal asymmetries and 344–5, 347
  and causal reasoning 366
  and fundamental dynamical laws 300–7, 315–16
  and memories 374–5
  and records 363, 366, 368, 371, 373–81, 387, 388, 392–3
Pearl, J. 4, 92, 93
  and causes as difference makers 203–4

on interventions 171
  and possible worlds 217–18
  and SE framework 202
Pearson, K. 21
perspectival realism 193
perspectives 180–1, 286
  asymmetries of 251–2
  nature of 250–3
perspectivalism 250–91
  causal asymmetries 261–5
  cost of 287–8
  location of 283–8
  perspective, nature of 250–3
  reversibility argument 269–74
  science and 253–4, 288–90
  and spatial asymmetries 255–62
  temporal asymmetries 252, 262–4, 266–9
PH, *see* past hypothesis
physical explanations 178
physicalism 173, 295–6
physics 68–70, 161
  acausality in 22–8
  folk model of 106–18
  Mass on the Dome 23–8, 39–40
Popper, K. 312–13
positivist scepticism 21
possible worlds 215–18, 312 n.
  and default worlds 216–18
  semantics of 205, 215–16
  and SE framework 215–17
pragmatics 199–200
pragmatism 177–87
Price, H. 150, 187, 305
  and agency theory 121, 122–9, 149
  on colour judgements 229
primitivism 149–50, 273
probabilistic causation 17
probability 100, 135–43, 283–4, 289, 299–302, 318–20
  and betting behaviour 135–9
  Boltzmann's probability assumption 299–300, 306
  and causation 281–2
  PROB 300, 304–7, 315–16, 338–9, 342–6, 347
probability-raising 330–1, 332
projectivism 245
  Blackburn and 224–5, 227
  causal/ethical 236–41

402 INDEX

projectivism (*cont.*)
  Hume and 225–30, 232–3, 236–8
  Stroud and 226–7, 236–7
Putnam, H. 53–4

quantum mechanics 54, 297
quantum theory 17, 18–19
  and earlier theories 28–9

Ramsey, F. P. 183, 281–2
ready conditions 363, 366–8, 369, 370, 375–7
ready states 371–2
realism
  causal 191–2, 243–5, 286–7, 288
  perspectival 193
  sceptical 224
reality 30–1
  representation of 178–9, 180–2
reason, beliefs and 239–40
reasoning 371
  backtracking 337–8
  causal 219–20, 231–5, 265, 290, 366
  counterfactual 365–6, 380–4
  inductive 246–7
  scientific 218–19
records 362–3, 369, 370–1
  and counterfactual reasoning 365–6, 380–4
  and inferences 366–7, 374
  and past hypothesis 363, 366, 368, 371, 373–81, 387, 388, 392–3
  and ready conditions 370
reduction relations, generative capacity of 28–32, 41
reference, causal theories of 62–3
Reichenbach, H. 299 n. 15
Reid, T. 121
relationships 58–63
  causal influences in 93–7
  and correlations 97–8
  macroscopic 98–102
  roulette wheel scenario 99–101
representation 187
  of reality 178–9, 180–2
reversibility argument 269–74
Rosen, D. 197
Russell, B. 14, 21, 37 n., 38, 164–5, 173, 175
  arguments against causation 156–61
  cause, notion of 45, 46–7, 49–51

  and determinism 48
  dispensability of causation 327
  eliminativism 177–8
  on free will 47, 49
  functional dependency of causation 351
  on fundamental laws 296–7
  misleading nature of word 'cause' 46, 48–9
  spurious causes argument 164
  status of causation 191

Salmon, W. 192, 197–8
sceptical realism 224
scepticism 13, 21–2
Schaffer, J. 82 n. 17
Schrödinger equation 69
scientific reasoning 218–19
SE, *see* structural equations
SM, *see* statistical mechanics
small worlds 4
Smith, S. 68
smoothness 115–16
Sober, E. 198, 218, 221–2
spatial asymmetries 255–62
speech, objective mode of 229
stable objects, behaviour of 107–10
Stalnaker, R. C. 187
static laws 296
statistical explanations 112–13
statistical mechanical probabilities 330
statistical mechanics
  coarse/fine graining in 172
  conditionals, and counterfactuals 316–24
  probability distribution 299 n. 15, 300–7, 316, 318–20, 323, 387–8
statistics 70
Stich, S. 287
stochastic dynamics 332–3
Strevens, M. 99, 115, 116 n.
Stroud, B. 225, 230, 238–41
  on causal realism 243
  on connectionless beings 242, 243, 246
  on necessity 238–9, 242–4
  and projectivism 226–7, 236–7
structural equations (SE) framework 202–9
  and causes as difference makers 209–15
  and possible worlds 215–17
Suppes, P. 51, 55–6, 59
systems, sensitive
  detector-like 111–12
  quasi-chancy 113–17

taste 236, 238, 239–40, 241
temporal asymmetries
  and causal perspectivism 252, 262–4, 266–9
  of counterfactuals 294–5, 310, 312, 313, 353
  of decision conditionals 317–18
  interventions and 279–80
  and miracles 310–12, 313, 314
thermodynamic asymmetries 334–5, 360
thermodynamic systems 359–64
thermodynamics, second law of 293–325
token causation 58–9
Tooley, M. 297 n. 8
total causation 73
truth conditions 386–7
  for counterfactuals 215–16, 218, 308–16, 360–1, 363–4, 389–90

Unger, P. 193

vacua 31–2
variables 165–9, 171–2, 367
  default values 208, 210–15
  exogenous 169, 203, 208, 210–15
  omissions 168–9
  parenthood relations 167–8, 203
  random 165–8
von Wright, G. H. 121

Williams, D. C. 275 n. 19
Woodward, J. 3, 73, 91, 94, 70 n. 5, 202
worlds 322
  counterfactual 380–4
  default 216–22
  miracle 354
  possible 205, 215–18, 312 n.
  small 4
  temporal asymmetries of counterfactuals in 313–16

The manufacturer's authorised representative in the EU for product safety is
Oxford University Press España S.A. of el Parque Empresarial San Fernando de
Henares, Avenida de Castilla, 2 – 28830 Madrid (www.oup.es/en or product.
safety@oup.com). OUP España S.A. also acts as importer into Spain of products
made by the manufacturer.

www.ingramcontent.com/pod-product-compliance
Ingram Content Group UK Ltd.
Pitfield, Milton Keynes, MK11 3LW, UK
UKHW022230230426
12048UKWH00016BA/1163